# MATHEMATICAL ESSAYS

# MATHEMATICAL ESSAYS

## In Honor of Su Buchin

*Edited by*

**C.C. Hsiung**
*Lehigh University*
*Bethlehem, PA 18015*

World Scientific

World Scientific Publishing Co. Pte. Ltd.
P O Box 128
Farrer Road
Singapore 9128

*The editor and publisher are indebted to the original authors and publishers of the various journals for their assistance and permission to reproduce the selected papers found in this volume.*



ISBN 9971-950-98-7

**Printed in Singapore by Kim Hup Lee Printing Co. Pte. Ltd.**

Professor Su Buchin

# PREFACE

This is a collection of research papers published in various mathematical journals by friends, colleagues and former students of Professor Buchin Su in honor of his 80th birthday and 50th year of educational work.

Professor Su was born in 1902 in Pingyang County, Zhejiang Province, People's Republic of China. He received the degree of Bachelor of Science in mathematics from Tôhoku University, Sendai, Japan in 1927, and the degree of Doctor of Science from the same university in 1931. After returning to China in 1931, he first taught at Zhejiang University in Hangzhou until 1952 when the whole College of Science of Zhejiang University was merged into Fudan University in Shanghai. During his 50 years of educational work besides teaching, he also has taken up various administrative positions serving as Chairman, Dean, Vice President and finally the President of Fudan University in 1978.

Professor Su was one of the pioneers of scientific research in China in the late twenties. He has made notable contributions to the development of differential geometry and has exercised great influence in the mathematical community. He has published over 150 research papers and 10 monographs and textbooks chiefly concerning affine and projective differential geometry, differential geometry of generalized spaces, and computational geometry. Professor Su was also an outstanding teacher attracting a large number of talented students to found a school of differential geometry starting at Zhejiang University.

Professor Su has dedicated his life to teaching and research, persisting in his scientific activity under adverse conditions even during wars and political turmoil.

As a former humble student of his, I present this collection to Professor Su. I respect him for his glorious durable dedication, thank him for his influence on my interest in differential geometry, and wish him many happy returns for his birthday.

*Bethlehem, May, 1983* C. C. Hsiung

# CONTENTS

# MATHEMATICAL ESSAYS

*Chin. Ann of Math.*
3 (4) 1982

# LOCAL ISOMETRIC IMBEDDING OF RIEMANNIAN MANIFOLDS $M^n$ INTO A SPACE OF CONSTANT CURVATURE $S^{n+1}$

Bai Zhengguo (Pa Chen-kuo)
(*Hangchow University*)

Dedicated to Professor Su Bu--chin on the Occasion of his 80th Birthday and his 50th Year of Educational Work

**1.** The problem of isometrically imbedding an $n$-dimensional Riemannian manifold $M^n$ into Euclidean space $E^{n+p}$ has received considerable attention. This problem, on which a series of results have been obtained both locally and globally, is still being studied now, thus as different sets of the necessary and sufficient conditions have been studied. These conditions are intrinsic and expressible by the Riemannian tensor of $M^n$. For example, Allendoerfer, C. B., obtained the conditions for an Einstein space of class one,[1] Jaak Vilms obtained the conditions for a Riemannian space of class one,[2] and so on. The enverloping space $E^{n+1}$ they discussed is Euclidean and the conditions obtained take different forms as the different mothods they used. These conditions are too complicated for applications. The purpose of this paper is to find a set of necessary and sufficient intrinsic conditions for the Riemannian manifold $M^n$ being a hypersurface of a space of constant curvature,and give some applications of these conditions. Since it is a problem of local imbedding, the following discussions are given on a coordinate neighborhood at a given point. Throughout this paper, the term Riemannian manifold or Riemannian space is understood to be a Riemannian manifold with a real fundamental quadratic form

$$\varphi=g_{\alpha\beta}dy^\alpha dy^\beta, \quad g=\det|g_{ij}|\neq 0,$$

as the basis of the metric, either positive definite or not, that is, the element of length $ds$ of the space is defined by

$$ds^2=eg_{\alpha\beta}dy^\alpha dy^\beta \quad (e=\pm 1).$$

Besides a set of necessary and suffiicient conditions of this manifold, we describe roughly the main results of this paper as follows: If any Riemannian manifold $M^n$ can be isometrically imbedded into two spaces $S^{n+1}$ of constant curvature, it can be isometrically imbedded into space $S^{n+1}$ of any constant curvature, and $M^n$ must be a conformally flat space of class one. Conversely, if conformally flat $M^n$ can be

Manuscript received April 13, 1981.

isometrically imbedded in one $S^{n+1}$ of constant curvature, it can be isometrically imbedded into an $S^{n+1}$ of any constant curvature. This class of spaces involves the spaces of constant curvature as its special class. In this sense, the former is an extension of the latter. we give in the end of this paper the characteristic Riemannian curvature form of this class of spaces.

**2.** The Gauss equations for a hypersurface $M^n$ of a space $S^{n+1}$ 1) of constant curvature $K_0$ is given by[3]

$$R_{ijkl}=e(b_{ik}b_{jl}-b_{il}b_{jk})+K_0(g_{ik}g_{jl}-g_{il}g_{jk}), \tag{1}$$

where $e=\pm 1$, $g_{ij}$ the Riemannian metric of $M^n$, $R_{ijkl}$ the Riemannian curvature tensor. If we put

$$T_{ijkl}=R_{ijkl}-K_0(g_{ik}g_{jl}-g_{il}g_{jk}), \tag{2}$$

the equation (1) becomes

$$T_{ijkl}=e(b_{ik}b_{jl}-b_{il}b_{jk}) \quad (e=\pm 1). \tag{3}$$

After a complicate calculation, we find from (3) the following identity

$$2eb_{il}b_j({}_qT_{mp})_{kh}=T_{hil}({}_qT_{mp})_{kj}+T_{ikl}({}_qT_{mp})_{hj}+T_{hkl}({}_qT_{mp})_{ij}, \tag{4}$$

where $b_j({}_qT_{mp})_{kh}$ indicates the sum of three terms obtained one by one by a cyclically permutation of $\begin{pmatrix} q & m & p \\ m & p & q \end{pmatrix}$ and so on. If the identity (4) is multiplied by $g^{hq}g^{pk}$ and summed for $h$, $q$, $p$, $k$, and put

$$T=R+n(n-1)K_0, \tag{5}$$

$$T_i^j=R_i^j+(n-1)K_0\delta_i^j, \quad T_{ij}=R_{ij}+(n-1)K_0g_{ij}, \tag{6}$$

$$T^p_{jkl}=R^p_{jkl}-K_0(\delta^p_k g_{jl}-\delta^p_l g_{jk}), \tag{7}$$

we have

$$2b_{jl}b_{ip}\left(T^p_m-\frac{1}{2}\delta^p_m T\right)=e\left(T_{jl}T_{im}-T_{ip}T^p_{jlm}+T_{hl}T^h_{mij}+T^p_{jlq}T^q_{ipm}-\frac{1}{2}T^q_{klm}T^k_{qij}\right). \tag{8}$$

If we put

$$P_{jlim}\equiv T_{jl}T_{im}-T_{ip}T^p_{jlm}+T_{pl}T^p_{mij}+T^p_{jlq}T^q_{ipm}-\frac{1}{2}T^q_{klm}T^k_{qij}, \tag{9}$$

the equation (8) becomes

$$2b_{jl}b_{ip}\left(T^p_m-\frac{1}{2}\delta^p_m T\right)=eP_{jlim} \tag{10}$$

and consequently we have

$$P_{jlim}P_{rshk}=b_{jl}b_{rs}(2b_{ip}T^p_m-b_{im}T)(2b_{hp}T^p_k-b_{hk}T).$$

By means of (10), this equation is reducible to

$$(2P_{iphk}T^p_m-P_{imhk}T)b_{jl}b_{rs}=eP_{jlim}P_{rshk}. \tag{11}$$

**Lemma 1.** *If*

$$H_{hijk}=a_{hj}a_{ik}-a_{hk}a_{ij} \quad (h,\ i,\ j,\ k=1,\ \cdots,\ n) \tag{12}$$

1) Both $S^{n+1}$ and $M^n$ are Riemannian spaces in the sense as stated in § 1, that is, their fundamental quadratic forms may be indefinite.

*and*

$$H_{hijk}+H_{hjki}+H_{hkij}=0 \tag{13}$$

*and if the rank r of the matrix* $(H_{hijk})$ *is* $\geqslant 3$, *we have* $a_{ji}=a_{ij}$.

*Proof* From (12) and (13) it follows that

$$a_{hj}(a_{ik}-a_{ki})+a_{hk}(a_{ji}-a_{ij})+a_{hi}(a_{kj}-a_{jk})=0. \tag{14}$$

Since the rank of the matrix $(a_{ij})$ is equal to that of $(H_{hijk})$, that is, equal to $r$, we suppose, without loss of generality, that $\det|a_{ij}|\neq 0$ $(i, j=1, \cdots, r)$, such that a set of $a^{hj}$ can be defined by the following equations

$$a^{hj}a_{hi}=\delta_i^j \quad (h, i, j=1, \cdots, r).$$

Firstly, if we take $h, i, j, k=1, \cdots, r$, in the equation (14) and contracting it by $a^{hj}$, we get

$$(r-2)(a_{ik}-a_{ki})=0$$

Since $r\geqslant 3$, it follows that $a_{ik}=a_{ki}$ $(i, k=1, \cdots, r)$.

Secondly, take in (14) $h, i, k=1, \cdots, r, j=r+1, \cdots, n$; we have

$$a_{hk}(a_{ji}-a_{ij})+a_{hi}(a_{kj}-a_{jk})=0.$$

Contracting it by $a^{hi}$

$$(r-1)(a_{kj}-a_{jk})=0,$$

and consequently

$$a_{kj}=a_{jk} \quad (k=1, \cdots, r, j=r+1, \cdots, n),$$

Finally, take in (14) $k, j=r+1, \cdots, n$; $h, i=1, \cdots, r$, we find

$$a_{kj}=a_{jk} \quad (k, j=r+1, \cdots, n).$$

Hence

$$a_{ij}=a_{ji} \quad (i, j=1, \cdots, n)$$

q. e. d.

The following lemma is due to Prof. Hu Hesheng.[5]

**Lemma 2.** *If a Riemannian manifold* $M^n$ *admits isometric imbedding into a space* $S^{n+1}$ *of constant curvature* $K_0$ *and if the rank of the matrix* $(T_{ijkl})$ *is* $\geqslant 4$, *the Codazzi equation is the algebraic consequence of the Gauss equation.*

Now we state the following

**Theorem 1.** *Let the rank of the matrix* $(T_{ijkl})$ *of a Riemannian manifold* $M^n$ *is* $\geqslant 4$, $T\neq 0$, $M^n$ *admits local isometric immedding into* $S^{n+1}$ *of constant curvature* $K_0$ *if and only if the following equations hold and are non trivial*

$$(2P_{hphk}T_k^p-P_{hkhk}T)T_{abcd}=P_{achk}P_{bdhk}-P_{adhk}P_{bchk}, \tag{15}$$

*for* $a, b, c, d=1, \cdots, n$, *and for a certain set of* $h, k$.

In order to prove this theorem, we state and prove the following

**Lemma 3.** *If a Riemannian manifold* $M^n$ *admits isometric imbedding into a space* $S^{n+1}$ *of constant curvature, and if the rank of the matrix* $(T_{ijkl})$ *is* $\geqslant 3$, $T\neq 0$, *there exists at least a set of* $h, k$ *such that*

$$2P_{hphk}T_k^p-P_{hkhk}T\neq 0. \tag{16}$$

If for any $h$, $k$

$$2P_{hphk}T^p_k - P_{hkhk}T = 0,$$

we have from (11)

$$P_{jlim} = 0. \tag{17}$$

From (10) it follows that

$$b_{ip}\left(T^p_m - \frac{1}{2}\delta^p_m T\right) = 0 \tag{18}$$

or

$$b_{jl}b_{ip}T^p_m - \frac{1}{2}b_{jl}b_{im}T = 0.$$

By means of this equation and the Gauss equation (3) we obtain

$$T_{ijpl}T^p_m - \frac{1}{2}TT_{ijml} = 0$$

or

$$T_{km}T^k_{lij} - \frac{1}{2}TT_{ijml} = 0.$$

Contracting it by $g^{li}$,

$$\left(T^k_j - \frac{1}{2}T\delta^k_j\right)T_{km} = 0. \tag{19}$$

Let the rank of the matrix $(T_{km})$ be $r$. By the theory of linear algebra there exists at a point $P$ of $M^n$ a coordinate system in which the matrix $(T_{km})$ takes the following form

$$\left(\begin{array}{c|c} \begin{matrix} T_{11} & \cdots & T_{1r} \\ \cdots & \cdots & \cdots \\ T_{r1} & \cdots & T_{rr} \end{matrix} & 0 \\ \hline 0 & 0 \end{array}\right)$$

In this coordinate system (19) may be written as

$$g^{kl}\left(T_{lj} - \frac{1}{2}Tg_{lj}\right)T_{km} = 0 \quad (j,\ k,\ m = 1, \cdots, r;\ l = 1, \cdots, n). \tag{20}$$

Since $\det|T_{ij}| \neq 0$ $(i, j = 1, \cdots, r)$, a set of $T^{mp}$ can be defined by

$$T_{km}T^{mp} = \delta^p_k. \tag{21}$$

Contracting (20) by $T^{mp}$

$$g^{pl}\left(T_{lj} - \frac{1}{2}Tg_{lj}\right) = 0,$$

and contracting again by $T^{jq}$,

$$g^{pq} = \frac{1}{2}TT^{pq} \quad (p,\ q = 1, \cdots, r).$$

Therefore

$$T = g^{ls}T_{ls} = g^{pq}T_{pq} = \frac{1}{2}TT^{pq}T_{pq} = \frac{r}{2}T.$$

Since $T\neq 0$, we have $r=2$. Then

$$T^p_m=g^{pl}T_{lm}=0 \quad (m\geqslant 3).$$

From (18) we find

$$b_{im}=0 \quad (m=3, \cdots, n;\ i=1, \cdots, n).$$

It means that the rank of the matrix $(b_{ij})$ is $\leqslant 2$. But it is known that the rank of the matrix $(T_{hijk})$ is equal to that of $(b_{ij})$,[4] and consequently the rank of $(T_{hijk})$ is $\leqslant 2$. This is a contradiction. Therefore there is at least a set of $h$, $k$ satisfying (16).

q. e. d.

We proceed to prove the Theorem 1.

Under the conditions stated in Lemma 3, it is seen that each equality of (15) is non trival. But the equalities (15) are the consequence of the conditions (11) and (3) which are the necessary conditions for $M^n$ to admit local isometric imbedding into $S^{n+1}$, and therefore (15) are also the necessary conditions.

We now prove that (15) are the sufficient conditions. According to (16), we can take $e=1$ or $-1$ such that

$$e(2P_{hphk}T^p_k-P_{hkhk}T)>0.$$

Define

$$b_{jl}=\frac{P_{jlhk}}{\sqrt{e(2P_{hphk}T^p_k-P_{hkhk}T)}}. \tag{22}$$

From (22) and (15) we have

$$T_{hijk}=e(b_{hj}b_{ik}-b_{hk}b_{ij}) \quad (e=\pm 1). \tag{23}$$

Since $T_{hijk}$ satisfies the conditions (13), and the rank of the matrix $(T_{hijk})$ is $\geqslant 4$, by Lemma 1 we have $b_{jl}=b_{lj}$, and cons equently the equations (23) can be taken as the Gauss equations of a hypersurface of $S^{n+1}$ of constant curvature $K_0$.

By Lemma 2, the Codazzi equations of this hypersurface are algebraic consequence of the Gauss equations. That is, $M^n$ admits local isometric imbedding into a space $S^{n+1}$ of constant curvature $K_0$.

q. e. d.

When $K_0=0$, we denote the corresponding $P_{hijk}$ by $Q_{hijk}$. As a consequence of Theorem 1 we have

**Corollary 1.** *Let the metric rank of a Riemannian manifold $V^n$, i. e., the rank of the matrix $(R_{hijk})$ is $\geqslant 4$, $R\neq 0$, a necessary and sufficient condition for $V^n$ to be of class one is as follows*

$$(2Q_{hphk}R^p_k-Q_{hkhk}R)R_{abcd}=Q_{achk}Q_{bdhk}-Q_{adhk}Q_{bchk}, \tag{24}$$

*where $a$, $b$, $c$, $d=1, \cdots, n$ and $h$, $k$ is a set of indices such that the equality is non trivial.*

**Corollary 2.** *If a Riemannian manifold $M^n$ admits isometric imbedding into a space $S^{n+1}$ of constant curvature $K_0$, then*

$$P_{jlhk}=P_{ljhk}, \tag{25}$$

$$P_{pqcd}P_{rsab}=P_{pqab}P_{rscd}, \tag{26}$$

$$2T_{jilp}\left(T^p_m-\frac{1}{2}T\delta^p_m\right)=P_{jilm}-P_{iljm},\tag{27}$$

$$2P_{ahcd}T^h_b-P_{abcd}T=2P_{chab}T^h_d-P_{cdab}T.\tag{28}$$

*Proof* From (11) we have (22), and since $b_{jl}=b_{lj}$ we obtain (25). Interchanging the indices $j$ and $r$, $l$ and $s$ in (11), we obtain (26). Substracting the equations (10) from the equation obtained by interchanging in (10) the indices $i$ and $j$, making use of (3), we get the equation (27). From (11) it follows that

$$2P_{ahcd}\left(T^h_b-\frac{1}{2}T\delta^h_b\right)T_{rasp}=P_{rsab}P_{qpcd}-P_{rpab}P_{qscd}.\tag{29}$$

From (26) it is seen that the right-hand member of (29) is invariant by interchanging the indices $a$ and $c$, $b$ and $d$, and consequently the equation (28) is easily obtained.

**3.** We give some applications of Theorem 1. After a direct computation we find

$$P_{jilm}=Q_{jilm}+(n-1)(n-2)K_0^2g_{jl}g_{im}+(n-1)K_0g_{im}R_{jl}+(n-3)K_0g_{jl}g_{im},\tag{30}$$

$$\begin{aligned}2T_{jilp}\left(T^p_m-\frac{1}{2}T\delta^l_m\right)=&2R_{jilp}R^p_m-RR_{jilm}+(n-1)(n-2)K_0^2(g_{jl}g_{im}-g_{il}g_{jm})\\&+[R(g_{jl}g_{im}-g_{jm}g_{il})-2(g_{jl}R_{im}-g_{il}R_{jm})\\&-(n-1)(n-2)R_{jilm}]K_0.\end{aligned}\tag{31}$$

Substituting (30) and (31), into (27) we have

$$2R_{jilp}R^p_m-RR_{jilm}=Q_{jilm}-Q_{iljm}+(n-1)(n-2)K_0C_{ijml},\tag{32}$$

where

$$\begin{aligned}C_{ijml}\equiv R_{ijml}&-\frac{1}{n-2}(g_{il}R_{jm}+g_{jm}R_{il}-g_{jl}R_{im}-g_{im}R_{jl})\\&-\frac{R}{(n-1)(n-2)}(g_{im}g_{jl}-g_{il}g_{jm})\end{aligned}\tag{33}$$

is the conformal curvature tensor of $M^n$.

**Theorem 2.** *If a Riemannian manifold $M^n$ ($n\geqslant4$) admits isometric imbedding into two $(n+1)$-dimensional spaces of constant curvatures $K_0$ and $K_1$ respectively, and $K_0\neq K_1$, $M^n$ is conformally flat.*

Since (32) is satisfied by $K_0$ and $K_1$, $K_0\neq K_1$, and $n\geqslant4$, we have consequently $C_{ijml}=0$, and $M^n$ is conformally flat.

**4.** When a Riemannian manifold $M^n$ is of constant curvature $a$, we have

$$R_{hijk}=a(g_{hj}g_{ik}-g_{hk}g_{ij}).\tag{34}$$

In this case (30) becomes

$$P_{jilm}=(n-1)(n-2)(a-K_0)^2g_{hl}g_{im},\tag{35}$$

and from (2), we have

$$T_{ijkl}=(a-K_0)(g_{ik}g_{jl}-g_{il}g_{jk}).$$

Substituting these equations into (15), we find that either the left-hand member or the right-hand member of (15) is equal to

$$(n-1)^2(n-2)^2(a-K_0)^4g^2_{hk}(g_{ac}g_{bd}-g_{ad}g_{bc})$$

which can not vanish identically for $n\geqslant4$, and $a-K_0\neq0$.

When $a-K_0\neq 0$, it is seen that the rank of the matrix $(T_{ijkl})$ is equal to that of the matrix $(g_{ij})$ and consequently the rank is $\geqq 4$. Since in this case

$$T=-n(n-1)(a-K_0)\neq 0,$$

we obtain by Theorem 1 the following

**Theorem 3.** *Any Riemannian manifold $M^n$ $(n\geqslant 4)$ of constant curvature a admits isometric imbedding into a space $S^{n+1}$ of any constant curvature $K_0(\neq a)$.*

From (11), we have (22) and consequently

$$e(2P_{hphk}T^p_k-P_{hkhk}T)>0 \quad (e=\pm 1)$$

is a necessary condition that $M^n$ admits isometric imbedding into $S^{n+1}$. When $M^n$ is of constant curvature $a$ and $S^{n+1}$ of constant curvature $K_0$, this condition becomes

$$e(2P_{hphk}T^p_k-P_{hkhk}T)=e(n-1)^2(n-2)^2g^2_{hk}(a-K_0)^3>0.$$

Moreover, if the fundamental quadratic form $g_{\alpha\beta}dy^\alpha dy^\beta$ of $S^{n+1}$ is positive definite, that is, $e=1$, from this condition it follows that $a-K_0>0$, and we have the following known result:

*Any Riemannian manifold $M^n$ $(n\geqslant 4)$ of constant curvature a does not admit isometric imbedding into a space $S^{n+1}$ of constant curvature $K_0$, if $K_0>a$ and the fundamental quadratic form of $S^{n+1}$ is positive definite.*

**5.** Substituting (2) and (30) into (15) we find that (15) is an algebraic equation in $K_0$ of the following form

$$(n-1)^2(n-2)^2g^2_{hk}C_{abcd}K^3_0+\cdots=0. \tag{36}$$

From Theorem 2, if $M^n$ $(n\geqslant 4)$ admits isometric imbedding into two $(n+1)$-dimensional spaces of constant curvature, $M^n$ is conformally flat, i. e., $C_{abcd}=0$, and (36) is a quadratic equation in $K_0$, we have the following result:

*If a Riemannian manifold $M^n$ admits isometric imbedding into three $(n+1)$-dimensional spaces of different constant curvatures, and if the rank of the matrix*

$$(R_{ijkl}-a(g_{ik}g_{jl}-g_{il}g_{jk}))$$

*is $\geqslant 4$, $R+n(n-1)a\neq 0$, where a is a constant, then $M^n$ admits isometric imbedding into a space of constant curvature a, and $M^n$ is of class one and conformally flat.*

**6.** From the foregoing results, a Riemannian manifold which is not conformally flat does not admit isometric imbedding into two spaces $S^{n+1}$ of different constant curvatures, and any $M^n$ of constant curvature admits isometric imbedding into a space $S^{n+1}$ of any constant curvature. It arises a question: does there exist a conformally flat $M^n$ which is not of constant curvature but admits isometric imbedding into two spaces of different constant curvature? The answer is affirmative. In fact, we can prove that if a conformally flat $M^n$ admits isometric imbedding into a space $S^{n+1}$ of constant curvature, it admits also isometric imbedding into a space $S^{n+1}$ of any constant curvature. In other words, this $M^n$ is of class one. Conversely, any conformally flat $M^n$ of class one admits isometric imbedding into a space $S^{n+1}$ of any constant curvature. According to Theorem 3, any Riemannian manifold $M^n$ $(n \geqslant 4)$ of constant curvature admits isometric imbedding into a space $S^{n+1}$ of any constant curvature. And any $M^n$ of constant curvature is conformally flat and of class one. So that the conformally flat $M^n$ of class one may be seen as an extension of the spaces of constant curvature.

We proceed to establish this conclusion.

It is well known that the manifold $M^n$ $(n \geqslant 4)$ is conformally flat if and only if $C_{ijkl}=0$ or by (33)

$$R_{ijkl}=g_{il}d_{jk}+g_{jk}d_{il}-g_{ik}d_{jl}-g_{jl}d_{ik}, \tag{37}$$

where

$$d_{jk}=\frac{1}{n-2}R_{jk}-\frac{R}{2(n-1)(n-2)}g_{jk}. \tag{38}$$

It is easy to show that (38) is algebraically consequence of (37).

If we put

$$\Delta \equiv g^{ij}d_{ij}, \quad d_i^t=g^{th}d_{hl}, \quad D_{ij}\equiv g^{pk}d_{pi}d_{kj}=d_i^k d_{kj}, \quad D\equiv g^{ij}D_{ij} \tag{39}$$

and contract (38) by $g^{jk}$, we obtain

$$R=2(n-1)\Delta, \quad R_{jk}=\Delta g_{jk}+(n-2)d_{jk}. \tag{40}$$

From (37) we have

$$R^h_{jkl}=g_{jk}d^h_l+\delta^h_l d_{jk}-d^h_k g_{jl}-\delta^h_k d_{jl} \tag{41}$$

and

$$R_{ih}R^h_{jkl}=\Delta R_{ijkl}+(n-2)(d_{il}d_{jk}-d_{ik}d_{jl}+g_{jk}D_{il}-g_{jl}D_{ik}), \tag{42}$$

$$R^q_{klm}R^k_{qij}=4(d_{im}d_{lj}-d_{il}d_{mj})+2g_{lj}D_{im}+2g_{im}D_{lj}-2g_{il}D_{mj}-2g_{mj}D_{il}, \tag{43}$$

$$R^p_{ilq}R^q_{jpm}=2g_{il}D_{mj}+2g_{mj}D_{il}-2d_{jl}d_{im}-(n-4)d_{il}d_{mj}-g_{im}D_{jl}$$
$$-g_{jl}D_{im}-\Delta(g_{il}d_{jm}+d_{il}g_{jm})-Dg_{il}g_{jm}. \tag{44}$$

Since $Q_{jlim}=P_{jlim}|_{K_0=0}$, we have from (9)

$$Q_{jlim}=R_{jl}R_{im}-R_{ip}R^p_{jlm}+R_{hl}R^h_{mij}+R^p_{jlq}R^q_{ipm}-\frac{1}{2}R^q_{klm}R^k_{qij}. \tag{45}$$

Substituting (40)—(44) into (45), we find

$$Q_{jlim}=\Delta^2 g_{jl}g_{im}+(n-3)\Delta(g_{jl}d_{im}+g_{im}d_{jl})$$
$$+(n-1)g_{im}D_{jl}-(n-3)g_{jl}D_{im}+(n-2)(n-3)d_{jl}d_{im}-Dg_{jl}g_{im}. \tag{46}$$

Substituting (46) into (30) and put.

$$\lambda_{ij}\equiv d_{ij}-\frac{\Delta}{n}g_{ij}, \quad \mu_{ij}\equiv D_{ij}-\frac{D}{n}g_{ij}, \tag{47}$$

we obtain

$$P_{jlim}=(n-1)(n-2)g_{im}g_{jl}K_0^2+\Big[(n-3)g_{jl}\lambda_{im}+(n-1)g_{im}\lambda_{jl}$$
$$+\frac{4(n-1)}{n}\Delta g_{im}g_{jl}\Big](n-2)K_0+(n-2)(n-3)\lambda_{jl}\lambda_{im}$$
$$+(n-1)g_{im}\mu_{jl}-(n-3)g_{jl}\mu_{im}+\frac{2(n-1)(n-3)}{n}\Delta(g_{jl}\lambda_{im}+g_{im}\lambda_{jl})$$
$$+(n-2)\Big(\frac{4n-3}{n^2}\Delta^2-\frac{D}{n}\Big)g_{im}g_{jl}. \tag{48}$$

Multiplying (48) by $g^{im}$ and summing for $i$, $m$, we have

$$g^{im}P_{jlim}=n(n-1)(n-2)g_{jl}K_0^2+(n\lambda_{jl}+4\Delta g_{jl})(n-1)(n-2)K_0$$
$$+n(n-1)\mu_{jl}+2(n-1)(n-3)\Delta\lambda_{jl}+(n-2)\left(\frac{4n-3}{n}\Delta^2-D\right)g_{jl}. \quad (49)$$

Multiplying (48) by $g^{jl}$ and summing for $j$, $l$ we obtain

$$g^{jl}P_{jlim}=n(n-1)(n-2)g_{im}K_0^2+[n(n-3)\lambda_{im}+4(n-1)\Delta g_{im}](n-2)K_0$$
$$-n(n-3)\mu_{im}+2(n-1)(n-3)\Delta\lambda_{im}+(n-2)\left(\frac{4n-3}{n}\Delta^2-D\right)g_{im}. \quad (50)$$

Moreover, we have

$$g^{jl}g^{im}P_{jlim}=n^2(n-1)(n-2)K_0^2+4n(n-1)(n-2)\Delta K_0$$
$$+(n-2)[(4n-3)\Delta^2-nD]. \quad (51)$$

From (26) it follows that

$$P_{jlpq}P_{abim}-P_{abpq}P_{jlim}=0.$$

Multiplying by $g^{pq}g^{ab}$ and summing for $p$, $q$, $a$, $b$, by means of (48)—(51) we find

$$(g^{pq}P_{jlpq})(g^{ab}P_{abim})-(g^{ab}g^{pq}P_{abpq})P_{jlim}$$
$$=-n(n-3)\{[n\Delta^2-(n-2)^2D]\lambda_{im}+(n-1)[n(n-2)K_0$$
$$+2(n-3)\Delta]\mu_{im}\}\lambda_{jl}+n(n-1)(n-3)\{[n(n-2)K_0$$
$$+2(n-1)\Delta]\lambda_{im}-n\mu_{im}\}\mu_{jl}=0. \quad (52)$$

Therefore there exists a function $\rho$ which is independent of the indices $i$ and $m$ such that

$$\mu_{jl}=\rho\lambda_{jl}. \quad (53)$$

When $\lambda_{jl}=d_{jl}-\frac{\Delta}{n}g_{jl}=0$, $M^n$ is a space of constant curvature, as follows easily from (38). In this case $\mu_{jl}=0$ and the equation (53) or (52) reduces to an identity. In the general case, substituting (53) into (52), we have

$$\rho^2-\frac{4}{n}\Delta\rho+\frac{1}{n-1}\left[\Delta^2-\frac{(n-2)^2}{n}D\right]=0. \quad (54)$$

(53) and (54) are the necessary conditions for a conformally flat $M^n$ which admits isometric imbedding into a space $S^{n+1}$ of constant curvature. The equation obtained by eliminating $\rho$ from (53) and (54) depends only upon the metric tensor $g_{ij}$ and hence is an intrinsic necessary condition.

Eliminating $\mu_{ij}$ and $D$ from (53), (54) and (48), we obtain

$$P_{jlim}=\frac{n-3}{n-2}\left\{(n-2)\lambda_{jl}+\left[(n-2)K_0+\frac{2(n-1)}{n}\Delta-\rho\right]g_{jl}\right\}$$
$$\cdot\left\{(n-2)\lambda_{im}+\frac{n-1}{n-3}\left[(n-2)K_0+\frac{2(n-3)}{n}\Delta+\rho\right]g_{im}\right\}. \quad (55)$$

By means of (2), (5)—(7) and (55), we have

$$2P_{hphk}T_k^p-P_{hkhk}T=-\frac{(n-3)^2}{n-2}\{(n-2)K_0+2(\Delta-\rho)\}$$
$$\cdot\left\{(n-2)\lambda_{hm}+\frac{n-1}{n-3}\left[(n-2)K_0+\frac{2(n-3)}{n}\Delta+\rho\right]g_{hk}\right\}^2. \quad (56)$$

In deriving (56) we have used the following relations

$$\lambda^j_m = d^j_m - \frac{\Delta}{n}\delta^j_m,$$
$$D_{im} = d_{ij}d^j_m = \lambda_{ij}\lambda^j_m + \frac{2\Delta}{n}\lambda_{im} + \frac{\Delta^2}{n^2}g_{im}, \tag{57}$$
$$\lambda_{ij}\lambda^j_m = \left(\rho - \frac{2\Delta}{n}\right)\lambda_{im} + \frac{n-1}{(n-2)^2}\left(\rho - \frac{2\Delta}{n}\right)^2 g_{im}.$$

Substituting (55), (56) and (2) into (15), we obtain

$$\begin{aligned}
&-(n-2)\{R_{jlim} - K_0(g_{ji}g_{lm} - g_{jm}g_{li})\}\{(n-2)K_0 + 2(\Delta-\rho)\} \\
&= (n-2)^2(\lambda_{ji}\lambda_{lm} - \lambda_{jm}\lambda_{li}) \\
&\quad + (n-2)\left[(n-2)K_0 + \frac{2(n-1)}{n}\Delta - \rho\right](g_{lm}\lambda_{ji} + g_{ji}\lambda_{lm} - g_{li}\lambda_{jm} - g_{jm}\lambda_{li}) \\
&\quad + \left[(n-2)K_0 + \frac{2(n-1)}{n}\Delta - \rho\right]^2 (g_{ji}g_{lm} - g_{jm}g_{li}).
\end{aligned} \tag{58}$$

By means of (33), (34) and $C_{jlim}=0$ the equation (58) or (15) becomes

$$\begin{aligned}
2(n-2)(\Delta-\rho)R_{jlim} &= (n-2)^2(\lambda_{jm}\lambda_{li} - \lambda_{ji}\lambda_{lm}) \\
&\quad + \left[\frac{2(n-1)}{n}\Delta - \rho\right]^2 (g_{jm}g_{li} - g_{ji}g_{lm}) \\
&\quad + (n-2)\left[\frac{2(n-1)}{n}\Delta - \rho\right](g_{jm}\lambda_{li} + g_{li}\lambda_{jm} - g_{ji}\lambda_{lm} - g_{lm}\lambda_{ji}).
\end{aligned} \tag{59}$$

Eliminating $g_{jm}\lambda_{li} + g_{li}\lambda_{jm} - g_{ji}\lambda_{lm} - g_{lm}\lambda_{ji}$ from (59) and the following equation

$$R_{jlim} = g_{jm}\lambda_{li} + g_{li}\lambda_{jm} - g_{ji}\lambda_{lm} - g_{lm}\lambda_{ji} + \frac{2\Delta}{n}(g_{jm}g_{li} - g_{ji}g_{lm}), \tag{37'}$$

which is equivalent to (37), We can write (59) as follows

$$\begin{aligned}
(n-2)\left(\frac{2}{n}\Delta - \rho\right)R_{jlim} &= (n-2)^2(\lambda_{jm}\lambda_{li} - \lambda_{ji}\lambda_{lm}) \\
&\quad + \left[\frac{2(n-1)}{n}\Delta - \rho\right]\left(\frac{2}{n}\Delta - \rho\right)(g_{jm}g_{li} - g_{ji}g_{lm}).
\end{aligned} \tag{59'}$$

By means of (47) and (54) we can also write (59′) either in the form

$$(n-2)(\rho-\Delta)R_{jlim} = (n-2)^2(d_{ji}d_{lm} - d_{jm}d_{li}) + (\rho-\Delta)^2(g_{ji}g_{lm} - g_{jm}g_{li}) \tag{59''}$$

or

$$R_{jlim} = \frac{1}{2(n-2)(\rho-\Delta)}\{(R_{ji} - \rho g_{ji})(R_{lm} - \rho g_{lm}) - (R_{jm} - \rho g_{jm})(R_{li} - \rho g_{li})\} \tag{59'''}$$

or

$$(n-2)(\rho-2\Delta)R_{jlim} = R_{ji}R_{lm} - R_{jm}R_{li} + \rho(\rho-2\Delta)(g_{ji}g_{lm} - g_{jm}g_{li}). \tag{59''''}$$

In other words, when $C_{jlim}=0$ the condition (15) is independent of $K_0$, we have

**Theorem 4.** *If a conformally flat $M^n$ $(n\geqslant 4)$ admits isometric imbedding into a space $S^{n+1}$ of constant curvature and if the rank of the matrix $(R_{ijkl} - a(g_{ik}g_{jl} - g_{il}g_{jk}))$ is $\geqslant 4$, $R + n(n-1)a \neq 0$, where $a$ is a constant, then $M^n$ admits also isometric imbedding into the space $S^{n+1}$ of constant curvature $a$. This is a characteristic property of the conformally flat manifold.*

By means of the foregoing results, we have also the following

**Theorem 5.** *If the rank of the matrix* $(T_{ijkl})$ *is* $\geqslant 4$ *and* $T \neq 0$, *then* (37) *and* (59″) *are the characteristic intrinsic conditions for a conformally flat Riemannian manifold* $M^n (n \geqslant 4)$ *to admit isometric imbedding into a space* $S^{n+1}$ *of constant curvature* $K_0$.

It is remarkable that the conditions (37) and (59″) are independent of $K_0$.

It is easily seen that the condition (54) is algebraically consequence of (37) and (59″).

From (54), it follows that

**Corollary 1.** *A necessary condition that a conformally flat Riemannian space* $M^n (n \geqslant 4)$ *is of class one is that*

$$nD \geqslant \Delta^2. \tag{60}$$

From (53) we obtain

**Corollary 2.** *A necessary condition that a conformally flat* $M^n (n \geqslant 4)$ *is of class one is that*

$$\begin{vmatrix} \lambda_{jl} & \mu_{jl} \\ \lambda_{im} & \mu_{im} \end{vmatrix} = 0. \tag{61}$$

## References

[1] Allendoerfer, C. B., Einstein spaces of class one, *Bull. Amer. Math. Soc.*, **43** (1937), 265—270.

[2] Jaak Vilms, Local isometric imbedding of Riemannian $n$-manifolds into Euclidean $(n+1)$-space, *J. Diff Geometry,* **12** (1977), 197—202.

[3] Eisenhart L. P., Riemannian Geometry, Princeton, 1949.

[4] Thomas T. Y., Riemannian spaces of class one and their characterization, *Acta Math.*, **67** (1936), 169—211.

[5] Hu Hesheng, On the deformations of a Riemannian metric $V_m$ in the space $S_{m+1}$ of constant curvature (in Chinese), *Acta Mathematica Sinica,* (1956), 320—331.

J. DIFFERENTIAL GEOMETRY
**16 (1981) 559–576**

# GEOMETRICAL CLASS AND DEGREE FOR SURFACES IN THREE-SPACE

THOMAS BANCHOFF & NICOLAAS H. KUIPER

*Dedicated to Professor Buchin Su on his 80th birthday*

## 1. Definitions and theorems

We consider immersions and embeddings $f\colon M \to E^3$ of a closed surface $M$ into euclidean three-space $E^3$, in the *smooth* ($C^\infty$) and the *polyhedral* (= piecewise linear = $PL$) category. Occasionally we will mention the topological category.

For any line $l$ in $E^3$ we can choose the parallel projection $\pi_l$ of $E^3$ into an orthogonal plane $\alpha_l$. We will use polar coordinates $(r, \theta)$ and euclidean coordinates $(u, v) = (r\cos\theta, r\sin\theta)$ around $(0,0) = l \cap \alpha_l$ in that plane.

The line $l$ is said to be *transversal* to $f$ in case there is a neighborhood $U_p \subset M$ for any point $p \in f^{-1}(l) = \{q \in M\colon f(q) \in l\}$, for which the projection

$$\pi_l \circ f\colon U_p \to \alpha_l$$

is an isomorphism (diffeomorphism, etc.) onto its image, which image we can assume to be a round disc $r < \delta$ for some $\delta > 0$. Then we define the *geometrical degree* $d = d(f)$ of the surface $f$,

$$(1.1) \qquad d(f) = \sup_{l\,\text{transversal}} d_l(f) \leqslant \infty,$$

to be the least upper bound of the number of points $d_l(f)$ of $f^{-1}(l)$ for all transversal lines $l$. The number $d_l(f)$ is constant on each component of the open dense subspace of $f$-transversal lines in the Grassmann manifold Gr of all lines in $E^3$. Any transversal line can be moved into the position of any other one in such a way that the number of points $d_l(f)$ changes by an even number at isolated times at which the line is not transversal. In $E^3$ we can moreover

Received December 17, 1981. The authors had the benefit of conversations with E. Calabi, Y. Kergosien, W. Kühnel, W. Pohl and R. Thom.

move $l$ away from $f(M)$, so that eventually $d_l(f)$ is zero. We easily conclude

**Lemma 1.** *The geometrical degree $d(f) \leqslant \infty$ is even for smooth or PL immersed closed surfaces $f\colon M \to E^3$. If $d(f) = 2$, then $f$ is an embedding of the 2-sphere onto a convex surface. The degree $d(f)$ is finite for a PL-surface.*

Observe also that if $h(x, y, z)$ is a polynomial of degree $m$ in euclidean coordinates $x, y, z$ for $E^3$, and $f\colon M \to E^3$ is an embedding into the real algebraic variety with equation $h(x, y, z) = 0$, then

(1.2) $$\text{geometrical degree } d(f) \leqslant \text{algebraic degree } m,$$

(over **C** the two degrees are equal).

In order to define the geometrical class of a surface $f$, we examine (and define) the "number of tangent planes $c_l(f)$" through a general transversal line $l$ as follows. The function $\tilde{\theta}_l = \tilde{\theta} = \theta \circ f$ modulo $\pi$, in terms of the polar coordinate $\theta$, is well defined on $M \setminus f^{-1}(l) = \{q \in M\colon f(q) \notin l\}$ with values in the circle $\mathbf{R}/\pi\mathbf{Z}$. It has no critical points near $f^{-1}(l)$ and it is constant on each plane through $l$. We call the transversal line $l$ *general* for a *smooth* surface in case the function $\tilde{\theta}_l$ is nondegenerate. We denote by $c_l^+(f)$ the number of critical points with relative extreme value (maximum or minimum), and by $c_l^-(f)$ the number of other critical values (saddle points), and we put $c_l(f) = c_l^+(f) + c_l^-(f)$.

We call the transversal line $l$ *general* for a *piecewise linear* surface (the pieces are the simplices of some triangulation of $M$) in case the function $\tilde{\theta}_l$ has only isolated critical points, say $c_l^+(f)$ with relative extreme values, $c_l^-(f)$ others, with $c_l(f) = c_l^+(f) + c_l^-(f)$ as "total number of tangent planes". But in the $PL$-case the level curve $\{q \in M\colon \tilde{\theta}_l(f(q)) = \gamma\}$ of a critical nonextreme value $\gamma$ is seen to consist near the isolated critical point $p$ of an even number $2(m+1)$ of straight line segments ($m \geqslant 1$) each ending at $p \in M$, and then that critical point must be counted with *multiplicity* $m$ in $c_l^-(f)$ in order to agree with the Morse theory to be applied later. Compare Kuiper [6].

In the $C^\infty$ and $PL$-categories the *general lines* form an open dense set in Gr, and $c_l^+(f)$, $c_l^-(f)$, $d_l(f)$ are constant on each component of this subspace.

**Definition.** The *geometrical class* of the immersion $f\colon M \to E$ is the least upper bound

(1.3) $$c = c(f) = \sup\nolimits_{l\,\text{general}} c_l(f)$$

We also put

(1.3)′ $$c^+(f) = \sup\nolimits_{l\,\text{general}} c_l^+(f),$$

(1.3)″ $$c^-(f) = \sup\nolimits_{l\,\text{general}} c_l^-(f).$$

Clearly

$$c^+(f) + c^-(f) \geq c(f) \geq c^{\pm}(f).$$

In §2 we prove the main theorem relating class and degree:

**Theorem 1.** *For any smooth or polyhedral immersion $f\colon M \to E^3$ of a closed surface, we have*

$$c(f) \geq d(f) - \chi(M) \tag{1.4}$$

*where $\chi(M)$ is the Euler characteristic of $M$.*

Our interest in the class of a surface originated in its relation to tightness. For a map of a *closed surface* into euclidean space tightness has a simple characterization. The map $f\colon M \to E^3$ is *tight* if the set

$$f^{-1}(h) = \{q \in M\colon f(q) \in h\}$$

is connected for any closed half space $h$ bounded by a plane. A smooth immersion $f$ in $E^3$ can be shown to be tight if and only if the total absolute curvature

$$\frac{1}{2\pi}\int |K\,d\sigma|,$$

which is $\geq 4 - \chi(M)$, attains its minimal value $4 - \chi(M)$. See [6]. We also prove, in §2,

**Theorem 2.** *The geometrical class of a TIGHT smooth or polyhedral immersion $f\colon M \to E^3$ of degree $d(f) \geq 4$ satisfies the equality*

$$c(f) = c^-(f) = d(f) - \chi(M). \tag{1.5}$$

By Lemma 1 the assumption $d(f) \geq 4$ only excludes convex surfaces ($S^2$). If $l$ is a general line which does not meet the convex hull $\mathcal{H}f(M)$ of $f(M)$, then tightness of $f$ is seen to imply $c_l^+(f) = 2$, as there could not be more than two "extreme" tangent planes through $l$. Then by the Morse equality $c_l^+(f) - c_l^-(f) = \chi(M)$, we obtain

$$c_l(f) = 4 - \chi(M). \tag{1.6}$$

Even so $c(f)$ can take bigger values as seen in the end of §4.

**Theorem 2a.** *For a tight smooth immersed surface $f\colon M \to E^3$, if the Euler characteristic $\chi(M) \neq 2$, then the geometrical class can be made as large as we please by a $C^2$-small perturbation, while preserving tightness.*

This answers problem 4 in [7] in the negative.

In the rest of the paper we are interested in immersions and embeddings $f$ for which $d(f)$ and $c(f)$ are small, and which are tight as well if possible. In the polyhedral case and for *orientable surfaces all wishes can be attained.*

**Theorem 3.** *There are tight polyhedral imbeddings $f\colon M \to E^3$ of geometrical degree* 4 *and class* $4 - \chi(M)$ *for any orientable surface $M$ of genus $g \geqslant 0$.*

*Construction.* For $g = 0$ we take the surface of a cube. For $g = 1$ we delete from the cube the convex hull of the union of two concentric equally large squares in opposite faces, and take the boundary of the remaining ring. See Fig. 1a.

For $g \geqslant 2$ we proceed as follows. See Fig. 1b. Let $B_1$ be the convex body which is a vertical cylinder with basis the two-disc of a regular $g + 1 - gon$, $g \geqslant 2$, and constant height. Consider in each vertical rectangular face a concentric rectangle $D_i$, $i = 0, 1, \cdots, g$, and let these rectangles all have the same size. The convex hull

$$B_2 = \mathcal{H}(D_0 \cup D_1 \cup \cdots \cup D_g)$$

is a second convex body. Then the boundary of the closure of $B_1 \setminus B_2$, $f(M) = M = \partial(\overline{B_1 \setminus B_2})$, is a polyhedral, tight(!) surface of genus $g$. Its geometrical degree is $d(f) = 4$, because $M$ is contained in $\partial B_1 \cup \partial B_2$, the union of two convex surfaces, and its class is $4 - \chi(M)$ by Theorem 2.

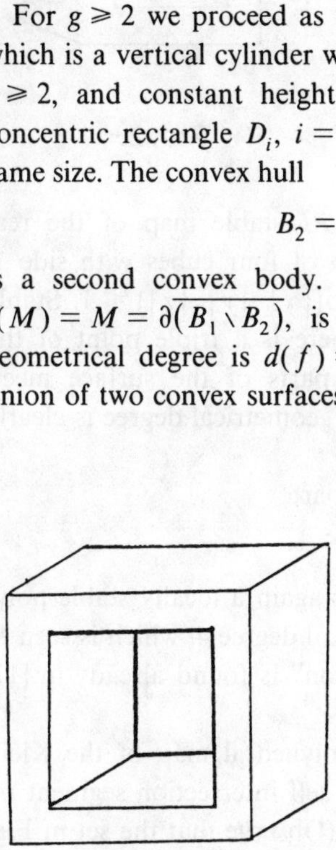

FIG. 1a

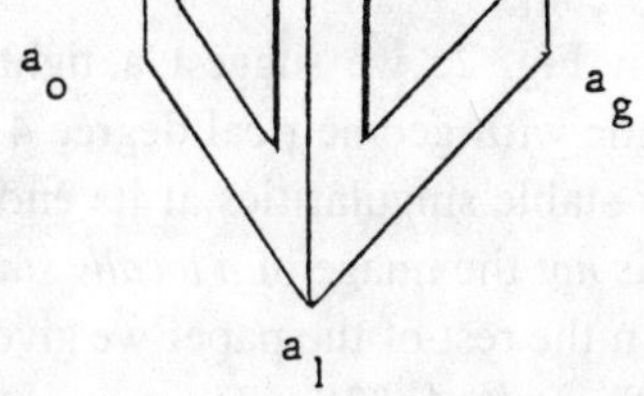

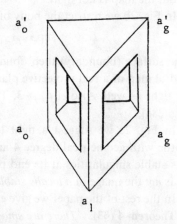

FIG. 1b

Nonorientable closed surfaces $M$ seem to have no immersions of gemetrical degree 4, and we can prove that for $\chi(M)$ odd. However if we allow "locally stable" maps (generalizing immersions), we do have examples for the projective plane and the Klein bottle. Without going into the definition we recall that a smooth or $PL$ map $f; \mathbf{R}^2 \to \mathbf{R}^3$ is stable but not an immersion at $0 \in \mathbf{R}^2$, if it is homeomorphic near $0 \in \mathbf{R}^2$ to a cone on a circle immersion with one transversal self intersection in a plane $\alpha$ not containing $f(0)$.

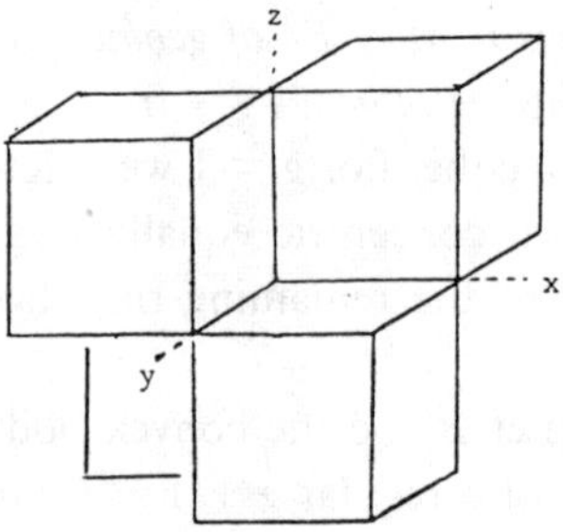


FIG. 2a

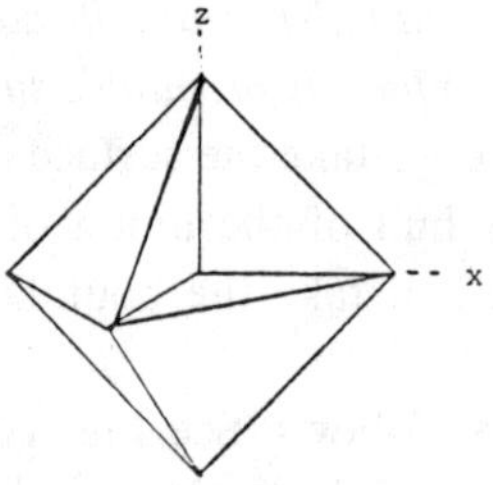


FIG. 2b

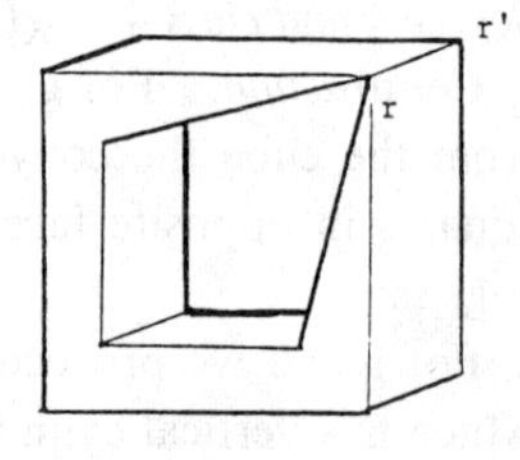


FIG. 2c

In Fig. 2a we have the *Petrobras surface* a *PL*-stable map of the real projective plane onto the boundary of the union of four cubes with side 1, placed as indicated in the cube with equation $\sup\{|x|, |y|, |z|\} \leqslant 1$. Stable singularities are at $(x, y, z) = (\pm 1, 0, 0)$, etc. There is a triple point of the surface at $(0, 0, 0)$ where the coordinate plane parts of the surface meet. Self-intersections are on the coordinate axes. The geometrical degree is clearly 4, but the map is not tight.

In Fig. 2b we restrict the body of Fig. 2a to the part

$$\sup_{e^2=f^2=1} |x + ey + fz| \leqslant 1.$$

The surface (boundary) then obtained determines again a locally stable polyhedral map of a real projective plane, of geometrical degree 4, which is seen to be tight as well. Also $c(f) = 3$. This "heptahedron" is found already in [11, Fig. 288].

In Fig. 2c we suggest a tight locally stable polyhedral map of the Klein bottle with geometrical degree 4 and class 4. The self intersection segment $rr'$ has stable singularities at its end points $r$ and $r'$. (Observe that the set in Fig. 2c is *not* the image of *a locally stable map of a torus*.)

In the rest of the paper we give examples for smooth surfaces.

**Theorem 4** (§3). *There are smooth (even locally algebraic) embeddings in* $E^3$ *of geometrical degree* 4 *for orientable closed surfaces of any genus* $g \geqslant 1$.

This was announced in [10] by R. Thom as an insight of E. Calabi.

We recall also in §3 that there is a real algebraic model (surface) of algebraic degree 4 for any orientable surface of genus $g \leqslant 7$.

**Theorem 5** (§4). *There are smooth (even locally algebraic) TIGHT embeddings in* $E^3$ *of geometrical degree* 6 *for orientable surfaces of any genus* $g \geqslant 2$.

**Theorem 6** (§5). *There is a smooth TIGHT embedding in* $E^3$ *of geometrical degree* 4 *for the orientable surface of genus* $g = 2$. *The given example is a component of an algebraic variety of algebraic degree* 5.

**Conjecture.** There is no tight smooth surface of geometrical degree 4 in $E^3$ of genus $g > 3$. Calabi constructs an example of genus 3.

We will make some comments about the robustness of the degree and of tightness under $C^4$-small perturbations of our examples in §3, 4, 5.

**Generalizations.** 1. Lemma 1 and Theorems 1 and 2 remain true for smooth and *PL* immersions of a closed surface $M$ into euclidean $N$-space $E^N$, $N \geqslant 3$, with suitable modifications in the definitions and proofs. In particular instead of lines $l$ one takes affine subspaces of codimension 2 in $E^N$.

2. Theorem 1 remains true for immersions of a closed surface $M$ into real projective $N$-space $\mathbf{R}P^N$, $N \geqslant 3$, with suitable modifications in the definitions. The degree $d(f)$ can be odd, for example $d(f) = 1$ for the straight projective plane $\mathbf{R}P^2$ in $\mathbf{R}P^N$. Surfaces of geometrical degree 2 in $\mathbf{R}P^3$ are either convex surfaces of algebraic quadratic hyperboloid surfaces. The last are homeomorphic to the torus. We intend to discuss immersions of nonorientable surfaces of small degree in another paper.

## 2. The relation between class and degree

We recall the Morse equality: If $\varphi: M \to \mathbf{R}$ is a nondegenerate smooth function on a closed manifold with $\mu_j$ critical points of index $j$, then

$$\sum_j (-1)^j \mu_j = \chi(M).$$

For a *surface* put $\mu_0 + \mu_2 = c^+$ and $\mu_1 = c^-$. Then

$$c^+ - c^- = \chi(M). \tag{2.1}$$

Suitably modified the same holds for continuous functions with isolated critical points on a closed surface, for example on a polyhedral surface. We now prove

**The litte Morse equality:** (2.1) *holds also for maps of a closed surface $M$ into the circle $S^1$ with all critical points isolated* (*see* [12]).

*Proof.* Let the map $\varphi: M \to S^1 (= \mathbf{R}/\pi\mathbf{Z}$ say) have $c(\varphi) = c^+(\varphi) + c^-(\varphi)$ isolated critical points, all other points assumed regular for $\varphi$. Let $[s - 2\varepsilon, s + 2\varepsilon]$, $\varepsilon > 0$, be an interval of regular values. $\varphi^{-1}(s)$ is a union of $k (\geqslant 0)$ circles, $\varphi^{-1}([s - 2, \varepsilon, s + 2\varepsilon])$ is a union of $k$ orientable bands in $M$. Delete these bands and close each of the $2k$ bounding circles of the rest of $M$ by a cone, so that a new abstract surface $M'$ is obtained. Extend the function $\varphi$ over the cones to obtain $\varphi'$ so that the only new critical points of $\varphi'$ are $k$ relative maxima with value $s - \varepsilon$ and $k$ relative minima with value $s + \varepsilon$, in the obvious way. The new function $\varphi'$ on the new manifold $M'$, for which $\chi(M') = \chi(M) + 2k$, has invariants:

$$c^+(\varphi') = c^+(\varphi) + 2k, \quad c^-(\varphi') = c^-(\varphi).$$

As $\varphi'$ maps into an interval (embeddable in **R**) the usual Morse equality holds (see [6]):

$$\chi(M) + 2k = \chi(M') = c^+(\varphi') - c^-(\varphi') = c^+(\varphi) + 2k - c^-(\varphi).$$

Hence the little Morse equality follows.

Next let $f\colon M \to E^3$ be a smooth or polyhedral immersion of a closed surface $M$ and $l$ a *general line*. We prove with $d_l = d_l(f)$, $c_l^- = c_l^-(f)$, etc.:

**Lemma 2.1.** $c_l^- = d_l - \chi(M) + c_l^+ \geqslant d_l - \chi(M)$.

*Proof*. See §1 for the meaning of round neighborhoods $U_p$, $p \in f^{-1}(l)$, and of $\tilde{\theta}_l = \theta \circ f$. Take out such round neighborhoods $U_1, \cdots, U_{d_l}$, $d_l = 2m$ from $M$, one for each point $p \in f^{-1}(l)$, and connect $\partial U_{2j-1}$ and $\partial U_{2j}$ by a cylinder to obtain a new abstract surface $M'$ with (clearly)

$$\chi(M') = \chi(M) - d_l.$$

Assume the discs $f(U_{2j-1})$ and $f(U_{2j})$ disjoint, and extend $f$ to get $f'$, mapping the cylinders onto tubes around $l$. Then $f'(M')$ avoids $l$, and $\tilde{\theta}_l' = \theta \circ f'$ maps $M'$ into the circle $\mathbf{R}/\pi\mathbf{Z}$. We may assume that the critical points of $\theta$ on $f'(M')$ are the same as those of $\theta$ on $f(M)$. Since $\theta_l'$ is defined at all points of $M'$, the little Morse equality holds and we have

$$c_l^+ - c_l^- = \chi(M') = \chi(M) - d_l.$$

Hence Lemma 2.1 follows.

*Another proof*. The above proof does not apply in case $d_l$ is odd, which may happen with immersions into $\mathbf{R}P^3$. Therefore we present a second proof. Replace $M$ by a new abstract surface $M'$, by taking out the open round disc $U_p$ for $p \in f^{-1}(l)$ again, and identifying diametrical points on $\partial U_p$. Two such diametrical points have the same value of $\bar{\theta}_l \circ f$ in $\mathbf{R}/\pi\mathbf{Z}$. The new manifold $M'$ (obtained by $d_l$ such "blowing ups", as they are called) has $\chi(M') = \chi(M) - d_l$, and the function $\tilde{\theta}_l \circ f$ has the same critical points as on $M$(!); Lemma 2.1 follows as before.

Theorem 1 follows from Lemma 2.1 by taking least upper bounds for general lines $l$:

$$c(f) \geqslant c^-(f) \geqslant d(f) - \chi(M).$$

*Proof of Theorem* 2. We now assume that $f$ is tight and $d(f) \geqslant 4$. For a general line $l$ which does not meet the convex hull $\mathcal{H}f(M)$ we observed already

$$c_l^+ = 2, c_l = 4 - \chi(M).$$

For a general line $l$ which does meet $\mathcal{H}f(M)$, clearly by definition of tightness, $c_l^+ = 0$ and, by Lemma 2.1

$$c_l = c_l^- = d_l - \chi(M).$$

Taking upper bounds for *all* general $l$ we get the equality of Theorem 2:

$$c(f) = c^-(f) = d(f) - \chi(M).$$

## 3. Smooth (even locally algebraic) orientable surfaces of geometrical degree 4

*A smooth two-sphere or revolution of degree* 4. Any tangent line $l$ in an asymptotic direction of a piece of smooth surface of negative Gauss curvature has in general local contact of order three, and a small perturbation of $l$ can give a contribution *three* to the degree $d_l$. Sweeping around such asymptotic tangents it seems hard to obtain $d(f) = 4$ for surfaces of high genus $g \geqslant 2$.

It is also interesting to observe what can happen if one rotates a convex curve $\gamma$ in a vertical plane around a disjoint vertical line in that plane. If $\gamma$ is a round circle we get the standard torus of algebraic and geometrical degree 4. If $\gamma$ is a square with two vertical sides, or a smooth approximation like the curve with equation

$$|x|^p + |y|^p = 1, \quad p \geqslant 4,$$

we get again an embedded torus with geometrical degree 4, as W. Kühnel observed for $p \geqslant 4$. But if we place the diagonal of the square, or a smooth approximation, parallel to the vertical axis, then the embedded torus of revolution so obtained has geometrical degree *at least six*. Look and see!

Even so, following a conviction of E. Calabi, we now proceed to get smooth surfaces of degree four of any genus $g$. We start from the following real algebraic variety $M_0$ of algebraic degree four in euclidean coordinates $x, y, z$ for $E^3$:

$$(3.1) \qquad (r^2-1)^2 + \varepsilon^4(1+\varepsilon)\left(\frac{z+1}{2}\right)^4 = \varepsilon^4, \qquad r^2 = x^2+y^2+z^2, \quad \varepsilon > 0 \text{ small.}$$

The planes $z$ = constant are said to be horizontal. $M_0$ is contained in the thick two-sphere

$$S\colon (r^2-1)^2 \leqslant \varepsilon^4,$$

and also in the convex domain

$$D: \left(\frac{z+1}{2}\right)^4 \leqslant \frac{1}{1+\varepsilon},$$

which for small $\varepsilon > 0$ cuts away a cap from the top of the thick two-sphere.

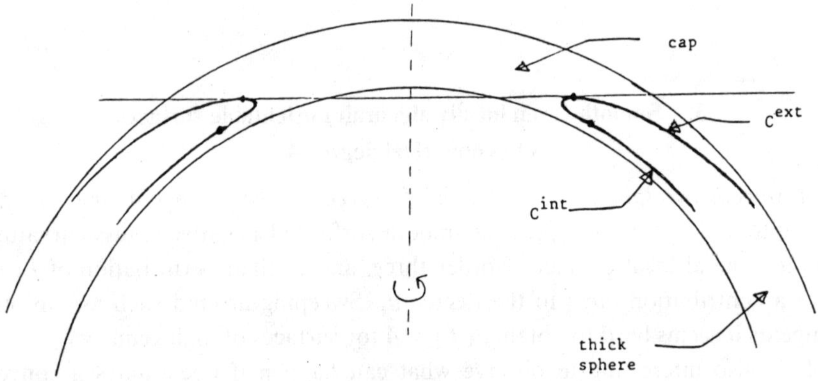


FIG. 3a

Half rays from the origin (0, 0, 0) give two or no intersection points with $M_0$. We have an embedded two-sphere, and a surface of revolution of algebraic and hence geometric degree 4. See Fig. 3a.

The surface has two large convex parts where $K \geqslant 0$; one exterior $C^{\text{ext}}$ and one interior $C^{\text{int}}$. The asymptotic curves in the complement are tangent to the boundary of $C^{\text{ext}}$ where the Gauss curvature is $K = 0$. They show a pattern as on the standard torus. It is not generic from the differential point of view as Thom and Kergosien observed in [3].

Near the boundary of $C^{\text{int}}$ the asymptotic curves end in cusps, orthogonal to that boundary. Now for small $\varepsilon$, all asymptotic tangents are nearly horizontal (!). That is why they cannot harm in the surface with respect to degree at points of $M_0$ which are far away. We use this in our constructions later. The degree is not influenced if we perturb only $C^{\text{ext}}$ and $C^{\text{int}}$ keeping them convex, and unchanged near their boundaries. We may do this in such a way that for some small $\delta > 0$ and $z \leqslant 1 - \delta$, the surface coincides there with parts of two concentric spheres with radii $r_1$ and $r_2$ close to 1 and $|r_2 - r_1|$ very small.

We call the part outside these two spherical parts a *Calabi handle* attached to the pair of spheres. For small $\varepsilon > 0$, it is as small as we please. The surface is clearly not tight, but is still an embedded 2-sphere of revolution of geometrical degree 4.

Finally for small $\varepsilon$ we can take two concentric spheres with radii $r_1$ and $r_2$, with $|r_2 - r_1|$ very small, and by attaching $g + 1$ disjoint Calabi handles we obtain a smooth orientable embedded surface of genus $g$, which is seen to be of geometrical degree 4.

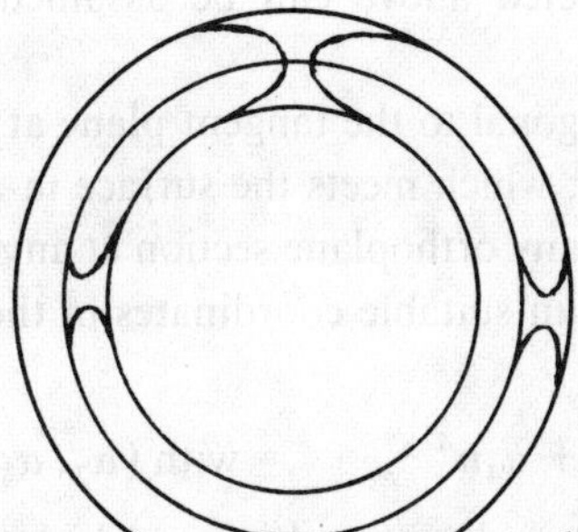

($g = 1$ Klein bottle)

FIG. 3b

*A smooth Klein bottle of geometrical degree six*. By taking three concentric spheres with radii very close to each other, we can connect by suitably attaching $g + 2$ Calabi handles and obtain a smooth immersion of geometrical degree six for any nonorientable surface of even Euler characteristic, $\chi(M) = -2(g-1)$.

*Real algebraic surfaces of algebraic degree* 4. These have been studied and classified by Gudkov [2], Kharlamov [4], [5]. The sum of the Betti numbers $s \leqslant 20$, but $\leqslant 16$ for connected surfaces, so that the genus is $g \leqslant 7$. See Thom [9] and Milnor [8] for earlier but weaker inequalities. We can obtain these surfaces with formulas like (3.1). Here are examples.

*A "regular" surface with the symmetry of the cube, of genus* 5 *and degree* 4 is given by

$$(3.2)\qquad \begin{aligned} &(r^2-1)^2 + \varepsilon^4(1+\varepsilon)(x^4+y^4+z^4) = \varepsilon^4, \\ &r^2 = x^2+y^2+z^2, \quad \varepsilon > 0 \text{ small.} \end{aligned}$$

Observe that the unit sphere $r = 1$ is interior to the surface $x^4 + y^4 + z^4 = 1$ and tangent at exactly 6 points where caps from the thick sphere are taken away.

*A "regular" surface with the same symmetry of genus* 7 *and degree* 4 *is given by*

$$(r^2-1)^2 + \varepsilon^4(1+\varepsilon)\cdot\frac{81}{84}\sum_{e^2=f^2=1}(x+ey+fz)^4 = \varepsilon^4,$$

where the analogous tangencies are at points $(x, y, z) = \frac{1}{\sqrt{3}}(\pm 1, \pm 1, \pm 1)$.

**Exercise.** Find a regular algebraic surface of algebraic degree 4 of genus 3, with the symmetry of the regular simplex.

*Algebraic surfaces of geometrical degree* 4 *of any genus g. $C^4$-robustness of geometrical degree*. Those smooth surfaces of geometrical degree 4 and genus $g \geqslant 2$ which we constructed above can be assumed to have the following property.

A plane which is orthogonal to the tangent plane at a point $p$ of a surface is called an orthoplane at $p$, which meets the surface in an *orthoplane section*. In our examples the 4-jet of any orthoplane section at any point $p$ of the surface is not flat. This means that in suitable coordinates in the orthoplane the section has an equation

$$v = \alpha_2 u^2 + \alpha_3 u^3 + \alpha_4 u^4 + \cdots, \quad \text{with } (\alpha_2, \alpha_3, \alpha_4) \neq (0,0,0).$$

This property is invariant under perturbations which are $C^4$-small: all derivatives of order 1, 2, 3, and 4 are small. This implies that locally the degree of our surfaces does not increase with $C^4$-small perburbations. Since in our examples, there is also no possible increase globally under such perturbations, the degree four remains unchanged. In other words, the *degree four of our surfaces* is a *$C^4$-robust property*. Then as we can $C^4$-approximate by (a component of) an algebraic variety, we find *(locally-)algebraic models of geometrical degree four, for orientable surfaces of any genus g*.

**Exercise (Conjecture).** If we choose $N$ sufficiently large, and then $\varepsilon > 0$ sufficiently small, the following equation defines an algebraic surface of genus $g$ and (probably) of geometrical degree 4:

$$(r^2 - 1)^2 + \varepsilon^4(1 + \varepsilon) \cdot \frac{1}{2} \sum_{j=0}^{g} \left(1 + \cos\frac{2\pi j}{g+1} + y \sin\frac{2\pi j}{g+1}\right)^N = \varepsilon^4.$$

## 4. TIGHT smooth (even locally algebraic) orientable surfaces of geometrical degree six and any genus $g$

We first construct *smooth surfaces*. Start from a round torus of degree 4 obtained by rotating a circle around a disjoint line in its plane. Any projective transform is also tight and of degree 4. Moreover the strictly convex part where $K > 0$ can be modified at will as long as it remains strictly convex, i.e., $K > 0$. Choose two points $p$ and $q$ on a round sphere. Place a projectively transformed standard torus such that its negative curvature-part approximates the line segment $pq$ and such that the two tangent planes along the points where $K = 0$ cut the sphere, and are parallel to and near to the tangent planes to the sphere at $p$ and $q$. With a modification we can arrange that the convex part of the surface contains almost all of $S^2$ except two small neighborhoods $U_p$ and $U_q$ of $p$ and $q$. We have attached one handle to the two-sphere $S^2$ near $p$ and $q$ so that the geometrical degree of the surface obtained is four. The new surface is tight: at each point where the Gauss curvature is $K > 0$, there is a globally supporting half-space $h$.

Next we attach $g$ such mutually disjoint thin handles near to chords $p_j q_j$, $j = 1, \cdots, g$, to the round sphere, and such that no straight line meets more than two handles. We get a surface $M_g$ of genus $g$, which is tight and whose geometrical degree is seen to be six. The geometrical class is then $6 - \chi(M_g)$ by Theorem 2. See Fig. 4.

*Robustness and algebraic examples*. The degree six of these surfaces is as before a $C^4$-robust property. Tightness on the other hand is not robust at all. However, if we can keep some high order jet at the union of the plane (!) sets of points where $K = 0$, fixed, then we have enough freedom for a $C^4$-small perturbation to obtain tight locally algebraic surfaces of geometrical degree six. Here is an explicit construction.

Let $f_1 = 0, \cdots, f_g = 0$ be the polynomial equations of the disjoint projective transforms of a standard torus with $f_j > 0$ representing the unbounded component of the complement for $j = 1, \cdots, g$. Each torus has two highly supporting half-spaces given by linear inequalities in the coordinates

$$h_i \geqslant 0, \quad i = 2j - 1, 2j; j = 1, \cdots, g.$$

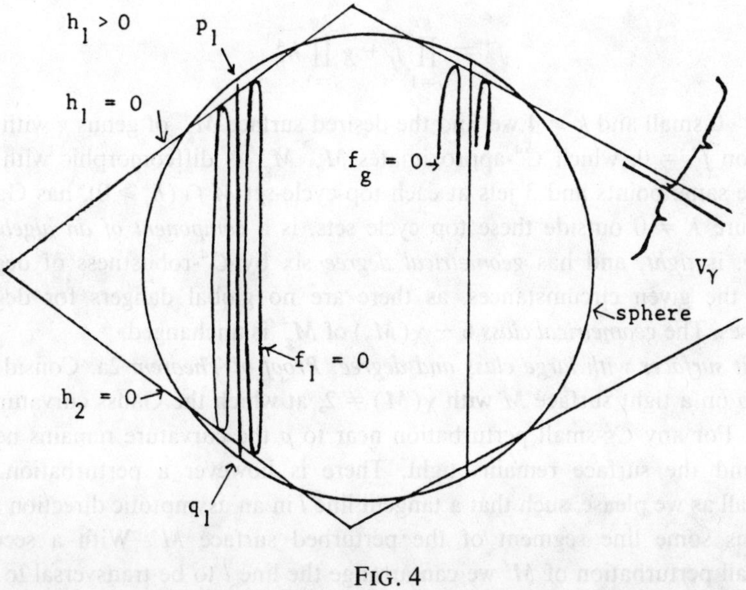


FIG. 4

The smooth surface we constructed has an equation $f = 0$ for some $C^\infty$-function $f$, for which we can assume by construction that it coincides with the polynomial function $\prod_{j=1}^{g} f_j$ *in the union* $U_\gamma$ of the two sets

$$\{p\colon \sup_j h_j(p) \geqslant 0\}, \quad \left\{q\colon \prod_{i=1}^{2g} h_i^4(p) \leqslant \gamma\right\}$$

for some small $\gamma > 0$. The complement $V_\gamma = E^3 \setminus U_\gamma$ of $U_\gamma$ in $E^3$ is homeomorphic to an open ball.

Define the $C^\infty$-function $\varphi\colon E^3 \to \mathbf{R}$ as zero on $U_\gamma$ and by

$$\varphi = \frac{f - \prod_{j=1}^{g} f_j}{\prod_{i=1}^{2g} h_i^4}$$

on $V_\gamma$. Let $g$ be a polynomial approximation of $\varphi$ on a large compact convex set $W$, which contains $V_\gamma$ and the surface $M$, such that the sum of the squares of all derivatives up to and including those or order $k$ obeys

$$\| g - \varphi \|_k^2 < \varepsilon, \quad \text{for some } \varepsilon > 0,$$

and at all points $p \in W$.

We had

$$f = \prod_{j=1}^{g} f_j + \varphi \cdot \prod_{i=1}^{2g} h_i^4,$$

and obtain the polynomial approximation on $W$

$$f^+ = \prod_{j=1}^{g} f_j + g \prod_{i=1}^{2g} h_i^4.$$

For $\varepsilon > 0$ small and $k \geqslant 4$ we find the desired surface $M_g^+$ of genus $g$ with the equation $f^+ = 0$, which $C^4$-approximates $M_g$. $M_g^+$ is diffeomorphic with $M$, has the same points and 3 jets at each top-cycle-set $M \cap (h_i \geqslant 0)$, has Gauss curvature $K \neq 0$ outside these top cycle sets, is a *component of an algebraic variety*, is *tight*, and has *geometrical degree* six by $C^4$-robustness of degree under the given circumstances, as there are no global dangers for degree increases. The *geometrical class* $6 - \chi(M_g)$ of $M_g^+$ is unchanged.

*Tight surfaces with large class and degree. Proof of Theorem* 2a. Consider a point $p$ on a tight surface $M$ with $\chi(M) \neq 2$, at which the Gauss curvature is $K < 0$. For any $C^2$-small perturbation near to $p$ the curvature remains negative, and the surface remains tight. There is however a perturbation, as $C^2$-small as we please, such that a tangent line $l$ in an asymptotic direction at $p$ contains some line segment of the perturbed surface $M'$. With a second $C^2$-small perturbation of $M'$ we can arrange the line $l$ to be transversal to the new surface $M''$ at as many points as we please. Then the degree and, by Theorem 1, also the class of our tight surface $M''$ are as large as we please. It could be made $\infty$.

## 5. A TIGHT smooth embedding with geometrical degree four of a surface of genus two

We construct a smooth surface $M = M_2$ of genus two with properties and symmetries as those of the example in Fig. 1b. It will turn out to be a component of the real algebraic variety of algebraic degree five with equation

$$\{(x, y, z): -1 \leqslant z \leqslant 1\} \subset \mathbf{R}^3, \tag{5.1}$$
$$2y(y^2 - 3x^2)(1 - z^2) + (x^2 + y^2)^2 = (9z^2 - 1)(1 - z^2).$$

*Construction and geometry of M.* Consider the function, on the $x$, $y$-plane,

$$\eta(x, y) = 2y(y^2 - 3x^2) + (x^2 + y^2)^2. \tag{5.2}$$

This function has the symmetry of the equilateral triangle, as $y(y^2 - 3x^2)$ is the real part of $(y + ix)^3$. We want the level curves of our surface $M$ to be similar to the level curves of $\eta$. The graph of the function $\eta$ as a monkey saddle at $x = y = 0$, and its asymptotic curves are nearly orthogonal to each other close to $(0, 0)$. Some levels are given in Fig. 5a.

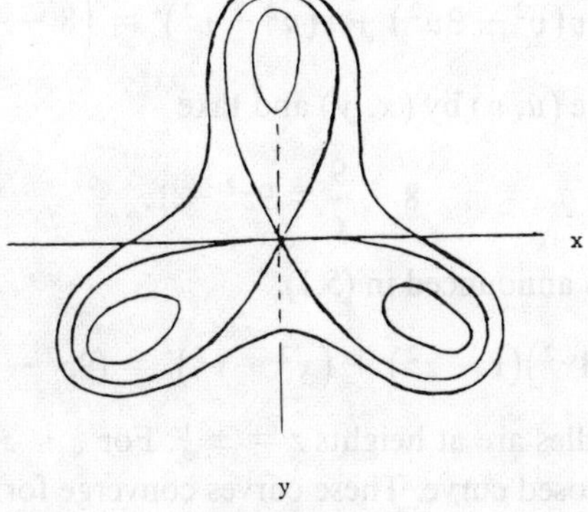


FIG. 5a

If we put $\eta = z^2 - 1$, we get monkey saddles at heights $z = +1$ and $-1$, and three vertical tubes in between. But this surface does not close either above or below. We proceed to change levels by similarities. As in Fig. 1b we do want three highly supporting vertical planes for tightness. To prepare this we determine double tangents $y = c$ for each level curve of $\eta$, and we arrange by a similarity $x = cu$, $y = cv$ that these level curves are tangent to the vertical plane $y = 1$. Then automatically by rotational-symmetry of order three, they are doubly tangent also to the other two corresponding vertical planes. Let us calculate the intersection of the plane $y = c$ and the level curve

$$2y(y^2 - 3x^2) + (x^2 + y^2)^2 = \eta = \text{constant}.$$

We get

$$x^4 + 2c^2x^2 + c^4 + 2c^3 - 6cx^2 - \eta = 0,$$

that is,

$$\left[x^2 + c(c-3)\right]^2 + c^2(8c - 9) - \eta = 0.$$

In order for $y = c$ to be a double tangent we must have

$$\eta = c^2(8c - 9). \tag{5.3}$$

For any $\eta$ we only want the largest value of $c$. Now we substitute $x = cu$ and $y = cv$ in

$$2y(y^2 - 3x^2) + (x^2 + y^2)^2 = c^2(8c - 9),$$

and find

$$2v(v^2 - 3u^2)c^3 + (u^2 + v^2)^2c^4 = c^2(8c - 9),$$

or

$$2v(v^2 - 3u^2) + c(u^2 + v^2)^2 = \left(8 - \frac{9}{c}\right).$$

Finally we replace $(u, v)$ by $(x, y)$ and take

$$8 - \frac{9}{c} = 9z^2 - 1.$$

We get the equation announced in (5.1),

$$2y(y^2 - 3x^2)(1 - z^2) + (x^2 + y^2)^2 = (9z^2 - 1)(1 - z^2).$$

The monkey saddles are at heights $z = \pm\frac{1}{3}$. For $\frac{1}{9} < z^2 < 1$ the level curves at height $z$ is one closed curve. These curves converge for $z \to \pm 1$ to the points $(x, y, z) = (0, 0, \pm 1)$. It then is geometrically seen that (5.1) defines a closed surface $M$ of genus 2.

The geometrical degree of $M$ is 4. Any line which meets $M$ transversally must meet it in an even number of points, the ends of intervals in which the line meets the body $B$ bounded by $M$. This even number is at most four since the algebraic degree of (5.1) is five. So *$M$ is of geometrical degree* 4.

Next we intersect the surface $M$ with the vertical plane $y = 1$. We find the equation

$$(x^2 + 3z^2 - 2)^2 = 0$$

of an ellipse $\mathcal{E}_1$ counted twice. It is the interaction of $M$ with the highly supporting vertical plane $y = 1$. Analogous ellipses $\mathcal{E}_2$ and $\mathcal{E}_3$ occur in the vertical planes obtained from $y = 1$ by rotation over $2\pi/3$ and $4\pi/3$ about the $z$-axis.

For making a nice model (Fig. 5b), we take $x$, $y$ and $3z$ as unit orthogonal coordinates. The union of the three ellipses $\mathcal{E}_1 \cup \mathcal{E}_2 \cup \mathcal{E}_3$ divides the surface $M$ into two open parts $M^+$ and $M^-$ each of which is orthogonally projected into the horizontal plane $z = 0$ with only folds along the intersection of $M$ with $z = 0$. The points of $M^-$ have $K < 0$ except at the monkey saddle points where $K = 0$. The points of $M^+$ are on the boundary of the convex hull $\mathcal{H}(M)$, and have $K > 0$ except at the top and bottom where $x = y = 0$.

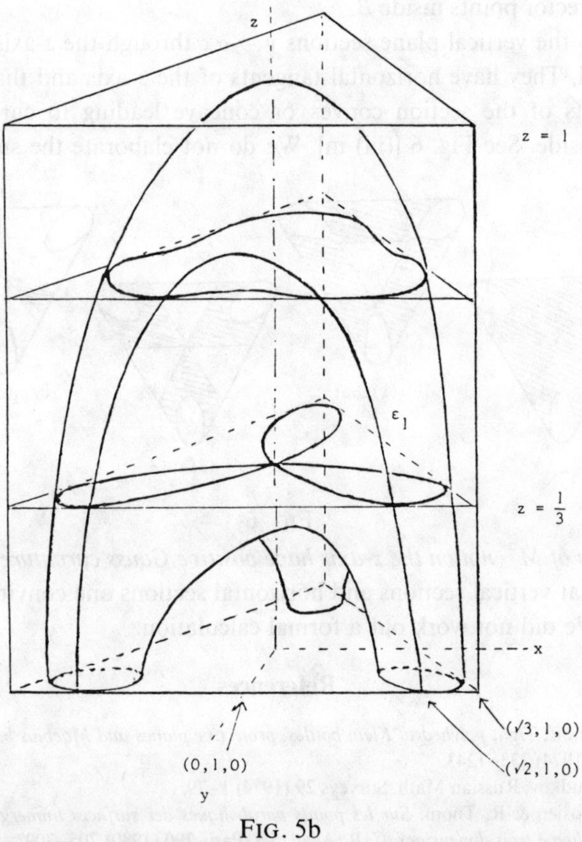


FIG. 5b

To check negative *Gauss curvature* $K < 0$, we study convenient plane sections through any point $p \in M^-$, one for which the curvature vector points inside the body $B$, and one for which the curvature vector points outside $B$. This suffices for $K(p)$ to be negative. Here are sections which do this for us in large parts of $M^-$.

(i) Take vertical planes $y = c$. By looking at horizontal sections (!) we see that such sections are at most tangent at the obvious points in the plane of symmetry $z = 0$: In between, the isotopy type of the section does not change. For $c < 1$, but close to 1, the section consists of two closed embedded curves one inside the other. As the geometrical degree of the union is 4, the inside curve is convex, and so its *curvature vector points outside B*. The same applies to the sections obtained from those discussed by rotation about the $z$-axis over $2\pi/3$ and $4\pi/3$. See Fig. 6 [(i) out], for the part of $M^-$ covered by this case.

(ii) For horizontal plane sections we get three ovals for $9z^2 - 1 < 0$, which can be shown to be convex. At the points covered (see Fig. 6 [(ii) in]) the curvature vector points inside $B$.

(iii) Also the vertical plane sections $y = \alpha x$ through the $z$-axis can be easily understood. They have horizontal tangents of the $z$-axis and that makes some of the parts of the section convex or concave leading to curvature vectors pointing inside. See Fig. 6 [(iii) in]. We do not elaborate the small remaining part of $M^-$.

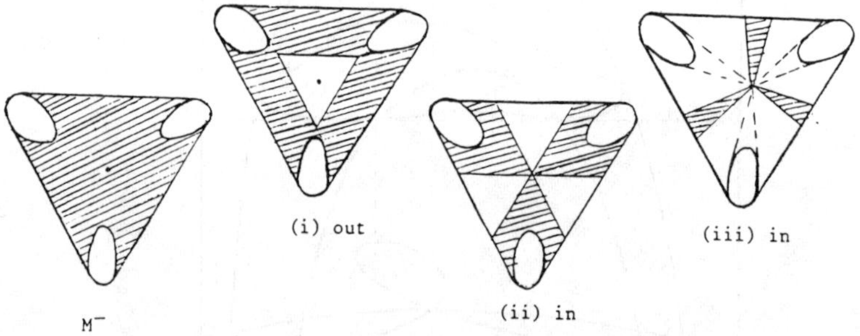


FIG. 6

*All points of $M^+$ not on the z-axis have positive Gauss curvature* $K > 0$. Again by looking at vertical sections and horizontal sections one convinces oneself of this fact. We did not work out a formal calculation.

## References

[1] T. Banchoff, *Tight polyhedral Klein bottles, projective planes and Moebius bands*, Math. Ann. **207** (1974) 233–1243.

[2] D. A. Gudkov, Russian Math. Surveys **29** (1974) 1–79.

[3] Y. Kergosien & R. Thom, *Sur les points paraboliques des surfaces immergées dans l'espace euclidien à trois dimensions*, C. R. Acad. Sci. Paris **290** (1980) 705–709.

[4] V. M. Kharlamov, *The maximum number of components of a surface of degree* 4 *in* $\mathbf{R}P^3$, Funkcional. Anal. i Prilŏzen **6** (1972) 101.

[5] _____, *Isotopic types of nonsingular surfaces of fourth degree in* $\mathbf{R}P^3$, Funkcional. Anal. i Prilŏzen **12** (1978) 86–87.

[6] N. H. Kuiper, *Morse relations for curvature and tightness*, Lecture Notes in Math. Vol. 209, Springer, Berlin, 1971, 77–89.

[7] _____, *Tight embeddings and maps. Submanifolds of geometrical class three in* $E^N$, The Chern Symposium 1979, Springer, Berlin, 97–145.

[8] J. Milnor, *On the Betti numbers of real varieties*, Proc. Amer. Math. Soc. **15** (1964) 275–280.

[9] R. Thom, *Sur l'homologie des variétés algébriques réelles*, Differential and Combinatorial Topology (A Sympos. in Honor of Marston Morse), Princeton University Press, Princeton, 1965, 255–265.

[10] _____, *Sur les variétés d'ordre fini*, Global Analysis (Papers in Honor of K. Kodaira), Princeton University Press, Princeton, 1969, 397–401.

[11] D. Hilbert & S. Cohn-Vossen, *Anschauliche Geometrie*, Springer, Berlin, 1932, Fig. 288.

[12] E. Pitcher, *Critical points of a map to a circle*, Proc. Nat. Acad. Sci. U.S.A. **25** (1939) 428–431.

Brown University
Institut Des Hautes Études Scientifiques, France

J. DIFFERENTIAL GEOMETRY
16 (1981) 137–145

# *CR*-SUBMANIFOLDS OF A COMPLEX SPACE FORM

AUREL BEJANCU, MASAHIRO KON & KENTARO YANO

*Dedicated to Professor Buchin Su on his 80th birthday*

## 0. Introduction

The *CR*-submanifolds of a Kaehlerian manifold have been defined by one of the present authors and studied by him [2], [3] and by B. Y. Chen [4].

The purpose of the present paper is to continue the study of *CR*-submanifolds, and in particular of those of a complex space form.

In §1 we first recall some fundamental formulas for submanifolds of a Kaehlerian manifold, and in particular for those of a complex space form, and then give the definitions of *CR*-submanifolds and generic submanifolds in our context. We also include Theorem 1 which seems to be fundamental in the study of *CR*-submanifolds.

In §2 we study the *f*-structures which a *CR*-submanifold and its normal bundle admit. We then prove Theorem 2 which characterizes generic submanifolds with parallel *f*-structure of a complex space form.

In §3 we derive an integral formula of Simons' type and applying it to prove Theorems 3, 4 and 5.

## 1. Preliminaries

Let $\overline{M}$ be a complex $m$-dimensional (real $2m$-dimensional) Kaehlerian manifold with almost complex structure $J$, and $M$ a real $n$-dimensional Riemannian manifold isometrically immersed in $\overline{M}$. We denote by $\langle\, ,\, \rangle$ the metric tensor field of $\overline{M}$ as well as that induced on $M$. Let $\overline{\nabla}$ (resp. $\nabla$) be the operator of covariant differentiation with respect to the Levi-Civita connection in $\overline{M}$ (resp. $M$). Then the Gauss and Weingarten formulas for $M$ are respectively written as

$$\overline{\nabla}_X Y = \nabla_X Y + B(X, Y), \quad \overline{\nabla}_X N = -A_N X + D_X N$$

Communicated May 5, 1979.

for any vector fields $X$, $Y$ tangent to $M$ and any vector field $N$ normal to $M$, where $D$ denotes the operator of covariant differentiation with respect to the linear connection induced in the normal bundle $T(M)^{\perp}$ of $M$. Both $A$ and $B$ are called the second fundamental forms of $M$ and are related by $\langle A_N X, Y\rangle = \langle B(X, Y), N\rangle$.

For any vector field $X$ tangent to $M$ we put

$$JX = PX + FX, \tag{1.1}$$

where $PX$ is the tangential part of $JX$, and $FX$ the normal part of $JX$. Then $P$ is an endomorphism of the tangent bundle $T(M)$ of $M$, and $F$ is a normal bundle valued 1-form on $T(M)$.

For any vector field $N$ normal to $M$ we put

$$JN = tN + fN, \tag{1.2}$$

where $tN$ is the tangential part of $JN$, and $fN$ the normal part of $JN$.

If the ambient manifold $\overline{M}$ is of constant holomorphic sectional curvature $c$, then $\overline{M}$ is called a complex space form, and will be denoted by $\overline{M}^m(c)$. Thus the Riemannian curvature tensor $\overline{R}$ of $\overline{M}^m(c)$ is given by

$$\begin{aligned}\overline{R}(X, Y)Z = \tfrac{1}{4}c[\langle Y, Z\rangle X - \langle X, Z\rangle Y + \langle JY, Z\rangle JX \\ -\langle JX, Z\rangle JY + 2\langle X, JY\rangle JZ]\end{aligned}$$

for any vector fields $X$, $Y$ and $Z$ of $\overline{M}^m(c)$. We denote by $R$ the Riemannian curvature tensor of $M$. Then we have

$$\begin{aligned}R(X, Y)Z = \tfrac{1}{4}c[\langle Y, Z\rangle X - \langle X, Z\rangle Y + \langle PY, Z\rangle PX - \langle PX, Z\rangle PY \\ +2\langle X, PY\rangle PZ] + A_{B(Y,Z)}X - A_{B(X,Z)}Y,\end{aligned} \tag{1.3}$$

$$\begin{aligned}&(\nabla_X B)(Y, Z) - (\nabla_Y B)(X, Z) \\ &= \tfrac{1}{4}c[\langle PY, Z\rangle FX - \langle PX, Z\rangle FY + 2\langle X, PY\rangle FZ]\end{aligned} \tag{1.4}$$

for any vector fields $X$, $Y$ and $Z$ tangent to $M$.

If the second fundamental form $B$ of $M$ satisfies the classical Codazzi equation $(\nabla_X B)(Y, Z) = (\nabla_Y B)(X, Z)$, then (1.4) implies (cf., [1, p. 434])

**Lemma 1.** *Let $M$ be an $n$-dimensional submanifold of a complex space form $\overline{M}^m(c)$, $c \neq 0$. If the second fundamental form of $M$ satisfies the classical Codazzi equation, then $M$ is holomorphic or anti-invariant.*

**Definition 1.** A submanifold $M$ of a Kaehlerian manifold $\overline{M}$ is called a $CR$-submanifold of $\overline{M}$ if there exists a differentiable distribution $\mathcal{D}: x \to \mathcal{D}_x \subset T_x(M)$ on $M$ satisfying the following conditions:

(i) $\mathcal{D}$ is holomorphic, i.e., $J\mathcal{D}_x = \mathcal{D}_x$ for each $x \in M$, and

(ii) the complementary orthogonal distribution $\mathcal{D}^{\perp}: x \to \mathcal{D}_x^{\perp} \subset T_x(M)$ is anti-invariant, i.e., $J\mathcal{D}_x^{\perp} \subset T_x(M)^{\perp}$ for each $x \in M$.

If dim $\mathfrak{D}_x^\perp = 0$ (resp. dim $\mathfrak{D}_x = 0$) for any $x \in M$, then the $CR$-submanifold is a holomorphic submanifold (resp. anti-invariant submanifold) of $\overline{M}$. If dim $\mathfrak{D}_x^\perp = \dim T_x(M)^\perp$ for any $x \in M$, then the $CR$-submanifold is a generic submanifold of $M$ (see [9]). It is clear that every real hypersurface of a Kaehlerian manifold is automatically generic submanifold. A $CR$-submanifold is called a proper $CR$-submanifold if it is neither a holomorphic submanifold nor an anti-invariant submanifold. From Lemma 1 we have

**Proposition 1.** *Let $M$ be a proper CR-submanifold of a complex space form $\overline{M}^m(c)$. If the second fundamental form of $M$ satisfies the classical Codazzi equation, then $c = 0$.*

A submanifold $M$ is said to be minimal if trace $B = 0$. If $B = 0$ identically, $M$ is called a totally geodesic submanifold.

**Definition 2.** A $CR$-submanifold $M$ of a Kaehlerian manifold $\overline{M}$ is said to be mixed totally geodesic if $B(X, Y) = 0$ for each $X \in \mathfrak{D}$ and $Y \in \mathfrak{D}^\perp$.

**Lemma 2.** *Let $M$ be a CR-submanifold of a Kaehlerian manifold $\overline{M}$. Then $M$ is mixed totally geodesic if and only if one of the following conditions is fulfilled*:

(i) $A_N X \in \mathfrak{D}$ *for any* $X \in \mathfrak{D}$ *and* $N \in T(M)^\perp$,

(ii) $A_N Y \in \mathfrak{D}^\perp$ *for any* $Y \in \mathfrak{D}^\perp$ *and* $N \in T(M)^\perp$.

The integrability of distributions $\mathfrak{D}$ and $\mathfrak{D}^\perp$ on a $CR$-submanifold $M$ is characterized by

**Theorem 1.** *Let $M$ be a CR-submanifold of a Kaehlerian manifold $\overline{M}$. Then we have*

(i) $\mathfrak{D}^\perp$ *is always involutive*, [4],

(ii) $\mathfrak{D}$ *is involutive if and only if the second fundamental form $B$ satisfies $B(PX, Y) = B(X, PY)$ for all $X, Y \in \mathfrak{D}$*, [2].

**Definition 3.** A $CR$-submanifold $M$ is said to be mixed foliate if it is mixed totally geodesic and $B(PX, Y) = B(X, PY)$ for all $X, Y \in \mathfrak{D}$.

Now, let $M^\perp$ be a leaf of anti-invariant distribution $\mathfrak{D}^\perp$ on $M$. then we have

**Proposition 2.** *A necessary and sufficient condition for the submanifold $M^\perp$ to be totally geodesic in $M$ is that*

$$B(X, Y) \in fT(M)^\perp \text{ for all } X \in \mathfrak{D}^\perp \text{ and } Y \in \mathfrak{D}.$$

*Proof.* For any vector fields $X$ and $Y$ tangent to $M$, (1.1) and Gauss and Weingarten formulas imply

$$tB(X, Y) = (\nabla_X P)Y - A_{FY}X, \tag{1.5}$$

where we have put $(\nabla_X P)Y = \nabla_X PY - P\nabla_X Y$.

Let $X, Z \in \mathfrak{D}^\perp$ and $Y \in \mathfrak{D}$. Then (1.5) implies that

$$\langle P\nabla_X Z, Y\rangle = -\langle A_{FZ}X, Y\rangle = -\langle B(X, Y), FZ\rangle,$$

which proves our assertion.

**Corollary 1.** *Let $M$ be a mixed totally geodesic CR-submanifold of a Kaehlerian manifold $\overline{M}$. Then each leaf of anti-invariant distribution $\mathfrak{D}^\perp$ is totally geodesic in $M$.*

**Corollary 2.** *A generic submanifold $M$ of a Kaehlerian manifold $\overline{M}$ is mixed totally geodesic if and only if each leaf of anti-invariant distribution is totally geodesic in $M$.*

**Lemma 3.** *Let $M$ be a mixed foliate CR-submanifold of a Kaehlerian manifold $\overline{M}$. Then we have*

$$A_N P + PA_N = 0$$

*for any vector field $N$ normal to $M$.*

*Proof.* From the assumption we have $B(X, PY) = B(PX, Y)$ for all $X, Y \in \mathfrak{D}$. On the other hand, we obtain $B(X, Y) = 0$ for $X \in \mathfrak{D}$ and $Y \in \mathfrak{D}^\perp$. Moreover, we see that $PX \in \mathfrak{D}$ for any vector field $X$ tangent to $M$. Consequently we can see that $B(X, PY) = B(PX, Y)$ for any vector fields $X, Y$ tangent to $M$, from which it follows that $A_N P + PA_N = 0$.

**Proposition 3.** *If $M$ is a mixed foliate proper CR-submanifold of a complex space form $\overline{M}^m(c)$, then we have $c \leqslant 0$.*

*Proof.* Let $X, Y \in \mathfrak{D}$ and $Z \in \mathfrak{D}^\perp$. Then we have

$$(\nabla_X B)(Y, Z) - (\nabla_Y B)(X, Z) = B(X, \nabla_Y Z) - B(Y, \nabla_X Z).$$

If we take a vector field $U$ normal to $M$ such that $Z = JU = tU$, we obtain that $\nabla_Y Z = -PA_U Y + tD_Y U$. Thus Lemma 3 implies that

$$(\nabla_X B)(Y, Z) - (\nabla_Y B)(X, Z) = B(PY, A_U X) + B(X, A_U PY).$$

Putting $X = PY$ and using (1.4) we see that $2B(PY, A_U PY) = -\frac{1}{2}c\langle PY, PY\rangle U$. Therefore we have

$$0 \leqslant 2\langle A_U PY, A_U PY\rangle = -\tfrac{1}{2}c\langle PY, PY\rangle\langle U, U\rangle, \tag{1.6}$$

which proves our assertion.

**Corollary 3.** *Let $M$ be a mixed foliate CR-submanifold of a complex space form $\overline{M}^m(c)$. If $c > 0$, then $M$ is a holomorphic submanifold or an anti-invariant submanifold of $\overline{M}^m(c)$.*

## 2. *f*-structure

Let $M$ be an $n$-dimensional $CR$-submanifold of a complex $m$-dimensional Kaehlerian manifold $\overline{M}$. Applying $J$ to both sides of (1.1) we have

$$-X = P^2 X + tFX,$$

from which it follows that $P^3X + PX = 0$ for any vector field $X$ tangent to $M$. Thus

$$P^3 + P = 0.$$

On the other hand, the rank of $P$ is equal to dim $\mathfrak{D}_x$ everywhere on $M$. Consequently, $P$ defines an $f$-structure on $M$ (see [7]).

Applying $J$ to both sides of (1.2) we obtain that

$$-N = FtN + f^2N,$$

so that $f^3N + fN = 0$ for any vector field $N$ normal to $M$, and the rank of $f$ is equal to dim $T_x(M)$ − dim $\mathfrak{D}_x$ everywhere on $M$. Thus $f$ defines an $f$-structure on the normal bundle of $M$.

**Definition 4.** If $\nabla_X P = 0$ for any vector field $X$ tangent to $M$, then the $f$-structure $P$ is said to be parallel.

**Proposition 4.** *Let $M$ be an $n$-dimensional generic submanifold of a complex $m$-dimensional Kaehlerian manifold $\overline{M}$. If the $f$-structure $P$ on $M$ is parallel, then $M$ is locally a Riemannian direct product $M^T \times M^\perp$, where $M^T$ is a totally geodesic complex submanifold of $\overline{M}$ of complex dimension $n - m$, and $M^\perp$ is an anti-invariant submanifold of $\overline{M}$ of real dimension $2m - n$.*

*Proof.* From the assumption and (1.5) we have $JB(X, Y) = tB(X, Y) = -A_{FY}X$. Thus $JB(X, PY) = 0$ and hence $B(X, PY) = 0$. On the other hand, we see that

$$fB(X, Y) = B(X, PY) + (\nabla_X F)Y. \tag{2.1}$$

Since $f = 0$, we have $(\nabla_X F)Y = -B(X, PY) = 0$.

Let $Y \in \mathfrak{D}^\perp$. Then we have that $P\nabla_X Y = \nabla_X PY - (\nabla_X P)Y = 0$ for any vector field $X$ tangent to $M$, so that the distribution $\mathfrak{D}^\perp$ is parallel. Similarly, the distribution $\mathfrak{D}$ is also parallel. Consequently, $M$ is locally a Riemannian direct product $M^T \times M^\perp$, where $M^T$ and $M^\perp$ are leaves of $\mathfrak{D}$ and $D^\perp$ respectively. From the constructions, $M^T$ is a complex submanifold of $\overline{M}$, and $M^\perp$ is an anti-invariant submanifold of $\overline{M}$. On the other hand, since $B(X, PY) = 0$ for any vector fields $X$ and $Y$ tangent to $M$, $M^T$ is totally geodesic in $\overline{M}$. Thus we have our assertion.

**Theorem 2.** *Let $M$ be an $n$-dimensional complete generic submanifold of a complex $m$-dimensional, simply connected complete complex space form $\overline{M}^m(c)$. If the $f$-structure $P$ on $M$ is parallel, then $M$ is an $m$-dimensional anti-invariant submanifold of $\overline{M}^m(c)$, or $c = 0$ and $M$ is $C^{n-m} \times M^{2m-n}$ of $C^m$, where $M^{2m-n}$ is an anti-invariant submanifold of $C^m$.*

*Proof.* First of all, we have

$$(\nabla_X B)(Y, PZ) = D_X(B(Y, PZ)) - B(\nabla_X Y, PZ) - B(Y, P\nabla_X Z) = 0,$$

which together with (1.4) implies

$$\tfrac{1}{4}c[\langle PY, PY\rangle FX - \langle PX, PY\rangle FY] = 0.$$

Thus we have $c = 0$ or $P = 0$. If $P = 0$, then $M$ is a real $m$-dimensional anti-invariant submanifold of $\overline{M}^m(c)$. If $c = 0$, then the ambient manifold $\overline{M}^m(c)$ is a complex number space $C^m$, and our assertion follows from Proposition 3.

**Proposition 5.** *Let $M$ be an $n$-dimensional complex mixed foliate proper generic submanifold of a simply connected complete complex space form $\overline{M}^m(c)$. If $c \geqslant 0$, then $c = 0$ and $M$ is $C^{n-m} \times M^{2m-\bar{n}}$ of $C^m$, where $M^{2m-n}$ is an anti-invariant submanifold of $C^m$.*

*Proof.* From Proposition 3 we see that $c = 0$ and hence $\overline{M}^m(c) = C^m$. Then (1.6) implies that $A_U X = 0$ for any $X \in \mathfrak{D}$. From this and (1.5) we see that $P$ is parallel. Thus theorem 2 proves our assertion.

## 3. An integral formula

First of all, we recall the formula of Simons' type for the second fundamental form [6].

Let $M$ be an $n$-dimensional minimal submanifold of an $m$-dimensional Riemannian manifold $\overline{M}$. Then the formula of Simons' type for the second fundamental form $A$ of $M$ is written as

$$\nabla^2 A = -A \circ \tilde{A} - \underset{\sim}{A} \circ A + \overline{R}(A) + \overline{R}', \tag{3.1}$$

where we have put $\tilde{A} = {}^t A \circ A$ and $\underset{\sim}{A} = \sum_{a=1}^{m-n} adA_a adA_a$ for a normal frame $\{V_a\}$, $a = 1, \cdots, m-n$, and $A_a = A_{V_a}$. For a frame $\{E_i\}$, $i = 1, \cdots, n$ of $M$, we put

$$\langle \overline{R}'^N(X), Y\rangle = \sum_{i=1}^{n} \left(\langle (\overline{\nabla}_X \overline{R})(E_i, Y)E_i, N\rangle + \langle (\overline{\nabla}_{E_i} \overline{R})(E_i, X)Y, N\rangle\right) \tag{3.2}$$

for any vector fields $X$, $Y$ tangent to $M$ and any vector field $N$ normal to $M$, $\overline{R}$ being the Riemannian curvature tensor of $\overline{M}$. Moreover, we put

$$\begin{aligned} &\langle \overline{R}(A)^N(X), Y\rangle \\ &\quad = \sum_{i=1}^{n} \big[2\langle \overline{R}(E_i, Y)B(X, E_i), N\rangle + 2\langle \overline{R}(E_i, X)B(Y, E_i), N\rangle \\ &\qquad - \langle A_N X, \overline{R}(E_i, Y)E_i\rangle - \langle A_N Y, \overline{R}(E_i, X)E_i\rangle \\ &\qquad + \langle \overline{R}(E_i, B(X, Y))E_i, N\rangle - 2\langle A_N E_i, \overline{R}(E_i, X)Y\rangle\big]. \end{aligned} \tag{3.3}$$

In the following, we assume that the ambient manifold $\overline{M}$ is a complex space form $\overline{M}^m(c)$. Since $\overline{M}^m(c)$ is locally symmetric, we have $\overline{R}' = 0$. A straightforward computation gives

$$
\begin{aligned}
\langle \overline{R}(A)^N(X), Y\rangle &= \tfrac{1}{4}cn\langle A_N X, Y\rangle - \tfrac{1}{2}c\langle A_{FY}X, tN\rangle - \tfrac{1}{2}c\langle A_{FX}Y, tN\rangle \\
&\quad + c\langle fB(X, PY), N\rangle + c\langle fB(Y, PX), N\rangle \\
&\quad + \tfrac{3}{4}c\langle PX, PA_N Y\rangle + \tfrac{3}{4}c\langle PY, PA_N X\rangle - \tfrac{3}{2}c\langle A_N PX, PY\rangle \qquad (3.4)\\
&\quad - \tfrac{1}{2}c\sum_{i=1}^{n}\Big[\langle A_{FE_i}E_i, X\rangle\langle FY, N\rangle + \langle A_{FE_i}E_i, Y\rangle\langle FX, N\rangle \\
&\quad + \tfrac{3}{2}\langle A_{FE_i}X, Y\rangle\langle FE_i, N\rangle\Big].
\end{aligned}
$$

We now prepare some lemmas for later use.

**Lemma 4**, [9]. *Let $M$ be a generic submanifold of a Kaehlerian manifold $\overline{M}$. Then we have*

$$A_{FX}Y = A_{FY}X$$

*for any vector fields $X$, $Y$.*

**Lemma 5.** *Let $M$ be a minimal CR-submanifold of a Kaehlerian manifold $\overline{M}$ with involutive distribution $\mathfrak{D}$. Then we have*

$$\sum B(E_\alpha, E_\alpha) = 0$$

*for a frame $\{E_\alpha\}$ of $\mathfrak{D}^\perp$.*

*Proof.* We take a frame $\{E_t, E_\alpha\}$ of $M$ such that $\{E_t\}$ and $\{E_\alpha\}$ are frames of $\mathfrak{D}$ and $D^\perp$ respectively. Since $\mathfrak{D}$ is involutive, we have that $\Sigma\, B(E_t, E_t) = 0$ so that $\Sigma\, B(E_\alpha, E_\alpha) = 0$.

We now define a vector field $H$ tangent to $M$ in the following way. Let $\{E_\alpha\}$ be a fame of $\mathfrak{D}^\perp$, and put $H = \Sigma_\alpha A_\alpha E_\alpha$. Then $H = \Sigma_i A_{fe_i} tfe_i$ for any frame $\{e_i\}$ of $M$ and $H$ is independent of the choice of a frame of $M$.

In the following we assume that $M$ is a generic minimal submanifold of $\overline{M}^m(c)$, $c > 0$, with the second fundamental form $B$ satisfying that $B(PX, Y) = B(X, PY)$ for all $X$, $Y \in \mathfrak{D}$, which implies that $\mathfrak{D}$ is involutive, and $H \in \mathfrak{D}^\perp$.

From (3.4), using Lemmas 4 and 5 we obtain

$$\langle \overline{R}(A), A\rangle \geqslant \tfrac{1}{4}(n+1)c\|A\|^2. \qquad (3.5)$$

On the other hand, we have [6]

$$\langle A \circ \tilde{A}, A\rangle + \langle \underset{\sim}{A} \circ A, A\rangle \leqslant \left(2 - \frac{1}{p}\right)\|A\|^4, \qquad (3.6)$$

where $p$ denotes the codimension of $M$, and $\|A\|$ is the length of the second fundamental form $A$ of $M$. Thus (3.1), (3.5) and (3.6) imply

$$-\langle \nabla^2 A, A\rangle \leqslant \left(2 - \frac{1}{p}\right)\|A\|^4 - \tfrac{1}{4}(n+1)c\|A\|^2. \tag{3.7}$$

If $M$ is compact orientable, then

$$\int_M \langle \nabla^2 A, A\rangle = -\int_M \langle \nabla A, \nabla A\rangle.$$

Therefore (3.7) implies the following.

**Theorem 3.** *Let $M$ be an $n$-dimensional compact orientable generic minimal submanifold of a complex space form $\overline{M}^m(c)$, $c > 0$. If $\mathfrak{D}$ is involutive and $H \in \mathfrak{D}^\perp$, then we have*

$$\int_M \langle \nabla A, \nabla A\rangle \leqslant \int_M \left\{\left(2 - \frac{1}{p}\right)\|A\|^4 - \tfrac{1}{4}(n+1)c\|A\|^2\right\}. \tag{3.8}$$

As the ambient manifold $\overline{M}^m(c)$ we take a complex projective space $CP^m$ with constant holomorphic sectional curvature 4. Then we have

**Theorem 4.** *Let $M$ be an $n$-dimensional compact orientable generic minimal submanifold of $CP^m$ with involutive distribution $\mathfrak{D}$. If $H \in \mathfrak{D}^\perp$ and $\|A\|^2 < (n+1)/(2 - 1/p)$, then $M$ is real projective space $RP^m$ and $n = m = p$.*

*Proof.* From (3.8) we see that $M$ is totally geodesic in $CP^m$. Thus $M$ is a complex or real projective space (see [1, Lemma 4]). Since $M$ is a generic submanifold, $M$ is a real projective space and anti-invariant in $CP^m$. Thus we have $n = m = p$ and $\dim \mathfrak{D} = 0$.

**Theorem 5.** *Let $M$ be an $n$-dimensional compact orientable generic minimal submanifold of $CP^m$. If $\mathfrak{D}$ is involutive, $H \in \mathfrak{D}^\perp$, and $\|A\|^2 = (n+1)/(2 - 1/p)$, then $M$ is $S^1 \times S^1$ in $CP^2$, and $n = m = p = 2$.*

*Proof.* From the assumption we have $\nabla A = 0$, and $M$ is an anti-invariant submanifold of $CP^m$, and hence $m = n = p$. Thus our assertion follows from [5, Theorem 3].

## Bibliography

[1] K. Abe, *Applications of Riccati type differential equation to Riemannian manifolds with totally geodesic distribution*, Tôhoku Math. J. **25** (1973) 425–444.

[2] A. Bejancu, *CR submanifolds of a Kaehler manifold*. I, Proc. Amer. Math. Soc. **69** (1978) 135–142.

[3] ______, *CR submanifold of a Kaehler manifold*. II, Trans. Amer. Math. Soc. **250** (1979) 333–345.

[4] B. Y. Chen, *On CR-submanifolds of a Kaehler manifold*. I, to appear in J. Differential Geometry.

[5] G. D. Ludden, M. Okumura & K. Yano, *A totally real surface in $CP^2$ that is not totally geodesic*, Proc. Amer. Math. Soc. **53** (1975) 186–190.

[6] J. Simons, *Minimal varieties in riemannian manifolds*, Ann. of Math. **88** (1968) 62–105.
[7] K. Yano, *On a structure defined by a tensor field f of type* (1, 1) *satisfying* $f^3 + f = 0$, Tensor, N. S., **14** (1963) 99–109.
[8] K. Yano & M. Kon, *Anti-invariant submanifolds*, Marcel Dekker, New York, 1976.
[9] ______, *Generic submanifolds*, Ann. Mat. Pura Appl. **123** (1980) 59–92.
[10] ______, *CR-sous-variétiés d'un espace projectif complexe*, C. R. Acad. Sci. Paris **288** (1979) 515–517.

IASI UNIVERSITY, RUMANIA
HIROSAKI UNIVERSITY, JAPAN
TOKYO INSTITUTE OF TECHNOLOGY, JAPAN

J. DIFFERENTIAL GEOMETRY
16 (1981) 537–550

# INTÉGRANDS DES NOMBRES CARACTÉRISTIQUES ET COURBURE : RIEN NE VA PLUS DÈS LA DIMENSION 6

JEAN PIERRE BOURGUIGNON & ALBERT POLOMBO

*Dédié au Professeur Buchin Su à l'occasion de son 80ième anniversaire*

## 1. Introduction, motivation

D'après la théorie de Chern-Weil (cf. [3]), les nombres caractéristiques d'une variété riemannienne compacte orientée $M$ de dimension $n = 2m$ s'expriment par des intégrales de polynôme de degré $m$ en son tenseur de courbure. Pour étudier le signe d'un nombre caractéristique, l'approche la plus simple consiste donc à regarder en chaque point le signe de l'intégrand correspondant.

En dimension 4 cette approche algébrique a permis à J. Thorpe puis à N. Hitchin d'obtenir le résultat suivant (cf. [14] et [5]) relatif aux métriques d'Einstein (i.e. celles dont la courbure de Ricci est un multiple de la métrique):

**Theoreme $E_4$.** *Sur une variété compacte orientée d'Einstein $M$ de dimension 4, la caractéristique d'Euler $\chi(M)$ et le premier nombre de Pontrjaguine $p_1(M)$ vérifient l'inégalité*

$$(E_4) \qquad |p_1(M)| \leq 2\chi(M),$$

*qui est conséquence, après intégration, de l'inégalité algébrique*

$$(AE_4) \qquad |p_1(R)| \leq 2\chi(R),$$

*où $R$ désigne le tenseur de courbure de la métrique; de plus l'égalité n'a lieu que si la courbure de Ricci est nulle.*

Il est établi dans [10] que cette inégalité est encore satisfaite si l'on impose différents types de pincement à la courbure sectionnelle ou à la courbure de Ricci, donc des conditions ouvertes sur le tenseur de courbure.

Le théorème $E_4$ contient le théorème $E_4$'suivant (qui fut ètabli antèrieurement par M. Berger [1]).

Received November 14, 1981.

**Théorème $E'_4$.** *Si M est une variété d'Einstein compacte de dimension* 4, *alors*

$$(E'_4) \qquad 0 \leqslant \chi(M),$$

*avec égalité si et seulement si M est plate. Cette inégalité est conséquence, après intégration, de l'inégalité ponctuelle*

$$(AE'_4) \qquad 0 \leqslant \chi(R),$$

*l'égalité ayant lieu si et seulement si $R = 0$.*

Cet article présente, dans une première partie, des contre-exemples à toutes les généralisations des inégalités ($AE_4$) et ($AE'_4$) aux dimensions supérieures. Nous montrons ainsi que, si des théorèmes ($E_{2m}$) et ($E'_{2m}$) existent, ils ne peuvent se démontrer par l'approche algébrique directe. En ce qui concerne l'articulation entre les théorèmes E et E′ nous montrons dans la section 1 que, pour les variétés riemanniennes de dimension $4k$, toute généralisation du théorème E implique que le théorème E′ est vrai à cause de la sensibilité différente au changement d'orientation de la caractéristique d'Euler et des nombres de Pontrjaguine.

Pour justifier la deuxième partie du titre, rappelons qu'en dimension 4 le résultat suivant dû à J. W. Milnor a aussi été obtenu par la méthode algébrique.

**Théorème $S'_4$.** *Si M est une variété riemannienne compacte de dimension* 4 *dont la courbure sectionnelle ne prend qu'un seul signe, alors*

$$(S'_4) \qquad 0 \leqslant \chi(M),$$

*inégalité qui est conséquence, après intégration, de l'inégalité ponctuelle*

$$(AS'_4) \qquad 0 \leqslant \chi(R).$$

R. Geroch a, le premier, donné dans [4] un exemple de tenseur de courbure en dimension 6 pour lequel ($AS'_6$) est violée (pour d'autres exemples, voir [2] et [7]).

Ainsi, pour les variétés riemanniennes, *dès la dimension* 6 *ni la condition d'Einstein ni le signe de la courbure sectionnelle ne permettent d'établir, par l'approche algébrique directe, des inégalités entre nombres caractéristiques.* A ce point il est intéressant de rappeler que dans [14] J. Thorpe avait donné une extension de l'inégalité ($AE_{2m}$) dans laquelle la condition d'Einstein était remplacée par la condition d'égalité de la courbure de Lipschitz-Killing d'un $2k$-plan et de celle de son orthogonal. Malheureusement si cette dernière condition se réduit bien à la condition d'Einstein en dimension 4 (voir [14]), elle devient extrêmement forte en dimension plus grande et n'est satisfaite par aucune classe géométriquement raisonnable de variétés riemanniennes (voir

[12] pour une discussion détaillée). De même B. Kostant avait établi que, si les valeurs propres de la courbure vue comme opérateur sur les 2-formes extérieures étaient non-négatives, alors ($AS'_{2m}$) était vraie (pour une preuve voir la section 1), mais cette hypothèse est, dès la dimension 4, plus forte que celle sur le signe de la courbure sectionnelle.

Fort de ces remarques, il est naturel de considérer des classes de variétés riemanniennes plus restreintes. *Nous consacrons la deuxième partie de cet article aux variétés kählériennes.* Pour ces variétés les nombres de Chern s'ajoutent aux autres nombres caractéristiques. De plus la condition d'Einstein implique que $c_1(R) = \frac{1}{2\pi}\rho J$ où $J$ désigne la forme de Kähler et $\rho$ la constante d'Einstein, relation qui permet de suspendre toute inégalité entre intégrands de nombres caractéristiques existant en une dimension aux dimensions supérieures. Cette remarque intervient dans le résultat suivant dû à S. T. Yau (cf. [16]):

**Théorème.** *Si $M$ est une variété complexe de dimension complexe $m$ dont la première classe de Chern est négative, alors*

$$m(-1)^m c_1^m(M) \leq 2(m+1)(-1)^m c_1^{m-2}(M)c_2(M).$$

Cette inégalité renforce l'inégalité ($E_4$) puisque, si on oublie la structure complexe, elle signifie que $|p_1(M)| \leq \chi(M)$. A. Beauville a remarqué que pour $m = 1, 2$ et 3, elle peut s'écrire

$$(KE_{2m}) \qquad \kappa^m(M) \leq (-1)^m(m+1)^m\tau(M),$$

où $\kappa$ désigne la classe du fibré canonique et $\tau$ le genre de Todd.

Nous montrons dans la section 6 que *l'inégalité* ($KE_8$) *ne peut être déduite par intégration sur $M$ de l'inégalité algébrique*

$$(AKE_8) \qquad \kappa^4(R) \leq 5^4\tau(R),$$

*où $R$ est un tenseur de courbure de Kähler-Einstein puisque nous donnons un contre-exemple à* ($AKE_8$).

Nous ne discuterons pas ici de généralisation au cas kählérien du théorème $S'_4$ si ce n'est pour remarquer que d'une part depuis la solution de laconjecture de Frankel (cf. [10] et [13]) la seule variété en jeu, lorsque la courbure est positive, est l'espace projectif $\mathbf{CP}^m$ et que d'autre part D. Johnson a établi le théorème $KS'_6$ (cf. [6]). Ce dernier résultat ne doit pas conduire à trop d'optimisme pour la généralisation à $KS'_{2m}$ puisque, du point de vue de l'algèbre de courbure, la première dimension "stable" est la dimension complexe 4.

L'organisation de l'article est la suivante: dans la section 2 nous développons la technique de l'algèbre de courbure (cf. [8]) qui semble un outil algébrique bien adapté pour discuter des relations entre intégrands caractéristiques et courbure. Dans le cadre de cet article elle peut paraître plus efficace à fournir des contre-exemples qu'à prouver des conjectures, mais elle nous semble cependant mériter un peu de publicité. Nous développons quelques remarques générales sur les inégalités entre nombres caractéristiques dans la section 3 (rôle de la caractéristique d'Euler, comportement par produit, cas de la dimension $4k+2$).

Dans la section 4 nous discutons de contre-exemples riemanniens en dimensions 6 et 8, le comportement multiplicatif de la caractéristique d'Euler nous obligeant à séparer ces deux cas.

L'article se termine par la section 5 où nous traitons des variétés kähleriennes. Diverses remarques utiles à des calculs généraux relatifs à de telles variétés s'y trouvent ainsi que des résultats de calculs très explicites sur un exemple particulier donnant le contre-exemple cherché.

Dans la préparation de cet article, qui s'est échelonnée sur une assez longue période, une aide précieuse nous a été apportée par le Centre de Calcul de l'Ecole Polytechnique (même si les calculs numériques ont presque disparu de cet article).

La mise au point finale a été faite lors d'un séjour à Stanford University que les auteurs remercient de son hospitalité.

## 2. L'algèbre de courbure

Soit $(V, \langle\, ,\rangle)$ un espace vectoriel euclidien de dimension $n$. L'espace vectoriel $KV = \oplus_k S^2\Lambda^k V$ est muni d'une loi de produit $\bullet$ définie pour $\alpha_1 \otimes \alpha_2$ dans $S^2\Lambda^p V$ et $\beta_1 \otimes \beta_2$ dans $S^2\Lambda^q V$ par

$$(\alpha_1 \otimes \alpha_2) \bullet (\beta_1 \otimes \beta_2) = \alpha_1 \wedge \beta_1 \bigcirc \alpha_2 \wedge \beta_2 + \alpha_1 \wedge \beta_2 \bigcirc \alpha_2 \wedge \beta_1,$$

qui en fait une algèbre, appelée *algèbre de courbure* (cf. [8]). Il est intéressant de noter que les espaces $\Lambda^{4k}V$ peuvent naturellement être considérés comme sous-espaces de $S^2\Lambda^{2k}V$.

Ainsi, en un point $x$ d'une variété riemannienne $(M, g)$, on peut prendre $(V, \langle\, ,\rangle) = (T_xM, g(x))$ où $T_xM$ est l'espace tangent en $x$. On remarque alors que la courbure de Ricci $r(x)$ appartient à $S^2V$ et la courbure de Riemann $R(x)$ à $S^2\Lambda^2V$. Via la métrique, $S^2\Lambda^2V$ sera considéré comme l'espace des endomorphismes symétriques de $\Lambda^2V$ muni de la métrique extension naturelle de celle de $V$.

Soit

$$R = \sum_i \lambda_i \omega_i \otimes \omega_i$$

la décomposition spectrale de l'endomorphisme $R$. La forme d'Euler de $M$ s'identifie, à un coefficient dépendant de $m$ près, à la puissance $m$-ième de $R$ dans l'algèbre de courbure, et peut donc s'écrire (cf. [8])

$$\chi(R) = \alpha_{2m} \sum \lambda_{i_1} \cdots \lambda_{i_m} \omega_{i_1} \wedge \cdots \wedge \omega_{i_m} \otimes \omega_{i_1} \wedge \cdots \wedge \omega_{i_m},$$

où $\alpha_{2m}$ est un coefficient universel dépendant de la dimension;

De même les formes de Pontrjaguine $p_k$ s'écrivent (cf [3], [14])

$$p_k(R) = \beta_{k,2m} \operatorname{proj}^{\perp}_{\Lambda^{4k}V}(R^{\cdot k}) \circ (R^{\cdot k}),$$

où $\beta_{k,2m}$ est un coefficient universel dépendant de $k$ et la dimension, $\operatorname{proj}^{\perp}_{\Lambda^{4k}V}$ l'opérateur de projection orthogonale sur $\Lambda^{4k}V$ vu comme sous-espace de $S^2\Lambda^{2k}V$ et $\circ$ la composition des endomorphismes de $\Lambda^{2k}V$.

On voit donc que le calcul d'une forme de Pontrjaguine d'ordre $k$ peut s'effectuer à partir des données spectrales de $R$ de la façon suivante:

(i) On calcule les produits extérieurs $k$ à $k$ des vecteurs propres de $R$. Il faut alors prendre garde que la décomposition de $R^{\cdot k}$ en somme d'endomorphismes de rang 1 obtenue à partir de ces produits n'est pas nécessairement sa décomposition spectrale, car les $2k$-formes ainsi obtenues n'ont aucune raison de former un système libre orthonormé ne serait-ce qu'à cause de leur nombre. *C'est précisément en raison de ce phénomène que les formes de Pontrjaguine rendent compte de l'irréductibilité de la métrique* (et par suite les classes de Pontrjaguine de celle de la variété).

(ii) On calcule le carré de composition de $R^{\cdot k}$. Plusieurs méthodes sont possibles: soit partir de la décomposition obtenue précédemment, soit décomposer les produits obtenus sur une base orthonormée de $\Lambda^{2k}V$. Nous préférons cette seconde méthode.

On remarquera que si

$$R^{\cdot k} = \sum \lambda_{i_1} \cdots \lambda_{i_k} \omega_{i_1} \wedge \cdots \wedge \omega_{i_k} \otimes \omega_{i_1} \wedge \cdots \wedge \omega_{i_k},$$

alors sa composante sur les $4k$-formes s'écrit

$$\operatorname{proj}^{\perp}_{\Lambda^{4k}V}(R^{\cdot k}) = \sum \lambda_{i_1} \cdots \lambda_{i_k} \omega_{i_1} \wedge \cdots \wedge \omega_{i_k} \wedge \omega_{i_1} \wedge \cdots \wedge \omega_{i_k}.$$

En particulier $R = \sum \lambda_i \omega_i \otimes \omega_i$ vérifie la première identité de Bianchi si et seulement si

$$\sum \lambda_i \omega_i \wedge \omega_i = 0.$$

On déduit ainsi de l'expression de $p_k$ donnée précédemment que, si $R^{\cdot k}$ admet une base de vecteurs propres décomposés, alors $p_l = 0$ pour tout $l > k$ (comparer avec [13]). Il serait donc intéressant de trouver des conditions sur $R$ impliquant que pour un $k$ donné, les vecteurs propres de $R^{\cdot k}$ sont des

$2k$-formes décomposables. Cette question ne semble pas facile à résoudre à cause de la complexité des relations de Grassmann traduisant la décomposabilité dans l'algèbre extérieure.

**Remarque.** Dans [15] J. Vilms introduit une notion de rang pour la coubure $R$ vue comme opérateur: la courbure est dite de rang $2r$ si $R$ applique les 2-formes de rang 2 sur les 2-formes de rang inférieur ou égal à $2r$ (le rang d'une 2-forme est celui de l'endomorphisme qui lui est associé via la métrique). Cette notion ne semble pas reliée à celle du rang des vecteurs propres de $R$: en effet $R$ peut avoir tous ses vecteurs propres de rang 2 sans que pour cela son rang au sens de J. Vilms soit égal à 2.

## 3. Quelques remarques sur les inégalités entre nombres caractéristiques

Nous discutons maintenant d'inégalités entre nombres caractéristiques de façon un peu générale. Supposons qu'existe une inégalité ($E_{4k}$) généralisant ($E_4$) et faisant intervenir les nombres caractéristiques de la variété de façon homogène (i.e. sans terme constant). On peut séparer l'inégalité en un terme contenant les nombres de Pontrjaguine, soit $B(p_1,\cdots,p_k)(M)$, et en un terme contenant la caractéristique d'Euler, soit $a_k\chi(M)$, de telle sorte que l'inégalité s'écrive

$$(E_{4k}) \qquad a_k\chi(M) + B_k(p_1,\cdots,p_k)(M) \geqslant 0.$$

Soit $M'$ la variété déduite de $M$ par changement d'orientation. En utilisant le comportement des nombres caractéristiques lors d'un changement d'orientation l'inégalité correspondant à $M'$ s'écrit

$$a_k\chi(M) - B_k(p_1,\cdots,p_k)(M) \geqslant 0.$$

Nous obtenons donc

$$a_k\chi(M) \geqslant 0.$$

Nous remarquons alors que $a_k$ ne peut être négatif car

$$M = S^2 \times \cdots \times S^2 (2k \text{ fois})$$

est une variété d'Einstein pour laquelle $\chi(M) = 2^{2k} > 0$. Si $a_k$ était nul, il faudrait qu'un nombre de Pontrjaguine soit nul pour toute variété de dimension $4k$ ayant une métrique d'Einstein. Nous savons que $a_1 \neq 0$ d'après ($E_4$). En dimension $4k$ on peut effectuer le raisonnement suivant: ordonnons les classes de Pontrjaguine et faisons de même pour les monômes présents dans $B_k(p_1,\cdots,p_k)$ suivant l'ordre lexicographique des ordres des classes de Pontrjaguine y figurant. Ainsi $p_3p_1^3$ et $p_2^2p_1^2$ se représentent respectivement par

(3, 1, 1, 1, 0, 0) et (2, 2, 1, 1, 0, 0) et par suite $p_3 p_1^3 \propto p_2^2 p_1^2$. Si $(\theta_1, \cdots, \theta_k)$ avec $k \geqslant \theta_1 \geqslant \theta_2 \geqslant \cdots \geqslant \theta_k \geqslant 0$ est la suite des ordres des classes de Pontrjaguine présentes dans la monôme minimal (pour l'ordre décrit ci-dessus) de l'expression de $B_k(p_1, \cdots, p_k)$, alors en considérant $M = \mathbf{H}P^{\theta_1} \times \cdots \times \mathbf{H}P^{\theta_k}$, où $\mathbf{H}P^r$ désigne l'espace projectif quaternionien de dimension $4r$, on obtient une contradiction. En effet sur $M$ il existe une métrique d'Einstein; par ailleurs $p_{\theta_1} \cdots p_{\theta_k}(M)$ est non nul et tout autre nombre de Pontrjaguine lui est supérieur pour l'ordre $\propto$ et par suite est nul. Nous avons donc prouvé la

**Proposition.** *S'il existe une inégalité homogène entre nombres caractéristiques vérifiée par toute variété compacte orientée de dimension* $4k$ *admettant une métrique d'Einstein, alors nécessairement l'inégalité*

$$(\mathrm{E}'_{4k}) \qquad 0 \leqslant \chi(M)$$

*est vérifiée par toute variété d'Einstein compacte de dimension* $4k$.

On est ainsi conduit à se demander si pour tout tenseur de courbure d'Einstein $R$ de dimension $2m$ on n'aurait pas

$$(\mathrm{AE}'_{2m}) \qquad 0 \leqslant \chi(R).$$

Nous allons montrer qu'en toute dimension paire supérieure à 4 il existe des contre-exemples à $(\mathrm{AE}'_{2m})$. Le cas des dimensions non multiples de 4 est à part dans la mesure où l'intégrand d'Euler est alors un polynôme homogène de degré impair en la courbure. Par suite, si l'intégrand d'Euler satisfait une inégalité, il devrait être nul pour tout tenseur de courbure à courbure de Ricci nulle, condition évidemment très forte. Une façon de raffiner cette discussion est de prendre en compte le signe de la constante d'Einstein (i.e. le facteur de proportionnalité entre la courbure de Ricci et la métrique qui par changement homothétique de la métrique peut toujours être pris égal à $-1, 0$ ou $1$). Lorsque la courbure de Ricci est définie, nous pouvons normaliser en la supposant positive. Toujours à cause de l'exemple constitué par les produits de sphères, il reste à trouver des tenseurs de courbure d'Einstein à constante positive et intégrand d'Euler négatif, le cas Ricci plat étant traité à part.

Pour produire des contre-exemples dans une dimension donnée, on pense de suite à prendre des produits de variétés de dimension plus basse, la caractéristique d'Euler se comportant multiplicativement ($\chi(M \times N) = \chi(M)\chi(N)$). Par contre on n'est sûr de l'existence de métriques d'Einstein sur le produit de deux variétés d'Einstein que si les constantes d'Einstein des deux facteurs ont le même signe, restriction qui est compatible avec la normalisation faite précédemment.

Si tout exemple de tenseur de courbure d'Einstein à constante d'Einstein positive et intégrand d'Euler négatif fournit des exemples analogues en dimension plus grande par produit avec des sphères $S^2$ de rayon convenablement ajusté, il n'en est pas de même d'un exemple de tenseur de courbure Ricci-plat en dimension 6 à intégrand d'Euler négatif qui ne peut se suspendre en dimension 8. En effet il faudrait faire un produit avec un facteur plat mais alors $\chi$ devient nul. Par contre dès la dimension 10 on peut effectuer un produit avec un tenseur de courbure à courbure de Ricci nulle, intégrand d'Euler positif et dimension 4. De tel tenseurs existent forcément puisque les surfaces $K3$ ont des métriques à courbure de Ricci nulle (cf. [16]) et que leur caractéristique d'Euler vaut 24.

## 4. Contre-exemples riemanniens en dimensions 6 et 8

Nous fournissons dans cette section des contre-exemples aux inégalités $(AE'_6)$ et $(AE'_8)$ ainsi qu'à leur cas d'égalité. Pour donner de tels contre-exemples, plusieurs approches sont possibles parmi lesquelles:

(1) partir de tenseurs de courbure suggérés par la géométrie;

(2) effectuer les calculs à partir d'objets algébriques choisis pour leur simplicité.

En utilisant l'approche (1), J. P. Bourguignon et H. Karcher construisent dans [2] un tenseur de courbure d'Einstein de dimension 6 à courbure sectionnelle positive dont l'intégrand d'Euler est négatif. Ils prennent une combinason linéaire des tenseurs de courbure des espaces symétriques irréductibles de dimension 6, soit

$$R = \lambda R_{S^6} + \mu R_{\mathbf{C}P^3} + \nu R_{\mathbf{R}G^{2,5}},$$

où $R_{S^6}$ désigne le tenseur de courbure de la sphère $S^6$, $R_{\mathbf{C}P^3}$ celui de l'espace projectif complexe $\mathbf{C}P^3$ et $R_{\mathbf{R}G^{2,5}}$ celui de la grassmannienne des 2-plans de $\mathbf{R}^5$. Si on remplace un des paramètres par la constante d'Einstein $\rho$ dans la formule donnant l'intégrand d'Euler, on obtient un polynôme du troisième degré en $\rho$ dont tous les coefficients ne sont pas identiquement nuls. Nous pouvons donc obtenir par exemple:

(i) des tenseurs de courbure d'Einstein de dimension 6 à intégrand d'Euler nul et constante d'Einstein non nulle (le polynôme en $\rho$ ne se réduit pas à son terme de plus haut degré);

(ii) un tenseur de courbure non nul de dimension 6 à courbure de Ricci nulle et à intégrand d'Euler nul (il suffit de faire $\rho = 0$ dans le polynôme et d'ajuster les paramètres $\lambda$ et $\mu$ pour que le terme constant du polynôme soit nul; ceci est possible puisque ce terme est un polynôme homogène de degré 3 en $\lambda$ et $\mu$), d'où un contre-exemple à une généralisation du cas d'égalité de ($\mathrm{AE}_4'$).

Une autre façon beaucoup plus implicite d'utiliser l'approche (1) consiste à prendre des métriques données sur des variétés par des méthodes globales. Ainsi la solution de la conjecture de Calabi (cf. [16]) permet d'affirmer l'existence de métriques de Kähler-Einstein sur les hypersurfaces complexes $M$ de $\mathbf{C}P^4$ à première classe de Chern négative ou nulle, i.e. celles dont le degré $d$ est supérieur ou égal à 5.

Il est facile de calculer la caractéristique d'Euler d'une telle hypersurface $M^d$ puisqu'elle s'identifie à $c_3(M^d)$: on a

$$\chi(M^d) = d^4 - 5d^3 + 10d^2 - 10d + 5.$$

On voit facilement que $\chi(M^d) \geqslant \chi(M^5) = 205$ pour $d \geqslant 5$.

Il existe donc un point $x$ de $M$ où l'intégrand d'Euler $\chi(R_d)$ est positif. En prenant le tenseur de courbure $-R_d(x)$, $d \geqslant 5$, on a donc un exemple de tenseur de courbure d'Einstein à constante d'Einstein positive et à intégrand d'Euler négatif. De plus pour $d = 5$, $-R_5$ est Ricci-plat et vérifie $\chi(-R_5) > 0$.

Nous passons maintenant à la dimension 8 où nous produisons un contre-exemple à ($\mathrm{AE}_8'$) en suivant la méthode (2).

Nous partons de la remarque suivante: si $z_1$ et $z_2$ sont deux tenseurs symétriques, alors la courbure de Ricci du tenseur de courbure $z_1 \bullet z_2$ vaut $\mathrm{trace}(z_1)z_2 + \mathrm{trace}(z_2)z_1 - z_1 \circ z_2 - z_2 \circ z_1$ où $\circ$ désigne la composition des endomorphismes.

Ainsi $z_1 \bullet z_2$ est un tenseur d'Einstein dès qu'il vérifie les propriétés $\mathrm{trace}(z_1) = \mathrm{trace}(z_2) = 0$ et $z_1 \circ z_2 = k\,\mathrm{Id}$ où $k$ est une constante.

Dans une base orthonormée $(e_i)$ de $V$ nous définissons deux tenseurs symétriques $z_1'$ et $z_2'$ par

$$z_1' = \begin{pmatrix} 2 & & & & & & & \\ & 2 & & & & & 0 & \\ & & 1 & & & & & \\ & & & -1 & & & & \\ & & & & -1 & & & \\ & 0 & & & & -1 & & \\ & & & & & & -1 & \\ & & & & & & & -1 \end{pmatrix}, \quad z_2' = \begin{pmatrix} \sqrt{3} & & & & \\ & \sqrt{3} & & & 0 \\ & & 0 & & \\ & 0 & & \ddots & \\ & & & & 0 \end{pmatrix}$$

de telle sorte que $z_1 = z_1' + z_2'$ et $z_2 = z_1' - z_2'$ vérifient les conditions mentionnées ci-dessus. La constante d'Einstein de $R$ vaut $16/7$ et les 2-vecteurs $e_i \wedge e_j$ forment une base de vecteurs propres de $R$ pour les valeurs propres

$a_{12} = 14$, $a_{13} = a_{23} = 4$, $a_{1\alpha} = a_{2\alpha} = -4$, $a_{3\alpha} = -2$ pour $4 \leqslant \alpha \leqslant 8$. On obtient ainsi

$$\chi(R) = -2^4 \cdot 3^2 \cdot 5^2,$$

d'où le contre-exemple cherché.

Par ailleurs $R - \frac{2}{7} R_{S^8}$ est à courbure de Ricci nulle et pour ce tenseur nous avons

$$\chi\left(R - \frac{2}{7}R_{S^8}\right) = -2^{10} \cdot 3 \cdot 5 \cdot 19 \cdot 7^{-3},$$

d'où un exemple de tenseur de courbure de dimension 8 à courbure de Ricci nulle et intégrand d'Euler négatif.

## 5. Contre-exemples kählériens en dimension 8

Nous nous proposons de construire un tenseur de courbure de Kähler-Einstein de dimension complexe 4 pour lequel

$$\kappa(R) > 5^4 \tau(R),$$

où $\kappa(R)$ désigne le produit de la constante d'Einstein par $2\pi$ et $\tau(R)$ l'intégrand du genre de Todd.

L'inégalité que nous voulons satisfaire s'exprime, en terme de classes de Chern, de la façon suivante:

$$c_1^4 > 5^3 \cdot 2^{-4} \cdot 3^{-2}\left(-c_4 + c_3 c_1 + 3c_2^2 + 4c_2 c_1^2 - c_1^4\right).$$

En utilisant les relations universelles entre classes de Pontrjaguine et classes de Chern, nous obtenons

$$p_1^2 = 4c_2^2 - 4c_2 c_1^2 + c_1^4,$$

$$p_2 = 2c_4 - 2c_1 c_3 + c_2^2.$$

L'inégalité précédente peut donc se réécrire sous la forme

$$3027c_1^4 > 7500 c_1^2 c_2 + 875 p_1^2 - 500 p_2.$$

Il est raisonnable de partir du tenseur de courbure de l'espace projectif complexe $\mathbf{C}P^4$ que nous notons $R_{\mathbf{C}P^4}$, puisque pour cet espace il y a égalité, et que, $\mathbf{C}P^4$ étant un espace symétrique, les intégrands des nombres caractéristiques sont constants.

L'espace euclidien $(V, \langle\,,\rangle)$ est muni d'une structure complexe compatible avec la métrique et notée $J$. Nous désignons par j l'élément de $S^2\Lambda^2 V$ défini pour $x$ et $y$ dans $V$ par

$$j(x \wedge y) = Jx \wedge Jy.$$

Pour être kählérien le tenseur de courbure $R$ doit vérifier la relation $R \circ j = R$. Ceci s'exprime encore, de façon équivalente, en disant que $R$ opère dans le sous-espace vectoriel $\Lambda^2_J V$ de dimension $m^2$ dont les éléments sont les 2-formes $\omega$ vérifiant $\omega \circ J = J \circ \omega$.

Soit $(e_1, e_2, e_3, e_4, Je_1, Je_2, Je_3, Je_4)$ une base orthonormée adaptée de $V$. Les vecteurs suivants forment une base orthonormée de $\Lambda^2_J V$ de vecteurs propres de $R_{\mathbf{C}P^4}$:

$$\sqrt{2}\,\omega_1 = e_1 \wedge e_2 + Je_1 \wedge Je_2, \quad \sqrt{2}\,\tilde{\omega}_1 = e_3 \wedge e_4 + Je_3 \wedge Je_4,$$
$$\sqrt{2}\,\omega_2 = e_1 \wedge e_3 + Je_1 \wedge Je_3, \quad \sqrt{2}\,\tilde{\omega}_2 = e_2 \wedge e_4 + Je_2 \wedge Je_4,$$
$$\sqrt{2}\,\omega_3 = e_1 \wedge e_4 + Je_1 \wedge Je_4, \quad \sqrt{2}\,\tilde{\omega}_3 = e_2 \wedge e_3 + Je_2 \wedge Je_3,$$
$$\sqrt{2}\,\pi_1 = e_1 \wedge Je_2 + e_2 \wedge Je_1, \quad \sqrt{2}\,\tilde{\pi}_1 = e_3 \wedge Je_4 + e_4 \wedge Je_3,$$
$$\sqrt{2}\,\pi_2 = e_1 \wedge Je_3 + e_3 \wedge Je_1, \quad \sqrt{2}\,\tilde{\pi}_2 = e_2 \wedge Je_4 + e_4 \wedge Je_2,$$
$$\sqrt{2}\,\pi_3 = e_1 \wedge Je_4 + e_4 \wedge Je_1, \quad \sqrt{2}\,\tilde{\pi}_3 = e_2 \wedge Je_3 + e_3 \wedge Je_2,$$
$$2\phi_0 = e_1 \wedge Je_1 + e_2 \wedge Je_2 + e_3 \wedge Je_3 + e_4 \wedge Je_4,$$
$$2\tilde{\phi}_0 = e_1 \wedge Je_1 + e_2 \wedge Je_2 - e_3 \wedge Je_3 - e_4 \wedge Je_4,$$
$$\sqrt{2}\,\phi_1 = e_1 \wedge Je_1 - e_2 \wedge Je_2,$$
$$\sqrt{2}\,\tilde{\phi}_1 = e_3 \wedge Je_3 - e_4 \wedge Je_4.$$

On a, pour $R_{\mathbf{C}P^4}$,

$$R\omega_i = 4\omega_i, \quad R\tilde{\omega}_i = 4\tilde{\omega}_i, \quad R\pi_i = 4\pi_i, \quad R\tilde{\pi}_i = 4\tilde{\pi}_i,$$
$$R\phi_0 = 20\phi_0, \quad R\tilde{\phi}_0 = 4\tilde{\phi}_0, \quad R\phi_1 = 4\phi_1, \quad R\tilde{\phi}_1 = 4\tilde{\phi}_1.$$

En fait, pour simplifier les calculs, on divisera toutes ces valeurs par 4.

On va chercher un tenseur de courbure $R$ admettant les vecteurs précédents comme vecteurs propres et ayant pour valeurs propres les nombres $\theta, \tilde{\theta}, p, \tilde{p}, f_0, f$, définis par

$$R\omega_i = \theta\omega_i, \quad R\tilde{\omega}_i = \tilde{\theta}\tilde{\omega}_i, \quad R\pi_i = p\pi_i, \quad R\tilde{\pi}_i = \tilde{p}\tilde{\pi}_i,$$
$$R\phi_0 = f_0\phi_0, \quad R\tilde{\phi}_0 = f\tilde{\phi}_0, \quad R\phi_1 = f\phi_1, \quad R\tilde{\phi}_1 = f\tilde{\phi}_1.$$

On obtient $R_{\mathbf{C}P^4}$ en faisant $\theta = \tilde{\theta} = p = \tilde{p} = f = 1, f_0 = 5$.

Pour ce tenseur de courbure on calculera successivement $p_2(R)$, $p_1^2(R)$, $c_1^2c_2(R)$ et $c_1^4(R)$ en tenant compte des relations imposées par la première identité de Bianchi, relations qui s'écrivent

$$\theta = \tfrac{1}{2}(f_0 - f - 2p),$$
$$\tilde{\theta} = \tfrac{1}{2}(f_0 - f - 2\tilde{p}).$$

Nous commençons par évaluer $p_2(R) = \beta_{2,8}\,\mathrm{proj}^{\perp}_{\Lambda^4 V}(R^{\cdot 2} \circ R^{\cdot 2})$. Il nous faut donc calculer $R^{\cdot 2}$, puis $R^{\cdot 2} \circ R^{\cdot 2}$. Le coefficient $\beta_{2,8}$ est obtenu en normalisant sur $\mathbf{C}P^4$ pour lequel $p_2(\mathbf{C}P^4) = 10$. Après un long calcul, nous obtenons

$$\begin{aligned} p_2(R) = 2^{-5}\Big\{ &\tfrac{1}{4}(f - f_0)(7f^3 - f^2 f_0 - 3f_0^2 f - 3f_0^3) + 8p\tilde{p}\theta\tilde{\theta} \\ &+ \tfrac{1}{2}(f - f_0)^2(p^2 + \tilde{p}^2 + \theta^2 + \tilde{\theta}^2) + 6(\theta^2\tilde{\theta}^2 + p^2\tilde{p}^2 \\ &+ \tilde{\theta}^2 p^2 + \theta^2\tilde{p}^2) + (f - f_0)\big[(3f + f_0)(p^2 + \tilde{p}^2 + \theta^2 + \tilde{\theta}^2) \\ &+ 2p(3\theta\tilde{\theta} + p\tilde{p}) + \tilde{p}(3\theta^2 + 3\tilde{\theta}^2 + p^2 + \tilde{p}^2)\big]\Big\} J^4, \end{aligned}$$

où $J^4 = J \wedge J \wedge J \wedge J$, et où nous avons effectué une nouvelle normalisation en prenant le volume de $\mathbf{C}P^4$ égal à 1.

Pour évaluer $p_1(R)$, nous partons de la relation

$$p_1(R) = \beta_{1,8}\,\mathrm{proj}^{\perp}_{\Lambda^4 V}(R \circ R).$$

Nous obtenons

$$\begin{aligned} p_1(R) &= \beta_{1,8}\Big\{ \theta^2 \sum_{i=1}^{3} \omega_i \wedge \omega_i + \tilde{\theta}^2 \sum_{i=1}^{3} \tilde{\omega}_i \wedge \tilde{\omega}_i + p^2 \sum_{i=1}^{3} \pi_i \wedge \pi_i + \tilde{p}^2 \sum_{i=1}^{3} \tilde{\pi}_i \wedge \tilde{\pi}_i \\ &\qquad + f_0\phi_0 \wedge \phi_0 + f(\tilde{\phi}_0 \wedge \tilde{\phi}_0 + \phi_1 \wedge \phi_1 + \tilde{\phi}_1 \wedge \tilde{\phi}_1)\Big\} \\ &= \beta_{1,8}\Big\{ (\theta^2 + p^2 - \tfrac{1}{2}f_0^2 + \tfrac{1}{2}f^2)(e_1 \wedge e_2 \wedge Je_1 \wedge Je_2 + e_1 \wedge e_3 \wedge Je_1 \\ &\qquad \wedge Je_3 + e_1 \wedge e_4 \wedge Je_1 \wedge Je_4) + (\tilde{\theta}^2 + \tilde{p}^2 - \tfrac{1}{2}f_0^2 + \tfrac{1}{2}f^2)(e_2 \wedge e_3 \\ &\qquad \wedge Je_2 \wedge Je_3 + e_2 \wedge e_4 \wedge Je_2 \wedge Je_4 + e_3 \wedge e_4 \wedge Je_3 \wedge Je_4)\Big\}. \end{aligned}$$

En normalisant sur $\mathbf{C}P^4$ nous obtenons

$$p_1^2(R) = \tfrac{1}{4}(\theta^2 + p^2 - \tfrac{1}{2}f_0^2 + \tfrac{1}{2}f^2)(\tilde{\theta}^2 + \tilde{p}^2 - \tfrac{1}{2}f_0^2 + \tfrac{1}{2}f^2)J^4.$$

Nous avons ensuite

$$c_2 = \tfrac{1}{2}(c_1^2 - p_1).$$

Or, lorsque l'on prend le volume de $\mathbf{C}P^4$ égal à 1, on a pour un tenseur de Kähler-Einstein $c_1(R) = f_0 J$. D'où

$$c_1^2 c_2(R) = \tfrac{1}{8}f_0^2(\theta^2 + \tilde{\theta}^2 + p^2 + \tilde{p}^2 + f^2 + 3f_0^2)J^4.$$

Posons

$$P(R) = 7500c_2^2(R) + 875p_1^2(R) - 500p_2(R) - 3027c_1^4(R).$$

Alors, en tenant compte des relations entre $p, \tilde{p}, \theta, \tilde{\theta}, f_0$ et $f$ exprimant la première identité de Bianchi, nous obtenons

$$P(R) = 5^3 \cdot 2^{-2}\{8,136f^4 + 2f^4 - 27f_0^3 f + 41 f_0^2 f^2 - f_0 f^3 + 12 p^2 \tilde{p}^2$$
$$+ (f - f_0)(6p\tilde{p}^2 - 2\tilde{p}^3) + 2(p^2 + \tilde{p}^2)(2f^2 + 27f_0^2 + ff_0)$$
$$+ (p + \tilde{p})(-27f_0^3 + 26f_0^2 f - f_0 f^2 + 2f^3)\}.$$

Nous pouvons trouver différents quadruplets $(p, \tilde{p}, f_0, f)$ rendant $P$ négatif. C'est le cas de $(2, -27, -1, -7)$ et $(-\frac{1}{2}, 39/2, 0, -1)$. Ce polynôme est aussi négatif dans les régions

$$p = 0, \quad \tilde{p} = 1, \quad -0,36365 \leqslant f_0 < 0, \quad f = 0,$$
$$p = 0, \quad \tilde{p} = 1, \quad f_0 = 0, \quad 0 < f \leqslant 0,3926.$$

Nous avons donc trouvé les exemples cherchés. On peut vérifier que pour le polynôme $P$ le point $(1, 1, 5, 1)$ correspondant à $R_{\mathbf{C}P^4}$ est un minimum relatif. Il serait intéressant d'établir cette propriété de minimum relatif pour tout l'espace des tenseurs de courbure de Kähler-Einstein car cela fournirait une autre preuve de l'isolation de la métrique de Fubini-Study parmi les métriques de Kähler-Einstein sur $\mathbf{C}P^4$.

On peut encore utiliser l'exemple précédent pour exhiber un tenseur de courbure de Kähler-Einstein non nul à courbure de Ricci nulle et intégrand d'Euler nul.

Pour avoir $c_1(R) = 0$, il faut prendre $f_0 = 0$. Nous avons par ailleurs

$$\chi = c_4 = \tfrac{1}{2}(p_2 - \tfrac{1}{4}p_1^2),$$

d'où

$$\chi(R) = 2^{-6}[\tfrac{7}{4}f^4 + 8p\tilde{p}\theta\tilde{\theta} + \tfrac{1}{2}f^2(p^2 + \tilde{p}^2 + \theta^2 + \tilde{\theta}^2) + 6(\theta^2\tilde{\theta}^2 + p^2\tilde{p}^2$$
$$+\tilde{\theta}^2 p^2 + \theta^2\tilde{p}^2) + f\{3f(p^2 + \tilde{p}^2 + \theta^2 + \tilde{\theta}^2)$$
$$+2p(3\theta\tilde{\theta} + p\tilde{p}) + \tilde{p}(3\theta^2 + 3\tilde{\theta}^2 + p^2 + \tilde{p}^2)\}]$$
$$-2^{-5}(\theta^2 + p^2 + \tfrac{1}{2}f^2)(\tilde{\theta}^2 + \tilde{p}^2 + \tfrac{1}{2}f^2).$$

On remarque que si $\tilde{\theta} = -\tilde{p}$, $\theta = p = f = 0$, alors $R$ est un tenseur de courbure kählérien non nécessairement nul et dont l'intégrand d'Euler est nul.

Bien sûr, par perturbation, en prenant par exemple $f = 2\varepsilon$, $p = \varepsilon$, $\theta = 0$, $\tilde{\theta} = 1 - \varepsilon$, $p = 1$, on a $\chi(R) \sim 2^{-3}\varepsilon$ lorsque $\varepsilon$ tend vers 0, d'où des exemples avec $\chi(R) < 0$.

## Références

[1] M. Berger, *Sur les variétés d'Einstein compactes*, C. R. III$^e$ Réunion Math. Expression Latine, Louvain Belgique, 1966, 35–55.

[2] J. P. Bourguignon & H. Karcher, *Curvature Operators: Pinching estimates and geometric examples*, Ann. Sci. École Norm. Sup. **11** (1978) 71–92.

[3] S. S. Chern, *On the curvature and characteristic classes of a Riemannian manifold*, Abh. Math. Sem. Univ. Hamburg **20** (1956) 117–126.

[4] R. Geroch, *Positive sectional curvature does not imply positive Gauss-Bonnet integrand*, Proc. Amer. Math. Soc. **54** (1976) 267–270.

[5] N. Hitchin, *Compact four-dimensional Einstein manifolds*, J. Differential Geometry **9** (1974) 435–441.

[6] D. L. Johnson, *Curvature and Euler characteristic for six-dimensional Kähler manifolds*, preprint.

[7] P. Klembeck, *On Geroch's counterexample to the algebraic Hopf conjecture*, Proc. Amer. Math. Soc. **59** (1976) 334–336.

[8] R. S. Kulkarni, *On the Bianchi identities*, Math. Ann. **199** (1972) 175–204.

[9] S. Mori, *Projective manifolds with ample tangent bundles*, Ann. of Math. **110** (1979) 590–606.

[10] A. Polombo, *Nombres caractéristiques d'une variété riemannienne de dimension* 4, J. Differential Geometry **13** (1978) 145–162.

[11] I. M. Singer & J. A. Thorpe, *The curvature of* 4*-dimensional Einstein spaces*, Global Analysis, Papers in Honor of K. Kodaira, Princeton University Press, Princeton, 1969, 355–365.

[12] Y. T. Siu & S. T. Yau, *Compact Kähler manifolds of positive bisectional curvature*, Invent. Math. **59** (1980) 189–204.

[13] A. Stehney, *Courbure d'ordre p et les classes de Pontrjagin*, J. Differential Geometry **8** (1973) 125–133.

[14] J. A. Thorpe, *Some remarks on the Gauss-Bonnet integral*, J. Math. Mech. **18** (1969) 779–786.

[15] J. Vilms, *On curvature operators of bounded rank*, Preprint, Colorado State University, Fort Collins.

[16] S. T. Yau, *On Calabi's conjecture and some new results in algebraic geometry*, Proc. Nat. Acad. Sci. U.S.A. **74** (1977) 1798–1799.

École polytechnique
Palaiseau Cedex, France

J. DIFFERENTIAL GEOMETRY
**16 (1981) 347–349**

# FOLIATIONS ON A SURFACE OF CONSTANT CURVATURE AND THE MODIFIED KORTEWEG–DE VRIES EQUATIONS

SHIING-SHEN CHERN & KETI TENENBLAT

*Dedicated to Professor Buchin Su on his 80th birthday*

ABSTRACT. The modified *KdV* equations are characterized as relations between local invariants of certain foliations on a surface of constant Gaussian curvature.

Consider a surface $M$, endowed with a $C^\infty$-Riemannian metric of constant Gaussian curvature $K$. Locally let $e_1, e_2$ be an orthonormal frame field and $\omega_1, \omega_2$ be its dual coframe field. Then the latter satisfy the structure equations

$$d\omega_1 = \omega_{12} \wedge \omega_2, \quad d\omega_2 = \omega_1 \wedge \omega_{12}, \quad d\omega_{12} = -K\omega_1 \wedge \omega_2, \tag{1}$$

where $\omega_{12}$ is the connection form (relative to the frame field). We write

$$\omega_{12} = p\omega_1 + q\omega_2, \tag{2}$$

$p, q$ being functions on $M$.

Given on $M$ a foliation by curves. Suppose that both $M$ and the foliation are oriented. At a point $x \in M$ we take $e_1$ to be tangent to the curve (or leaf) of the foliation through $x$. Since $M$ is oriented, this determines $e_2$. The local invariants of the foliation are functions of $p, q$ and their successive covariant derivatives. If the foliation is unoriented, then the local invariants are those which remain invariant under the change $e_1 \to -e_1$.

Under this choice of the frame field the foliation is defined by

$$\omega_2 = 0, \tag{3}$$

and $\omega_1$ is the element of arc on the leaves. It follows that $p$ is the geodesic curvature of the leaves.

We coordinatize $M$ by the coordinates $x, t$, such that

$$\omega_2 = Bdt, \quad \omega_1 = \eta dx + Adt, \quad \omega_{12} = udx + Cdt, \tag{4}$$

Received September 22, 1981. The first author is supported partially by NSF Grant MCS-8023356.

where $A, B, C, u$ are functions of $x, t$, and $\eta$ ($\neq 0$) is a constant. Thus the leaves are given by $t = \text{const}$, and $\eta x$ and $u/\eta$ are respectiviely the arc length and the geodesic curvature of the leaves. Substituting (4) into (1), we get

$$A_x = uB, \quad B_x = \eta C - uA, \quad C_x - u_t = -K\eta B. \tag{5}$$

Elimination of $B$ and $C$ gives

$$u_t = \left(\frac{A'_x}{u}\right)_{xx} + (uA')_x + \eta^2 K \frac{A'_x}{u}, \tag{6}$$

where

$$A' = A/\eta. \tag{7}$$

By choosing

$$A' = -K\eta^2 + \frac{1}{2}u^2, \tag{8}$$

we get

$$u_t = u_{xxx} + \frac{3}{2}u^2 u_x, \tag{9}$$

which is the modified Korteweg–de Vries (= $MKdV$) equation.

Condition (8) on the foliation can be expressed in terms of the invariants $p, q$ as follows: By (2) and (4) we have

$$u = \eta p, \quad C = Ap + Bq. \tag{10}$$

If we eliminate $B, C$ in the second equation by using (5), it can be written

$$\eta q = \left(\log \frac{A'_x}{u}\right)_x = (\log p_x)_x. \tag{11}$$

Introducing the covariant derivatives of $p$ by

$$dp = p_1\omega_1 + p_2\omega_2, \quad dp_1 = p_{11}\omega_1 + p_{12}\omega_2, \tag{12}$$

we have

$$p_x = p_1\eta, \quad p_{xx} = p_{11}\eta^2. \tag{13}$$

Hence condition (11) can be written

$$q = (\log p_1)_1. \tag{14}$$

A foliation will be called a $K$-foliation, if (14) is satisfied. We state our result in

**Theorem.** *The geodesic curvature of the leaves of a K-foliation satisfies, relative to the coordinates $x, t$ described above, an MKdV equation.*

The above argument can be generalized to $MKdV$ equations of higher order. The corresponding foliations are characterized by expressing $q$ as a function of $p, p_1, p_{11}, p_{111}, \cdots$.

Is there a similar geometrical interpretation of the *KdV*-equation itself, which is

$$(15) \qquad u_t = u_{xxx} + uu_x?$$

We do not have a simple answer to this question. Unlike the *MKdV*-equation, the sign of the last term is immaterial, because it reverses when $u$ is replaced by $-u$. It is therefore of interest to know that by a different foliation and a different coordinate system one can be led to a *MKdV*-equation (9) where the last term has a negative sign.

For this purpose we put

$$(16) \qquad \omega_2 = Bdt, \quad \omega_1 = vdx + Edt, \quad \omega_{12} = \lambda dx + Fdt,$$

where $\lambda$ is a parameter. Substitution into (1) gives

$$(17) \qquad F_x = -KvB, \quad B_x = -\lambda E + vF, \quad E_x - v_t = \lambda B.$$

Suppose $K \neq 0$, we get, by eliminating $B$, $E$,

$$(18) \qquad v_t = \left(\frac{F'_x}{Kv}\right)_{xx} + (vF')_x + \frac{\lambda^2}{Kv}F'_x,$$

where

$$(19) \qquad F = F'\lambda.$$

The choice

$$(20) \qquad F' = \frac{K}{2}v^2 - \lambda^2$$

reduces (18) into

$$(21) \qquad v_t = v_{xxx} + \frac{3}{2}Kv^2v_x.$$

Here the sign of the second term depends on the sign of $K$.

It can be proved that the choice (20) corresponds to a foliation which is characterized by

$$(22) \qquad q = \frac{p_{11}}{p_1} - 3\frac{p_1}{p} = \left(\log\frac{p_1}{p^3}\right)_1.$$

## References

[1] S. S. Chern & C. K. Peng, *Lie groups and KdV equations*, Manuscripta Math. **28** (1979) 207–217.

UNIVERSITY OF CALIFORNIA, BERKELEY
UNIVERSIDADE DE BRASILIA, BRASIL

*Chin. Ann. of Math.*
3 (4) 1982

# SINGULAR INTEGRALS IN SEVERAL COMPLEX VARIABLES (I)——HENKIN INTEGRALS OF STRICTLY PSEUDOCONVEX DOMAIN

GONG SHENG
(*University of Science and Technology of China, Institute of Applied Mathematics, Academia Sinica*)

SHI JIHUAI
(*University of Science and Technology of China*)

Dedicated to Professor Su Bu-chin on the Occasion of his 80th Birthday and his 50th Year of Educational Work

## § 0. Introduction

### 0.1. Cauchy-Fantappiè kernels

In the case of one complex variable, the Cauchy kernel gives rise to singular integrals on boundaries of arbitrary smooth domains. Cauchy kernel defines an analytic function $\mathbf{H}u$ on plane domain $\Omega$ by

$$\mathbf{H}u(w)=\int_{z\in b\Omega} H(w,\ z)u(z)d\sigma_z,\ w\in\Omega,$$

where $H(w, z)$ is the Cauchy kernel $[2\pi i(z-w)]^{-1}$, $d\sigma_z$ is Lebesgue measure on $b\Omega$. There is the famous Plemelj formula

$$\mathbf{H}u(w)=\frac{1}{2}u(w)+\text{p. v.}\int_{z\in b\Omega} H(w,\ z)u(z)d\sigma_z,\ w\in b\Omega, \tag{0.1}$$

as $w$ approaches $b\Omega$, where p. v. $\int_{z\in b\Omega}$ is defined by

$$\lim_{\substack{\varepsilon\to 0\\ \varepsilon>0}}\int_{\substack{z\in b\Omega\\ |z-w|>\varepsilon}} H(w,\ z)u(z)d\sigma_z. \tag{0.2}$$

An important point is that the deleted neighborhood around $w$ in (0.2), i. e. $\{z\in b\Omega,\ |z-w|<\varepsilon\}$ is symmetric. If the deleted neighborhood around $w$ is not symmetric, then the limit might fail to exist or the number $\frac{1}{2}$ in (0.1) might have to be modified. There is a corresponding Plemelj formula too as $w$ approaches $b\Omega$ from the exterior of $\Omega$. According to (0.1) and the corresponding formula, we may establish the theory of singular integrals and singular integral equations of one variable. For

Manuscript received April 24, 1981.

example, we can obtain the famous Poincaré-Bertrand formula: If $\varphi \in \mathrm{Lip}\alpha$, $(0<\alpha \leqslant 1)$, then

$$\mathbf{H}^2\varphi = \frac{1}{4}\varphi. \tag{0.3}$$

Here $\mathbf{H}$ is the Cauchy singular integral operator defined by (0.2). Using these formulas, we can solve the singular integral equations with Cauchy kernel, Hilbert kernel and some important boundary value problems. There are many works about this topic([1, 2]).

In several complex variables, there is no perfect analogue of the Cauchy kernel, what come closest to it are certain Cauchy-Fantappié (C-F) kernels

$$K(w, z) = \frac{c_n}{g^n}\,\omega \wedge dz_1 \wedge \cdots \wedge dz_n,\quad w \in \Omega,\ z \in b\Omega, \tag{0.4}$$

where

$$c_n = (-1)^{\frac{n(n-1)}{2}}(n-1)!\,(2\pi i)^{-n}, \tag{0.5}$$

$$g(w, z) = \sum_{i=1}^{n}(z_i - w_i)\,g_i(w, z), \tag{0.6}$$

$$\omega = g_1\bar\partial g_2 \wedge \cdots \wedge \bar\partial g_n + \cdots + (-1)^{n-1} g_n \wedge \bar\partial g_1 \wedge \cdots \wedge \bar\partial g_{n-1}. \tag{0.7}$$

These kernels are different from the Cauchy kernel of one complex variable, they depend on the domain $\Omega$. But they can be constructed for a wide class of $\Omega$ (smooth, bounded strictly pseudoconvex domains). The very important one of these domains is the ball. Henkin, Ramirez, Stein and Kerzman have constructed some important C-F kernels which are analytic when $w \in \Omega$. These kernels are called Henkin-Ramirez (H-R) kernel or Stein-Kerzman (S-K) kernel. It is well known that the requirement of the analyticity of the C-F kernels is very important.

### 0.2. Cauchy integral on the sphere

Singular integrals on the boundaries of special domains in $\mathbf{C}^n$ (such as the ball) first appeared and were studied in connection with Szegö kernel. Generally speaking, the Cauchy kernel is different from the Szegö kernel. C-F kernels are in the spirit of the Cauchy kernel and not of the Szegö kernel. However, if $\Omega$ is a ball in $\mathbf{C}^n$, the two kernels coincide. In 1965, Kung, S. and Sun, C. K.[3] studied the Cauchy integral on the sphere in $\mathbf{C}^n$. They proved: If $z = (z_1, \cdots, z_n) \in \mathbf{C}^n$, $z\bar z' < 1$, $u\bar u' = 1$, $f(u) \in \mathrm{Lip}\,p$ $(0<p\leqslant 1)$, then when $z$ approaches $v$ $(v\bar v' = 1)$ along the nontangential direction, we have

$$\lim_{z\to v} \frac{1}{\omega_{2n-1}}\int_{u\bar u'=1} \frac{f(u)\dot u}{(1-z\bar u')^n} = \frac{1}{2}f(v) + \text{p. v.}\,\frac{1}{\omega_{2n-1}}\int_{u\bar u'=1}\frac{f(u)\dot u}{(1-v\bar u')^n}, \tag{0.8}$$

where $\omega_{2n-1}$ is the volume of $u\bar u' = 1$. The principal value is defined by

$$\text{p. v.}\,\frac{1}{\omega_{2n-1}}\int_{u\bar u'=1}\frac{f(u)\dot u}{(1-v\bar u')^n} = \lim_{\varepsilon\to 0}\frac{1}{\omega_{2n-1}}\int_{\substack{u\bar u'=1\\ |1-v\bar u'|>\varepsilon}}\frac{f(u)\dot u}{(1-v\bar u')^n}.$$

After that, Koranyi and Vagi[4] obtained the same results by generalized Cayley transformation, this is equivalent to the Plemelj formula of generalized upper half plane

$$D=\left\{z=(z_1, \cdots, z_n):\ \operatorname{Im} z_1-\sum_{k=2}^{n}|z_k|^2>0\right\}.$$

On the other hand, using the Plemelj formulas of the sphere, we may obtain the Plemelj formulas of the matrix hyperbolic space and the Lie sphere hyperbolic space[5,6]. Shi Jihuai[7] also proved the following result

$$\lim_{z\to v}\frac{1}{\omega_{2n-1}}\int_{u\bar{u}'=1}\frac{f(u)\dot{u}}{(1-z\bar{u}')^n}=2^{n-2}f(v)+\text{p. v. }\frac{1}{\omega_{2n-1}}\int_{u\bar{u}'=1(\mathrm{Im})}\frac{f(u)\dot{u}}{(1-v\bar{u}')^n},$$

where p. v. $\frac{1}{\omega_{2n-1}}\int_{u\bar{u}'=1(\mathrm{Im})}$ is defined by $\lim_{\varepsilon\to 0}\frac{1}{\omega_{2n-1}}\int_{\substack{u\bar{u}'=1\\ |\mathrm{Im}(v\bar{u}')|>\varepsilon}}$.

If $u\bar{u}'=1$, $v\bar{v}'=1$, we have the following singular kernels

$$H(v, u)=(1-v\bar{u}')^{-n} \quad \text{(Cauchy kernel)},$$

$$B(v, u)=H(v, u)+H(u, v)-1 \quad \text{(B kernel)},$$

$$h(v, u)=\frac{1}{i}(H(v, u)-H(u, v)) \quad \text{(Hilbert kernel)}.$$

By these singular kernels, we may get the following singular integral operators on sphere

$$\mathbf{H}\varphi=2\omega_{2n-1}^{-1}\int_{u\bar{u}'=1}\varphi(u)H(v, u)\dot{u},$$

$$\mathbf{B}\varphi=2\omega_{2n-1}^{-1}\int_{u\bar{u}'=1}\varphi(u)B(v, u)\dot{u},$$

$$\mathbf{h}\varphi=2\omega_{2n-1}^{-1}\int_{u\bar{u}'=1}\varphi(u)h(v, u)\dot{u},$$

where the principal value is defined by $\lim_{\substack{\varepsilon\to 0\\ \varepsilon>0}}\int_{\substack{u\bar{u}'=1\\ |1-v\bar{u}'|>\varepsilon}}$. If $\varphi\in \mathrm{Lip}\,p$ $(0<p\leqslant 1)$, these singular integrals exist. From them we may also obtain the generalized Poincaré-Bertrand formula ($\mathbf{H}^2=I$), the generalized Hilbert formula of Hilbert kernel and solve the linear singular integral equations or the systems of linear singular integral equation with Cauchy kernel, $B$ kernel and Hilbert kernel, etc[8,9]. It is worth mentioning that $B$ kernel does not appear in one varirable. This shows the difference between one and several variables.

**0.3. Main results of this paper**

Pseudoconvex domains are the domains of holomorphy. For the strictly pseudoconvex domains, we can construct the analytic Henkin-Ramirez (H-R) kernel and Stein-Kerzman (S-K) kernel.

Alt[13] in 1974, Kerzman and Stein[14] in 1978 proved respectively the plemelj formula for the H-R kernel and S-K kernel. They proved: If $\Omega$ is $C^\infty$ smooth, bounded strictly pseudoconvex domain, $u\in C^\infty(b\Omega)$, $H(w, z)$ is H-R kernel or S-K kernel

$$\mathbf{H}u(w)=\int_{z\in b\Omega} H(w, z)u(z)d\sigma_z, \ w\in\Omega.$$

Then when $w$ approaches $b\Omega$ along the nontangential direction

$$\mathbf{H}u(w)=\frac{1}{2}u(w)+\text{p. v.}\int_{z\in b\Omega} H(w, z)u(z)d\sigma_z \quad (w\in b\Omega) \tag{0.9}$$

holds, where p. v. $\int_{z\in b\Omega}$ is defined by $\lim\limits_{\varepsilon\to 0}\int_{z\in b\Omega-B(w,\varepsilon)}$, and

$$B(w, \varepsilon)=\{z\in b\Omega, \ |g(z, w)|<\varepsilon\}.$$

In the section 1 of this paper, we consider that the deleted neighborhood $B(w, \varepsilon)$ around $w$ in the boundary of a strictly pseudoconvex domain is a more general form instead of above form, a corresponding plemelj formula is obtained. This shows the essential difference between one and several variables: There is only one method to define the principal value of the Cauchy integral in the former, but there are infinite many methods in the latter, even if the deleted part is symmetry. Basing upon these results, we discuss the singular integral on the boundary of a strictly pseudoconvex domain.

In the sections 2 and 3, we discuss some special cases of $B(w, \varepsilon)$ and the theory of the corresponding singular integral equations. In the section 2, we consider the situation that the deleted neighborhood $B(w, \varepsilon)$ is an "ellipse" or a "belt" when the "ellipse" is a "disc", we obtain the results of Alt and Kerzman-Stein. We also discover that there is a method of deletion, such that the term $u(w)$ does not appear in the Plemelj formula, in other words, when $w$ approaches the boundary from the interior, the Cauchy integral approaches a special principal value. In the section 3, we consider the situation that the deleted neighborhood is a "rectangle". All these show the variety of the definition of the principal value of the Cauchy integral.

Part results of this paper were announced in [10].

## § 1. Cauchy integrals on a strictly pseudoconvex domain

### 1.1 Henkin-Ramirez kernel and Stein-Kerzman kernel

Suppose $\Omega$ is a smooth, bounded strictly pseudoconvex domain, $\lambda$ is a real function, $\lambda\in C^\infty(\overline{\Omega})$, $\lambda(z)<0$, if $z\in\Omega$; $\lambda(z)=0$, if $z\in b\Omega$; $\lambda(z)>0$, if $z\bar{\in}\Omega$; grad $\lambda(z)\neq 0$, if $z\in b\Omega$ and

$$\left(\frac{\partial^2\lambda}{\partial z_i\partial \bar{z}_i}\right)\geqslant CI, \tag{1.1.1}$$

where the constant $C>0$ is independent of $z\in\overline{\Omega}$, $I$ is the identity matrix. For $z\in b\Omega$, $w\in\Omega$ and $w$ near $z$, set

$$g_i^L(w, z)=\frac{\partial\lambda}{\partial z_i}(z)+\frac{1}{2}\sum_{j=1}^{n}\frac{\partial^2\lambda}{\partial z_i\partial z_j}(z)(w_j-z_j), \tag{1.1.2}$$

$$g^L(w, z) = \sum_{i=1}^{n}(z_i - w_i) g_i^L(w, z).$$

Extend the local $g^L(w, z)$ to a global $g(w, z)$: for $w$ near $z$, set

$$g(w, z) = g^L(w, z)\phi(w, z),$$

where $\phi(w, z)$ is a locally defined function and is holomorphic in $w$ for $w$ near $z$. Furthermore $\phi(z, z) \neq 0$. The global $g_i(w, z)$ are obtained from the division problem

$$g(w, z) = \sum (z_i - w_i) g_i(w, z). \tag{1.1.3}$$

It may be proved that there are the functions $g_i(w, z)$ which are holomorphic in $w$ such that (1.1.3) holds. Substitute such $g_i(w, z)$ into $g_i(w, z)$ which are needed in the C-F kernel(0.6) of $\Omega$ and substitute $g_i(w, z)$ of (1.1.3) into $g$ of (0.4), we obtain the C-F kernel $H(w, z)$ of $\Omega$, this is just the Henkin-Ramirez (H-R) kernel. As for the details of proof, see [11, 15].

Stein and kerzman[14] constructed another kernel $E+C$, called Stein-Kerzman (S-K) kernel, where $E(w, z)\, d\sigma_z$ is a C-F form (0.4), in which $g_i(w, z)$ are totally explicit, but are holomorphic only when $w$ is close to $z \in b\Omega$, namely as $w$ near $z$, $g_i(w, z) = g_i^L(w, z)$ and $C(w, z)$ is the correction term, such that $H(w, z) = E(w, z) + C(w, z)$ is holomorphic in $w$ globally. Furthermore

$$C(w, z) \in C^\infty(U(\overline{\Omega}) \times V(b\Omega)),$$

where $U(\overline{\Omega})$ and $V(b\Omega)$ are the neighborhoods of $\overline{\Omega}$ and $b\Omega$ respectively. That is to say, $C(w, z)$ is infinitely smooth even when $w=z$. $C(w, z)$ is a solution of a $\bar\partial$ problem. $C(w, z)$ is not a C-F form, but $E+C$ can reproduce holomorphic functions. As for the details, see [14].

It is known that for H-R kernel and S-K kernel, we both have

$$\int_{z\in b\Omega} |\mathrm{H}(w, z)|\, d\sigma_z = \infty, \quad w \in b\Omega, \tag{1.1.4}$$

where $d\sigma_z$ is the Lebesgue element of area on $b\Omega$.

Fix $w \in b\Omega$, without loss of generality, take $w=0$, then we have the following

**Theorem 1.1.**[12,14] *Suppose $\Omega$ is a smooth, bounded strictly pseudoconvex domoin, $0 \in b\Omega$, $H(w, z)$ is the H-R kernel or S-K kernel. Then near $z=0$, there is a holomorphic local change of variables, such that $b\Omega$ and $H(0, z)$ have the following forms: $z \in b\Omega$ if and only if $z_n = t + i\tau$ satisfies*

$$\tau = |\xi|^2 + \varepsilon(\xi, t), \tag{1.1.5}$$

*where $z = (z_1, \cdots, z_n)$, $\xi = (z_1, \cdots, z_{n-1})$, the error term $\varepsilon$ is of third order, namely*

$$|\varepsilon(\xi, t)| = O(\rho^3), \tag{1.1.6}$$

*where $\rho^4 = |\xi|^4 + t^2$. For $H(0, z)$, we have*

$$H(0, z) = \frac{\gamma_n}{(|\xi|^2 + it)^n} + \phi(z), \tag{1.1.7}$$

*where $\phi(z)$ is absolutely integrable in $z \in b\Omega$ and $\gamma_n$ is a constant.*

Obviously $(\xi, t) \in \mathbf{C}^{n-1} \times \mathbf{R}$ and $\tau = |\xi|^2$ is "Heisenberg group" surface $S_{n-1}$.

(1.1.5) shows that near $z=0$ the $b\Omega$ may be approximated by the "Heisenberg group" surface $S_{n-1}$ and the $\varepsilon(\xi, t)$ of (1.1.6) is the error of approximation.

It is easy to know[12,13]

$$\int_{\rho<1} |(|\xi|^2+it)|^{-n} d\mu=\infty, \tag{1.1.8}$$

$$\int_{0<a<\rho<b} |(|\xi|^2+it)^{-n} d\mu=0, \tag{1.1.9}$$

where $d\mu$ is the usual Lebesgue measure $dx_1 dy_1 \cdots dx_{n-1} dy_{n-1} dt$ $(z_j=x_j+iy_j)$. In fact, we may prove

$$\frac{1}{g^n(0, z)}=\frac{1}{(|\xi|^2+it)^n}(1+O(\rho)). \tag{1.1.10}$$

## 1.2. General plemelj formula

We first prove the following general Plemelj formula.

**Theorem 1.2.** *Suppose $\Omega$ is a smooth, bounded strictly pseudoconvex domain, $H(w, z)$ is the H-R kernel or S-K kernel, $u\in C^\infty(b\Omega)$,*

$$\mathbf{H}u(w)\equiv\int_{z\in b\Omega} H(w, z)u(z)d\sigma_z, \quad w\in\Omega \tag{1.2.1}$$

*is the Cauchy integral, where $d\sigma_z$ is the Lebesgue element of area on $b\Omega$. Then $\mathbf{H}u$ is holomorphic in $\Omega$ and admits a continous extension up to $\overline{\Omega}$. When $w\in b\Omega$, define*

$$\text{p. v.}\int_{z\in b\Omega} H(w, z)u(z)d\sigma_z\equiv\lim_{\substack{\varepsilon\to0\\ \varepsilon>0}}\int_{z\in b\Omega-D(w,\varepsilon)\cap b\Omega} H(w, z)u(z)d\sigma_z, \tag{1.2.2}$$

*here $D(w, \varepsilon)$ is the neighborhood around $w$ and contracts to the point $w$ as $\varepsilon\to0$. If*

$$\lim_{\varepsilon\to0}\lim_{\delta\to0}\int_{z\in D(w,\varepsilon)\cap b\Omega} H(w+\delta\nu, z)d\sigma_z=a \tag{1.2.3}$$

*exists, where $a$ is a constant and $\nu$ is the inner normal to $b\Omega$ at $w$, $\delta>0$. Denote the value of (1.2.2) by $\mathbf{H}_a u$ and*

$$\mathbf{H}u(w)=\lim_{w_0\to w}\mathbf{H}u(w_0),$$

*where $w_0$ approaches $w\in b\Omega$ along the nontangential direction from the interior of $\Omega$. Then we have the Plemelj formula*

$$\mathbf{H}u(w)=au(w)+\mathbf{H}_a u(w). \tag{1.2.4}$$

Obviously, this deduces to the Theorem of Alt[13] and Kerzman-Stein[14] when $D(w, \varepsilon)$is $\{z\in b\Omega, |g(z, w)|<\varepsilon\}$. In this case, $a=\dfrac{1}{2}$ and Plemelj formula becomes

$$\mathbf{H}u(w)=\frac{1}{2}u(w)+\mathrm{H}_{\frac{1}{2}}u(w). \tag{1.2.5}$$

We now prove Theorem 1.2.

Keep $w\in b\Omega$ fixed, say $w=0$. Then the condition (1.2.3) becomes

$$\lim_{\varepsilon\to0}\lim_{\delta\to0}\int_{z\in D(0,\varepsilon)\cap b\Omega} H(0+\delta\nu, z)d\sigma_z=a, \tag{1.2.6}$$

where $\nu$ is the inner normal to $b\Omega$ at $w=0$. Write

$$\lim_{\varepsilon\to 0}\lim_{\delta\to 0}\int_{z\in b\Omega-D(0,\varepsilon)\cap b\Omega} H(0+\delta\nu,\ z)u(z)d\sigma_z$$
$$=\lim_{\varepsilon\to 0}\lim_{\delta\to 0}\int_{z\in b\Omega-D(0,\varepsilon)\cap b\Omega} H(0+\delta\nu,\ z)[u(z)-u(0)]d\sigma_z$$
$$+u(0)\lim_{\varepsilon\to 0}\lim_{\delta\to 0}\int_{z\in b\Omega-D(0,\varepsilon)\cap b\Omega} H(0+\delta\nu,\ z)d\sigma_z,$$

since $u\in C^\infty(b\Omega)$, the above first integral exists[13], and the second integral is equal to $u(0)(1-a)$ by (1.2.6). Therefore, for $0\in b\Omega$, we have

$$\text{p. v.}\int_{z\in b\Omega} H(0,\ z)u(z)d\sigma_z=\lim_{\substack{\varepsilon\to 0\\ \varepsilon>0}}\int_{z\in b\Omega-D(0,\varepsilon)\cap b\Omega} H(0,\ z)u(z)d\sigma_z$$
$$=\int_{z\in b\Omega} H(0,\ z)[u(z)-u(0)]d\sigma_z+u(0)(1-a).$$

Hence, for $w\in b\Omega$, we also have

$$\text{p. v.}\int_{z\in b\Omega} H(w,\ z)u(z)d\sigma_z=\int_{z\in b\Omega} H(w,\ z)[u(z)-u(w)]d\sigma_z+u(w)(1-a). \tag{1.2.7}$$

Let $w_0\in\Omega$

$$\int_{z\in b\Omega} H(w_0,\ z)u(z)d\sigma_z=\int_{z\in b\Omega} H(w_0,\ z)[u(z)-u(w)]d\sigma_z$$
$$+u(w)\int_{z\in b\Omega} H(w_0,\ z)d\sigma_z=I_1+I_2.$$

When $w_0$ approaches $w$ along the nontangential direction, by (1.2.7) we obtain

$$\lim_{w_0\to w} I_1=\int_{z\in b\Omega} H(w,\ z)[u(z)-u(w)]d\sigma_z$$
$$=\text{p. v.}\int_{z\in b\Omega} H(w,\ z)u(z)d\sigma_z-u(w)(1-a),$$

hence

$$\mathbf{H}u(w)=\text{p. v.}\int_{z\in b\Omega} H(w,\ z)u(z)d\sigma_z-u(w)(1-a)+u(w)$$
$$=\text{p. v.}\int_{z\in b\Omega} H(w,\ z)u(z)d\sigma_z+au(w).$$

This ends the proof.

Notice the condition (1.2.3) By Theorem 1.1, there is a holomorphic local change of variables, such that $b\Omega$ and $H(0,\ z)$ can be represented by (1.1.5) and (1.1.7) respectively near $0\in b\Omega$ and (1.2.6) beocmes[13, 14]

$$\lim_{\varepsilon\to 0}\lim_{\delta\to 0}\left\{\int_{z\in D(0,\varepsilon)} \gamma_n(|\xi|^2+\delta+it)^{-n}\,d\mu+\int_{z\in D(0,\varepsilon)} \phi_\delta(z)d\mu\right\},$$

where $\phi_\delta(z)$ is absolutely integrable in $z\in b\Omega$ and $\lim_{\delta\to 0}\phi_\delta(z)=\phi(z)$. Since $D(0,\ \varepsilon)$ contracts to a point as $\varepsilon\to 0$, obviously

$$\lim_{\varepsilon\to 0}\lim_{\delta\to 0}\int_{z\in D(0,\varepsilon)} \phi_\delta(z)d\mu=0,$$

therefore the condition (1.2.3) becomes

$$\lim_{\varepsilon\to 0}\lim_{\delta\to 0}\int_{z\in D(0,\varepsilon)}\gamma_n(|\xi|^2+\delta+it)^{-n}\,d\mu=a.$$

Consider generalized Cayley transformation $T$

$$z_1=\frac{u_1}{1+u_n},\ \cdots,\ z_{n-1}=\frac{u_{n-1}}{1+u_n},\ z_n=i\,\frac{1-u_n}{1+u_n}.$$

Since $z_n=t+i\tau$, hence

$$t=\frac{2\operatorname{Im}u_n}{|1+u_n|^2},\quad \tau=\frac{1-|u_n|^2}{|1+u_n|^2},$$

$T$ transforms $|\xi|^2+it+\delta$ into

$$\frac{|u_1|^2+\cdots+|u_{n-1}|^2+2i\operatorname{Im}u_n}{|1+u_n|^2}+\delta.$$

Since $T$ maps upper half plane $\tau>0$, $\tau=|\xi|^2$ onto $u\bar{u}'=1$, the above can also be written as

$$\frac{1+2i\operatorname{Im}u_n-|u_n|^2}{|1+u_n|^2}+\delta=\frac{(1+u_n)(1-\bar{u}_n)}{|1+u_n|^2}+\delta=\frac{1-\bar{u}_n}{1+\bar{u}_n}+\delta=\frac{(1+\delta)-(1-\delta)\bar{u}_n}{1+\bar{u}_n}$$

and $d\mu=\omega_{2n-1}^{-1}(1+\bar{u}_n)^{-n}\dot{u}$.

So the condition (1.2.6) becomes

$$\lim_{\varepsilon\to 0}\lim_{\delta\to 0}\frac{1}{\omega_{2n-1}}\int_{z\in D(p_n,\varepsilon)}\frac{\dot{u}}{(1+\delta)^n\left(1-\dfrac{1-\delta}{1+\delta}\bar{u}_n\right)^n}=\lim_{\varepsilon\to 0}\lim_{\rho\to 1}\frac{1}{\omega_{2n-1}}\int_{z\in D(p_n,\varepsilon)}\frac{\dot{u}}{(1-\rho\bar{u}_n)^n}$$

$$=\lim_{\varepsilon\to 0}\lim_{\rho\to 1}\frac{1}{\omega_{2n-1}}\int_{z\in D(p_n,\varepsilon)}\frac{\dot{u}}{(1-\rho p_n\bar{u}')^n}=a, \tag{1.2.8}$$

where $p_n=(0,\cdots,0,1)$, $D(p_n,\varepsilon)=T(D(0,\varepsilon))$ is the neighborhood around $p_n$, $\rho=\dfrac{1-\delta}{1+\delta}$.

Let us observe the $D(w,\varepsilon)$ in detail. In Theorem 1.1, when changing the variables, we set $g(z,0)=-it+\tau=-iz_n$. Hence, if the definition of $D(w,\varepsilon)$ depends on $g(z,w)$, $D(w,\varepsilon)$ may be written as $D(w,\varepsilon;\operatorname{Re}g,\operatorname{Im}g)$ it becomes $D(0,\varepsilon,\operatorname{Re}g(z,0),\operatorname{Im}g(z,0))$, i. e. $D(0,\varepsilon,\tau,-t)$ after the change of variables and becomes

$$D\left(p_n,\ \varepsilon,\ \frac{1-|u_n|^2}{|1+u_n|^2},\ \frac{-2\operatorname{Im}u_n}{|1+u_n|^2}\right)$$

under the generalized transformation of Cayley.

For example

$$D(w,\varepsilon,\operatorname{Re}g,\operatorname{Im}g)=\{\alpha^2(\operatorname{Re}g)^2+\beta^2(\operatorname{Im}g)^2\leqslant\varepsilon^2\}$$

may be reduced to

$$\{\alpha^2(1-|u_n|^2)^2+4\beta^2(\operatorname{Im}u_n)^2\leqslant\varepsilon^2\};$$

$$D(w,\varepsilon,\operatorname{Re}g,\operatorname{Im}g)=\{|\operatorname{Re}g|<\alpha\varepsilon,\ |\operatorname{Im}g|<\beta\varepsilon,\ \alpha>0,\ \beta>0\}$$

may be reduced to

$$\{1-|u_n|^2<\alpha\varepsilon,\ 2|\operatorname{Im}u_n|<\beta\varepsilon,\ \alpha>0,\ \beta>0\}.$$

In the sections 2 and 3, we shall give some concrete neighborhoods $D(w,\varepsilon)$; calculate the value of $a$ in (1.2.8), one can discover some interesting results.

## 1.3. Singular integrals on $b\Omega$

When $w\in b\Omega$, $\int_{z\in b\Omega} H(w, z)u(z)d\sigma_z$ is a singular integral.

Let $z\in b\Omega$, $\varphi(z)\in \mathrm{Lip}\,p$, $0<p\leqslant 1$, $v$, $w\in b\Omega$, set

$$\varphi_1(w)=\int_{b\Omega(a)} H(w, z)\varphi(z)d\sigma_z,\quad \varphi_2(v)=\int_{b\Omega(b)} H(v, z)\varphi_1(z)d\sigma_z,$$

where $\int_{b\Omega(a)}$ stands for the Cauchy principal value, the deleted neighborhood is $D_a(w, \varepsilon)$, such that the corresponding value of (1.2.3) is $a$.

Consider the Cauchy integrals

$$f(\eta)=\int_{b\Omega} H(\eta, z)\varphi(z)d\sigma_z,\ f_1(\eta)=\int_{b\Omega} H(\eta, z)\varphi_1(z)d\sigma_z,\ \eta\in\Omega,$$

when $\eta$ approaches $\zeta\in b\Omega$ along the nontangential direction, by (1.2.4)

$$f(\zeta)=\int_{b\Omega(a)} H(\zeta, z)\varphi(z)d\sigma_z+a\varphi(\zeta),\quad f_1(\zeta)=\int_{b\Omega(b)} H(\zeta, z)\varphi_1(z)d\sigma_z+b\varphi_1(\zeta).$$

By the definition of $\varphi_1(\zeta)$, $\varphi_2(\zeta)$, we have

$$\varphi_1(\zeta)=f(\zeta)-a\varphi(\zeta),\ \varphi_2(\zeta)=f_1(\zeta)-b\varphi_1(\zeta),\ \zeta\in b\Omega.$$

Substitute $\varphi_1(\zeta)$ into the expression of $f_1(\eta)$

$$f_1(\eta)=\int_{b\Omega(b)} H(\eta, z)f(z)d\sigma_z-a\int_{b\Omega(b)} H(\eta, z)\varphi(z)d\sigma_z=(1-a)f(\eta),$$

therefore

$$\varphi_2(\zeta)=(1-a)f(\zeta)-b[f(\zeta)-a\varphi(\zeta)]=(1-a-b)f(\zeta)+ab\varphi(\zeta).$$

For $a$, if there is a neighborhood $D_b(w, \varepsilon)$ around $w$, such that the corresponding value of (1.2.3) is $b$ and $1-a-b=0$, then $\varphi_2(\zeta)=ab\varphi(\zeta)$, that is to say

$$\mathbf{H}_b\mathbf{H}_a=abI.$$

where $I$ stands for the identity operator. In the same manner, we can also prove

$$\mathbf{H}_a\mathbf{H}_b=abI.$$

Thus we obtain the generalized Poincarè-Bertrand formula.

**Theorem 1.3.** *Let $\Omega$ be a smooth, bounded strictly pseudoconvex domain of $\mathbf{C}^n$, $H(w, z)$ is H-R kernel or S-K kernel, $\varphi\in C^\infty(b\Omega)$. The singular integral operator $\mathbf{H}_a$ is defined by (1.2.2). If for $a$, there is another definition of Cauchy principal value, such that the value of (1.2.3) is $b$, and $1-a-b=0$. Suppose the corresponding singular integral operator is $\mathbf{H}_b$, then*

$$\mathbf{H}_a\mathbf{H}_b=\mathbf{H}_b\mathbf{H}_a=abI, \tag{1.3.1}$$

*where $I$ stands for the identity operator.*

If $ab\neq 0$, (1.3.1) may also be written as: denote $\frac{1}{a}\mathbf{H}_a\varphi=\psi$, then $\frac{1}{b}\mathbf{H}_b\psi=\varphi$; or denote $\frac{1}{b}\mathrm{H}_b\varphi=\psi$, then $\frac{1}{a}\mathrm{H}_a\psi=\varphi$.

It is easy to know by the operator theory, if $\lambda$ is not a characteristic value of the singular integral operator $\mathbf{H}_a$, $\mathbf{K}$ is a continuous operator, $f\in C^\infty(b\Omega)$, then the singular integral equation on $b\Omega$

$$(-\lambda I+\mathbf{H}_a+\mathbf{K})\varphi=\psi \tag{1.3.2}$$

has exactly one solution, namely it can be normalized to a Fredholm integral equation. When $D(w, \varepsilon)=\{|g(w, z)|<\varepsilon\}$, $\Omega$ is a ball, this is just the theory of the singular integral equation on a sphere[8, 9].

When $z\in b\Omega$, $w\in b\Omega$, the H-R kernel or S-K kernel $H(w, z)$ of $\Omega$ are both singular kernels. For the case of sphere, one can define $B$ kernel and Hilbert kernel from $H(w, z)$

$$B(w, z)=H(w, z)+H(z, w)-1,$$

$$h(w, z)=\frac{1}{i}[H(w, z)-H(z, w)].$$

From these kernels, we also have

$$\mathbf{B}_a\varphi=\int_{b\Omega(a)} B(w, z)\varphi(z)d\sigma_z,$$

$$\mathbf{h}_a\varphi=\int_{b\Omega(a)} h(w, z)\varphi(z)d\sigma_z,$$

where the integrals denote the Cauchy principal value, the deleted neighborhood around $w$ is $D_a(w, \varepsilon)$ such that the value of (1.2.3) is $a$.

Obviously, if $\mathbf{H}_a$ has an inverse operator, then $\mathbf{B}_a$ and $\mathbf{h}_a$ also have the inverse operators. Therefore, we can also solve the singular integral equations on $b\Omega$

$$(-\mu I+\mathbf{B}_a+\mathbf{K})\varphi=f \tag{1.3.3}$$

and

$$(-\nu I+\mathbf{h}_a+\mathbf{K})\varphi=f, \tag{1.3.4}$$

where $\mu$, $\nu$ are not the characteristic value of $\mathbf{B}_a$ and $\mathbf{h}_a$ respectively, $\mathbf{K}$ is a continuous operator, $f\in C^\infty(b\Omega)$.

As in [8, 9], one can solve the systems of singular integral equations similar to (1.3.2), (1.3.3), (1.3.4) and when $D(w, \varepsilon)=\{|g(w, \varepsilon)|<\varepsilon\}$, $\Omega$ is a ball, this is just the theory of singular integral equations discussed in [8].

## § 2. The case that the neighborhood is an "ellipse"

### 2.1. Plemelj formula

In this section, we shall consider some special neighborhoods $D(w, \varepsilon)$, such that the concrete Plemelj formulas are obtained from the general Plemelj formula. We first consider the case that the neighborhood $D_e(w, \varepsilon)$ is an "ellipse":

$$\{z\in b\Omega,\ \alpha^2(\operatorname{Re} g)^2+\beta^2(\operatorname{Im} g)^2\leqslant\varepsilon^2\},$$

where $\alpha\geqslant 0$, $\beta\geqslant 0$, $\alpha+\beta\neq 0$. When $\alpha=\beta$, the "ellipse" becomes a, "disc", this is just the case discussed by Alt[13] and Kerzman-Stein[14]. In this case, we shall prove that (1.2.3) becomes

$$\lim_{s\to 0}\lim_{\delta\to 0}\int_{z\in D_s(w,s)} H(w+\delta\nu,\ z)\,d\sigma_s=\frac{1}{2}\left(\frac{2\beta}{\alpha+\beta}\right)^{n-1}. \tag{2.1.1}$$

That is to say, the value of $a$ in Plemelj formula (1.2.4) is

$$a=\frac{1}{2}\left(\frac{2\beta}{\alpha+\beta}\right)^{n-1}.$$

When $\alpha=\beta$, this is just the Theorem of Alt[13] and Kerzman-Stein[14]. Therefore, this result has generalized the works of Alt and Kerzman-Stein.

Another case $\beta=0$ must be noted. In this case, $a=0$, it follows that the Plemelj formula becomes vary simple

$$\mathbf{H}u(w)=\mathbf{H}_0u(w).$$

In other words, when $w_0$ approaches $w\in b\Omega$ along the nontangential direction from the interior of $\Omega$

$$\lim_{w_0\to w}\int_{z\in b\Omega} H(w,\ z)u(z)\,d\sigma_z=\lim_{\substack{s\to 0\\ s>0}}\int_{z\in b\Omega-D(w,s)\cap b\Omega} H(w,\ z)u(z)\,d\sigma_z.$$

Now $D(w,\ z)$ is a "belt"

$$\{z\in b\Omega,\ |\mathrm{Re}\, g|<s\}.$$

Namely, one can find a manner to define the principal value of Cauchy integral, such that this principal value is just the limit value of the Cauchy integral $\mathbf{H}u(w_0)$ when $w_0$ approaches $w\in b\Omega$ along the nontangential direction from the interior of $\Omega$. It is impossible in one variable that the limit value of the Cauchy integral on $b\Omega$ can be represented by a certain principal value of the Cauchy integral on $b\Omega$.

We now discuss the conditions under which the singular integral operator $\mathbf{H}_a$ has an inverse operator. It has been pointed out in the section 1.3 that if one can find an operator $\mathbf{H}_b$, such that $1-a-b=0$, $ab\neq 0$, then $\frac{1}{b}\mathbf{H}_b$ is the inverse operator of $\frac{1}{a}\mathbf{H}_a$. Obviously, if

$$\frac{1}{2}\left(\frac{2\beta}{\alpha+\beta}\right)^{n-1}<1,$$

i. e.

$$\frac{\alpha}{\beta}>2^{\frac{n-2}{n-1}}-1, \tag{2.1.2}$$

take $\quad D_s(w,\ s)=\{z\in b\Omega,\ \alpha'^2(\mathrm{Re}\,g)^2+\beta'^2(\mathrm{Im}\,g)^2\leqslant s^2\}$,

where $\alpha'\geqslant 0$, $\beta'\geqslant 0$, $\alpha'+\beta'\neq 0$ and

$$\frac{\alpha'}{\beta'}=\left[2^{2-n}-\left(\frac{\alpha+\beta}{\beta}\right)^{1-n}\right]^{-\frac{1}{n-1}}-1,$$

if denote

$$\frac{1}{2}\left(\frac{2\beta'}{\alpha'+\beta'}\right)^{n-1}=b,$$

then $\frac{1}{ab}\mathbf{H}_b$ is the inverse operator of $\mathbf{H}_a$. That is to say, under the condition of (2.1.2), $\mathbf{H}_a$ has an inverse operator. In the same way, we may set up the theory of singnlar integral equations.

**2.2. The proof of (2.1.1)**

To prnve (2.1.1), it is enough to prove that (1.2.8) holds by § 1.2. Namely

$$\lim_{\varepsilon\to 0}\lim_{\rho\to 1}\frac{1}{\omega_{2n-1}}\int_{u\in D(p_n,\varepsilon)}\frac{\dot{u}}{(1-\rho p_n\bar{u}')^n}=\lim_{\varepsilon\to 0}\lim_{\rho\to 1}\frac{1}{\omega_{2n-1}}\int_{u\in D(p_n,\varepsilon)}\frac{\dot{u}}{(1-\rho\bar{u}_n)^n}$$

$$=\frac{1}{2}\left(\frac{2\beta}{\alpha+\beta}\right)^{n-1},$$

where $p_n=(0, \cdots, 0, 1)$, $0<\rho<1$, and

$$D(p_n, \varepsilon)=\{u\bar{u}'=1,\ \alpha^2(1-|u_n|^2)^2+4\beta^2(\operatorname{Im}u_n)^2\leqslant\varepsilon^2\}.$$

Let
$$\tilde{D}(p_n, \varepsilon)=\{u\bar{u}'=1,\ \alpha^2(1-|u_n|^2)^2+4\beta^2(\operatorname{Im}u_n)^2>\varepsilon^2\}.$$

Since
$$\frac{1}{\omega_{2n-1}}\int_{u\in D(p_n,\varepsilon)}\frac{\dot{u}}{(1-\rho\bar{u}_n)^n}+\frac{1}{\omega_{2n-1}}\int_{u\in \tilde{D}(p_n,\varepsilon)}\frac{\dot{u}}{(1-\rho\bar{u}_n)^n}=1,$$

we only need to prove

$$\lim_{\varepsilon\to 0}\lim_{\rho\to 1}\frac{1}{\omega_{2n-1}}\int_{u\in \tilde{D}(p_n,\varepsilon)}\frac{\dot{u}}{(1-\rho\bar{u}_n)^n}=\lim_{\varepsilon\to 0}\frac{1}{\omega_{2n-1}}\int_{u\in \tilde{D}(p_n,\varepsilon)}\frac{\dot{u}}{(1-\bar{u}_n)^n}$$

$$=1-\frac{1}{2}\left(\frac{2\beta}{\alpha+\beta}\right)^{n-1}, \tag{2.2.1}$$

where $\alpha\geqslant 0$, $\beta\geqslant 0$, $\alpha+\beta\neq 0$.

when $\alpha=0$, $\beta>0$, (2.2.1) has been proved in [7].

We now consider the case $\alpha>0$, $\beta>0$.

As has been discussed in [3, 7], let

$$\bar{u}_n=re^{i\theta},\ v=(u_1, \cdots, u_{n-1}),$$

then $\tilde{D}(p_n, \varepsilon)$ may be witten in form

$$\begin{cases} v\bar{v}'=1-r^2,\\ \alpha^2(1-r^2)^2+4\beta^2r^2\sin^2\theta>\varepsilon^2. \end{cases}$$

Set
$$c=\arcsin\frac{\sqrt{\varepsilon^2-\alpha^2(1-r^2)^2}}{2\beta r},$$

we have

$$\frac{1}{\omega_{2n-1}}\int_{u\in\tilde{D}(p_n,\varepsilon)}\frac{\dot{u}}{(1-\bar{u}_n)^n}=\frac{1}{\omega_{2n-1}}\int_{v\bar{v}'<\frac{\varepsilon}{\alpha}}\dot{v}\left\{\int_{-(\pi-c)}^{-c}\frac{d\theta}{(1-re^{i\theta})^n}+\int_c^{\pi-c}\frac{d\theta}{(1-re^{i\theta})^n}\right\}$$

$$+\frac{1}{\omega_{2n-1}}\int_{v\bar{v}'>\frac{\varepsilon}{\alpha}}\dot{v}\int_{-\pi}^{\pi}\frac{d\theta}{(1-re^{i\theta})^n}=I_1+I_2. \tag{2.2.2}$$

It is known by [3, 7]

$$\int_{-(\pi-c)}^{-c}\frac{d\theta}{(1-re^{i\theta})^n}+\int_c^{\pi-c}\frac{d\theta}{(1-re^{i\theta})^n}$$

$$=2\operatorname{Im}\left\{\sum_{k=1}^{n-1}\frac{1}{k(1+re^{-ic})^k}-\sum_{k=1}^{n-1}\frac{1}{k(1-re^{ic})^k}+\log\frac{1-re^{ic}}{1+re^{-ic}}\right\}+2(\pi-2c),$$

hence
$$I_1=\frac{2}{\omega_{2n-1}}\operatorname{Im}\left\{\sum_{k=1}^{n-1}J_k-\sum_{k=1}^{n-1}H_k+J_0-H_0\right\}+\frac{2}{\omega_{2n-1}}\int_{v\bar{v}'<\frac{\varepsilon}{\alpha}}(\pi-2c)\dot{v},$$

where $J_0=\displaystyle\int_{v\bar{v}'<\frac{\varepsilon}{\alpha}}\log\frac{1}{1+re^{-ic}}\dot{v}$, $J_k=\displaystyle\int_{vv'<\frac{\varepsilon}{\alpha}}\frac{\dot{v}}{k(1+re^{-ic})^k}$, $k=1, 2, \cdots, n-1$,

$$H_0=\int_{v\bar{v}'<\frac{\varepsilon}{\alpha}}\log\frac{1}{1-re^{ic}}\dot{v},\quad H_k=\int_{v\bar{v}'<\frac{\varepsilon}{\alpha}}\frac{\dot{v}}{k(1-re^{ic})^k},\ k=1,2,\cdots,n-1.$$

Since $\sin c=\dfrac{\sqrt{\varepsilon^2-\alpha^2(1-r^2)^2}}{2\beta r},\quad \cos c=\dfrac{\sqrt{4\beta^2r^2-\varepsilon^2+\alpha^2(1-r^2)^2}}{2\beta r},$

so
$$1+re^{-ic}=1+\frac{\sqrt{4\beta^2(1-s^2)+\alpha^2s^4-\varepsilon^2}}{2\beta}-i\frac{\sqrt{\varepsilon^2-\alpha^2s^4}}{2\beta},$$

here $s^2=1-r^2$. Let $v=(x_1, x_2, \cdots, x_{2n-2})$, adopt the spherical polar coorolinates

$$x_1=s\cos\varphi_1,\ x_2=s\sin\varphi_1\cos\varphi_2,\ \cdots,\ x_{2n-2}=s\sin\varphi_1\sin\varphi_2\cdots\sin\varphi_{2n-3},$$

then
$$\dot{v}=s^{2n-3}\sin^{2n-4}\varphi_1\sin^{2n-5}\varphi_2\cdots\sin\varphi_{2n-4}ds\,d\varphi_1\cdots d\varphi_{2n-3}.$$

It follows that

$$J_k=\frac{2\pi^{n-1}}{k\Gamma(n-1)}\int_0^{\sqrt{\frac{\varepsilon}{\alpha}}}\left\{1+\frac{\sqrt{4\beta^2(1-s^2)+\alpha^2s^4-\varepsilon^2}}{2\beta}-i\,\frac{\sqrt{\varepsilon^2-\alpha^2s^4}}{2\beta}\right\}^{-k}s^{2n-3}ds.$$

Let $\eta=\dfrac{\varepsilon}{\alpha}$, $s=\sqrt{\eta}\,t$, then $J_k$ is equal to

$$\frac{2\pi^{n-1}(2\beta)^k}{k\Gamma(n-1)}\eta^{n-1}\int_0^1\frac{t^{2n-3}\,dt}{[2\beta+\sqrt{4\beta^2(1-\eta t^2)+\alpha^2\eta^2(t^4-1)}-i\sqrt{\alpha^2\eta^2(1-t^4)}]^k},$$

since the absolute value of the integrand is bounded, hence

$$\lim_{\varepsilon\to0}J_k=0,\ k=1,2,\cdots,n-1.$$

We may prove $\lim\limits_{\varepsilon\to0}J_0=0$ in the same manner.

As we do for $J_k$, $H_k$ may be written as

$$H_k=\frac{2\pi^{n-1}(2\beta)^k}{k\Gamma(n-1)}\int_0^1\frac{\eta^{n-1}t^{2n-3}\,dt}{[2\beta-\sqrt{4\beta^2(1-\eta t^2)+\alpha^2\eta^2(t^4-1)}-i\sqrt{\alpha^2\eta^2(1-t^4)}]^k}.$$

When $k<n-1$, the absolute value of the integrand is not greater than

$$\frac{M}{\beta^2t^4+\alpha^2(1-t^4)-1},$$

where $M$ is an absolute constant, it is integrable in $[0, 1]$, so

$$\lim_{\varepsilon\to0}H_k=0,\ k=1,2,\cdots,n-2.$$

When $k=n-1$, we have

$$H_{n-1}=\frac{2(2\beta\pi)^{n-1}}{\Gamma(n)}\int_0^1\frac{\eta^{n-1}t^{2n-3}\,dt}{[2\beta-\sqrt{4\beta^2(1-\eta t^2)+\alpha^2\eta^2(t^4-1)}-i\sqrt{\alpha^2\eta^2(1-t^4)}]^{n-1}}$$
$$=\frac{2(2\beta\pi)^{n-1}}{\Gamma(n)}\int_0^1\frac{t^{2n-3}\,dt}{[\beta t^2-i\alpha\sqrt{1-t^4}+O(\eta)]^{n-1}}.$$

Set $\gamma=\dfrac{\alpha}{\beta}$, then

$$\lim_{\varepsilon\to0}H_{n-1}=\frac{2^n\pi^{n-1}}{\Gamma(n)}\int_0^1\frac{t^{2n-3}\,dt}{(t^2-i\gamma\sqrt{1-t^4})^{n-1}}.$$

Let $t^2=\cos\theta$

$$\lim_{\varepsilon\to 0} H_{n-1}=\frac{(2\pi)^{n-1}}{\Gamma(n)}\int_0^{\frac{\pi}{2}}\frac{\cos^{n-2}\theta\sin\theta\,d\theta}{(\cos\theta-i\gamma\sin\theta)^{n-1}}$$

$$=\frac{(2\pi)^{n-1}}{\Gamma(n)(1+\gamma)^{n-1}}\int_0^{\frac{\pi}{2}}\frac{(e^{2i\theta}+1)^{n-2}(e^{2i\theta}-1)}{\left(1+\frac{1-\gamma}{1+\gamma}e^{2i\theta}\right)^{n-1}}\,d\theta$$

$$=\frac{(2\pi)^{n-1}}{\Gamma(n)(1+\gamma)^{n-1}}\int_0^{\frac{\pi}{2}}\Bigg\{\frac{1}{i}\sum_{p=0}^{n-2}\sum_{\substack{q=0\\ p+q\neq 0}}^{\infty}(-1)^q C_p^{n-2}C_q^{n+q-2}\left(\frac{1-\gamma}{1+\gamma}\right)^q \times\left[e^{2i\theta(p+q+1)}-e^{2i\theta(p+q)}\right]+\frac{1}{i}\left(e^{2i\theta}-1\right)\Bigg\}\,d\theta$$

$$=-\frac{(2\pi)^{n-1}}{2\Gamma(n)(1+\gamma)^{n-1}}\sum_{p=0}^{n-2}\sum_{\substack{p\neq 0\\ p+q\neq 0}}^{\infty}(-1)^q C_p^{n-2}C_q^{n+q-2}\left(\frac{1-\gamma}{1+\gamma}\right)^q \times\left[\frac{(-1)^{p+q+1}-1}{p+q+1}-\frac{(-1)^{p+q}-1}{p+q}\right]+\frac{(2\pi)^{n-1}}{\Gamma(n)(1+\gamma)^{n-1}}\left(1-\frac{\pi}{2i}\right).$$

It follows that

$$\lim_{\varepsilon\to 0}\operatorname{Im}(H_{n-1})=\operatorname{Im}(\lim_{\varepsilon\to 0}H_{n-1})=\frac{2^{n-2}\pi^n}{\Gamma(n)(1+\gamma)^{n-1}}.$$

We now calculate $H_0$

$$H_0=\int_{v\bar{v}'<\frac{\varepsilon}{\alpha}}\log\frac{1}{1-re^{ic}}\dot{v}=-\frac{2\pi^{n-1}}{\Gamma(n-1)}\int_0^{\sqrt{\frac{\varepsilon}{\alpha}}}s^{2n-3}\log(1-re^{ic})\,ds.$$

Since $\qquad \operatorname{Im}\log(1-re^{ic})=\arg(1-re^{ic})=O(1),$

hence when $n>1$

$$\operatorname{Im}(H_0)=O\left(\int_0^{\sqrt{\frac{\varepsilon}{\alpha}}}s^{2n-3}\,ds\right)=O(\varepsilon^{n-1})=o(1)\quad(\varepsilon\to 0).$$

By the same reason, when $n>1$, we have

$$\int_{v\bar{v}'<\frac{\varepsilon}{\alpha}}(\pi-2c)\dot{v}=o(1).$$

Sum up the above results, we obtain

$$\lim_{\varepsilon\to 0}I_1=-\frac{1}{2}\left(\frac{2}{1+\gamma}\right)^{n-1}=-\frac{1}{2}\left(\frac{2\beta}{\alpha+\beta}\right)^{n-1}.$$

As to $I_2$, by virtue of

$$\int_{-\pi}^{\pi}\frac{d\theta}{(1-re^{i\theta})^n}=2\operatorname{Im}\left\{\sum_{k=1}^{n-1}\frac{1}{k(1+r)^k}+\sum_{k=1}^{n-1}\frac{1}{k(1-r)^k}\right\}+2\pi=2\pi,$$

$$\lim_{\varepsilon\to 0}I_2=\frac{2\pi}{\omega_{2n-1}}\int_{0<v\bar{v}'<1}\dot{v}=1.$$

Substitute thses results into (2.2.2), we obtain (2.2.1) immediately.

Finally, we consider the case $\alpha>0$, $\beta=0$.

(2.2.1) now becomes

$$\lim_{\varepsilon\to 0}\frac{1}{\omega_{2n-1}}\int_{u\in\tilde{D}(p_n,\varepsilon)}\frac{\dot{u}}{(1-u_n)^n}=1,$$

where $\qquad \tilde{D}(p_n,\varepsilon)=\left\{u\bar{u}'=1,\ 1-|u_n|^2>\frac{\varepsilon}{\alpha}\right\}.$

using the above transformations, one may write $\tilde{D}(p_n, \varepsilon)$ as

$$\begin{cases} vv'=1-r^2, \\ 1-r^2>\dfrac{\varepsilon}{\alpha}, \end{cases}$$

then
$$\frac{1}{\omega_{2n-1}}\int_{u\in\tilde{D}(p_n,\varepsilon)}\frac{\dot{u}}{(1-\bar{u}_n)^n}=\frac{1}{\omega_{2n-1}}\int_{\frac{\varepsilon}{\alpha}<v\bar{v}'<1}\dot{v}\int_{-\pi}^{\pi}\frac{d\theta}{(1-re^{i\theta})^n}.$$

This is just the integral $I_2$ discussed before, but it is known $\lim\limits_{\varepsilon\to 0} I_2=1$, so (2.2.1) holds when $\alpha>0$, $\beta=0$.

## § 3. The case that the neighborhood is a "rectangle"

### 3.1. Plemelj formula

Now let the neighborhood $D_R(w, \varepsilon)$ around $w$ be a "rectangle"

$$\{z\in b\Omega,\ |\mathrm{Re}\, g|<\alpha\varepsilon,\ |\mathrm{Im}\, g|<\beta\},$$

here $\alpha>0$, $\beta>0$. In this case, we shall prove that (1.2.3) becomes

$$\lim_{\varepsilon\to 0}\lim_{\delta\to 0}\int_{z\in D_R(w,\varepsilon)} H(w+\delta\nu,\ z)d\sigma_z=\frac{2^{n-1}}{\pi}\left\{\frac{\pi}{2}-h_n\left(\mathrm{arc\ tg}\ \frac{\beta}{\alpha}\right)\right\}, \qquad (3.1.1)$$

where
$$h_n(x)=\int_0^x \cos^{n-2} t\,\frac{\sin(n-1)t}{\sin t}\,dt,$$

and the value of $a$ in Plemelj formula (1.2.4) is

$$a=\frac{2^{n-1}}{\pi}\left\{\frac{\pi}{2}-h_n\left(\mathrm{arc\ tg}\ \frac{\beta}{\alpha}\right)\right\}. \qquad (3.1.2)$$

In particular when $\alpha=\beta$, i. e. the rectangle becomes a square,

$$a=\frac{2^{n-1}}{\pi}\left\{\frac{\pi}{2}-h_n\left(\frac{\pi}{4}\right)\right\}.$$

When $\alpha=\infty$, $D_R(w, \varepsilon)$ becomes

$$\{z\in b\Omega,\ |\mathrm{Im}\, g|<\beta\varepsilon\},$$

this is just the case discussed in the preceding section. By virture of $h_n(0)=0$, so $a=2^{n-2}$, this coincides with the result of the preceding section.

When $\beta=\infty$, $D_R(w, \varepsilon)$ becomes

$$\{z\in b\Omega,\ |\mathrm{Re}\, g|<\alpha\varepsilon\},$$

this is the case discussed in the preceding section too. We shall see in § 3.2

$$h_n\left(\mathrm{arc\ tg}\ \frac{\beta}{\alpha}\right)=\mathrm{arc\ tg}\ \frac{\beta}{\alpha}+\sum_{k=1}^{n-2}\frac{\alpha^k}{k(\alpha^2+\beta^2)^{k/2}}\sin\left(k\ \mathrm{arc\ tg}\ \frac{\beta}{\alpha}\right),$$

therefore, when $\beta\to\infty$ arc tg $\frac{\beta}{\alpha}\to\frac{\pi}{2}$, so $h_n\left(\frac{\pi}{2}\right)=\frac{\pi}{2}$. It follows that $a=0$, this coincides with the result of the preceding section too.

As the preceding section, for the singular integal operator $\mathbf{H}_a$, if one can find an operator $\mathbf{H}_b$, such that $1-a-b=0$, $ab\neq 0$ then $\frac{1}{b}\,\mathbf{H}_b$ is the inverse of $\frac{1}{a}\,\mathbf{H}_a$.

It is not hard to prove $h_n(x)\geqslant 0$ for arbitrary positive integer $n$ and $x\in\left[0, \frac{\pi}{2}\right]$, and

$$\max_{0<x<\frac{\pi}{2}} h_n(x)=h_n\left(\frac{\pi}{n-1}\right).$$

If $\alpha$, $\beta$ satisfy

$$(1-2^{1-n})\pi-h_n\left(\frac{\pi}{n-1}\right)\leqslant h_n\left(\operatorname{arc\,tg}\frac{\beta}{\alpha}\right)\leqslant(1-2^{1-n})\pi,$$

then $\mathbf{H}_a$ ($a$ is defined by (3.1.2)) has an inverse operator. That is to say, we only choose $D_R(w, \varepsilon)$ as

$$\{z\in b\Omega,\ |\operatorname{Re} g|<\alpha'\varepsilon,\ |\operatorname{Im} g|<\beta'\varepsilon\},$$

where $\alpha'>0$, $\beta'>0$, such that

$$h_n\left(\operatorname{arc\,tg}\frac{\beta'}{\alpha'}\right)=(1-2^{1-n})\pi-h_n\left(\operatorname{arc\,tg}\frac{\beta}{\alpha}\right).$$

Set

$$b=\frac{2^{n-1}}{\pi}\left\{\frac{\pi}{2}-h_n\left(\operatorname{arc\,tg}\frac{\beta'}{\alpha'}\right)\right\},$$

then $\frac{1}{b}\mathbf{H}_b$ is the inverse of $\frac{1}{a}\mathbf{H}_a$. It follows that under the condition (3.1.3), $\mathbf{H}_a$ is invertible. We may also discuss the theory of the singular integral equations in the same way.

**3.2. The proof of (3.1.1)**

To prove (3.1.1), by the argument of § 1.2, we only need to prove that (1.2.8) holds, i. e.

$$\lim_{\varepsilon\to 0}\lim_{\rho\to 1}\frac{1}{\omega_{2n-1}}\int_{u\in D_R(p_n,\varepsilon)}\frac{u}{(1-\rho\bar{u}_n)^n}=\frac{2^{n-1}}{\pi}\left\{\frac{\pi}{2}-h_n\left(\operatorname{arc\,tg}\frac{\beta}{\alpha}\right)\right\},$$

where $p_n=(0, \cdots, 0, 1)$, $\alpha>0$, $\beta>0$

$$D_R(p_n, \varepsilon)=\{u\bar{u}'=1,\ 1-|u_n|^2<\alpha\varepsilon,\ 2|\operatorname{Im} u_n|<\beta\varepsilon\}.$$

Let

$$\tilde{D}_R(p_n, \varepsilon)=\{u\bar{u}'=1\}-D_R(p_n, \varepsilon),$$
$$M_\varepsilon=\{u\bar{u}'=1,\ 1-|u_n|^2>\alpha\varepsilon\},$$
$$N_\varepsilon=\{u\bar{u}'=1,\ 2|\operatorname{Im} u_n|>\beta\varepsilon\},$$
$$Q_\varepsilon=\{u\bar{u}'=1,\ 1-|u_n|^2>\alpha\varepsilon,\ 2|\operatorname{Im} u_n|>\beta\varepsilon\}.$$

Thus the result which needs to be proved is equivalent to

$$\lim_{\varepsilon\to 0}\frac{1}{\omega_{2n-1}}\int_{\tilde{D}_R(p_n,\varepsilon)}\frac{\dot{u}}{(1-\bar{u}_n)^n}=1-\frac{2^{n-1}}{\pi}\left\{\frac{\pi}{2}-h_n\left(\operatorname{arc\,tg}\frac{\beta}{\alpha}\right)\right\}. \tag{3.2.1}$$

But

$$\frac{1}{\omega_{2n-1}}\int_{\tilde{D}_R(p_n,\varepsilon)}\frac{\dot{u}}{(1-\bar{u}_n)^n}=\left\{\frac{1}{\omega_{2n-1}}\int_{M_\varepsilon}+\frac{1}{\omega_{2n-1}}\int_{N_\varepsilon}-\frac{1}{\omega_{2n-1}}\int_{Q_\varepsilon}\right\}\frac{\dot{u}}{(1-\bar{u}_n)^n}. \tag{3.2.2}$$

It is known by § 2.2

$$\lim_{\varepsilon\to 0}\frac{1}{\omega_{2n-1}}\int_{M_\varepsilon}\frac{\dot{u}}{(1-\bar{u}_n)^n}=1,\ \lim_{\varepsilon\to 0}\frac{1}{\omega_{2n-1}}\int_{N_\varepsilon}\frac{\dot{u}}{(1-\bar{u}_n)^n}=1-2^{n-2}. \tag{3.2.3}$$

so we only need to calculate

$$\lim_{\varepsilon\to 0}\frac{1}{\omega_{2n-1}}\int_{Q_\varepsilon}\frac{\dot{u}}{(1-\bar{u}_n)^n}.$$

As the argument in § 2.2, set

$$\bar{u}_n=re^{i\theta},\ v=(u_1,\ \cdots,\ u_{n-1}),$$

then $Q_\varepsilon$ may be written as

$$\begin{cases} v\bar{v}'=1-r^2, \\ 1-r^2>\alpha\varepsilon,\ 2r|\sin\theta|>\beta\varepsilon, \end{cases}$$

$$\frac{1}{\omega_{2n-1}}\int_{Q_\varepsilon}\frac{\dot{u}}{(1-\bar{u}_n)^n}=\frac{1}{\omega_{2n-1}}\int_{P_\varepsilon}\dot{v}\left\{\int_{-(\pi-c)}^{-c}+\int_{c}^{\pi-c}\right\}\frac{d\theta}{(1-re^{i\theta})^n},$$

where $c=\arcsin\dfrac{\beta\varepsilon}{2r}$ and

$$P_\varepsilon=\left\{v=(u_1,\ \cdots,\ u_{n-1}),\ \alpha\varepsilon<v\bar{v}'<1-\left(\frac{\beta\varepsilon}{2}\right)^2\right\}.$$

By [3, 7], the above equality may be written as

$$\frac{1}{\omega_{2n-1}}\int_{Q_\varepsilon}\frac{\dot{u}}{(1-\bar{u}_n)^n}=\frac{2}{\omega_{2n-1}}\operatorname{Im}\left\{\sum_{k=1}^{n-1}J_k-\sum_{k=1}^{n-1}H_k+J_0-H_0\right\}+\frac{1}{\omega_{2n-1}}\int_{P_\varepsilon}(\pi-2c)\dot{v},$$

here
$$J_0=\int_{P_\varepsilon}\log\frac{1}{1+re^{-ic}}\dot{v},\ J_k=\int_{P_\varepsilon}\frac{\dot{v}}{k(1+re^{-ic})^k},\ k=1,\ 2,\ \cdots,\ n-1,$$

$$H_0=\int_{P_\varepsilon}\log\frac{1}{1-re^{ic}}\dot{v},\ H_k=\int_{P_\varepsilon}\frac{\dot{v}}{k(1-re^{ic})^k},\ k=1,\ 2,\ \cdots,\ n-1.$$

Use the spherical polar coordinates

$$\int_{P_\varepsilon}c\dot{v}=\int_{P_\varepsilon}\arcsin\frac{\beta\varepsilon}{2\sqrt{1-v\bar{v}'}}\dot{v}=\frac{2\pi^{n-1}}{\Gamma(n-1)}\int_{\sqrt{\alpha\varepsilon}}^{\sqrt{1-\left(\frac{\beta\varepsilon}{2}\right)^2}}s^{2n-3}\arcsin\frac{\beta\varepsilon}{2\sqrt{1-s^2}}ds$$

$$=\frac{2\pi^{n-1}}{\Gamma(n-1)}\left\{\int_0^1\left[1-\left(\frac{\beta\varepsilon}{2}\right)^2\right]^{n-1}\arcsin\frac{\beta\varepsilon}{2\sqrt{1-\left[1-\left(\frac{\beta\varepsilon}{2}\right)^2\right]t^2}}dt\right.$$

$$\left.-\int_0^1(\alpha\varepsilon)^{n-1}\arcsin\frac{\beta\varepsilon}{2\sqrt{1-\alpha\varepsilon t^2}}dt\right\}.$$

Since the integrands are bounded,

$$\lim_{\varepsilon\to 0}\int_{P_\varepsilon}c\dot{v}=0.$$

It follows that

$$\lim_{\varepsilon\to 0}\frac{2}{\omega_{2n-1}}\int_{P_\varepsilon}(\pi-2c)\dot{v}=1.$$

By virtue of
$$1+re^{-ic}=1+\frac{1}{2}\sqrt{4r^2-\beta^2\varepsilon^2}-i\frac{1}{2}\beta\varepsilon,$$

hence

$$J_k=\frac{2\pi^{n-1}}{k\Gamma(n-1)}\int_{\sqrt{\alpha\varepsilon}}^{\sqrt{1-\left(\frac{\beta\varepsilon}{2}\right)^2}}s^{2n-3}\left[1+\frac{1}{2}\sqrt{4(1-s^2)-\beta^2\varepsilon^2}-\frac{i}{2}\beta\varepsilon\right]^{-k}ds$$

$$=\frac{2\pi^{n-1}}{k\Gamma(n-1)}\int_0^1\left[1-\left(\frac{\beta\varepsilon}{2}\right)^2\right]^{n-1}t^{2n-3}\left[1+\frac{1}{2}\sqrt{4\left[1-\left(1-\frac{1}{4}\beta^2\varepsilon^2\right)t^2\right]-\beta^2\varepsilon^2}\right.$$

$$\left.-\frac{i}{2}\beta\varepsilon\right]^{-k}dt-\frac{2\pi^{n-1}}{k\Gamma(n-1)}\int_0^1(\alpha\varepsilon)^{n-1}t^{2n-3}\left[1+\frac{1}{2}\sqrt{4(1-\alpha\varepsilon t^2)-\beta^2\varepsilon^2}-\frac{i}{2}\beta\varepsilon\right]^{-k}dt,$$

the integrands of these two integrals are all bounded, so

$$\lim_{\varepsilon\to0} J_k=\frac{2\pi^{n-1}}{k\Gamma(n-1)}\int_0^1\frac{t^{2n-3}}{(1+\sqrt{1-t^2})^k}\,dt,$$

it is a real number, hence

$$\lim_{\varepsilon\to0}\operatorname{Im}(J_k)=0,\ k=1,\ 2,\ \cdots,\ n-1.$$

We may prove in the same way

$$\lim_{\varepsilon\to0}\operatorname{Im}(J_0)=0,\ \lim_{\varepsilon\to0}\operatorname{Im}(H_0)=0.$$

As the calculation of $J_k$, $H_k$ may be written as

$$H_k=\frac{2\pi^{n-1}}{k\Gamma(n-1)}\left\{\int_0^1\left(1-\frac14\beta^2\varepsilon^2\right)^{n-1}t^{2n-3}\left[1-\sqrt{1-\left(1-\frac14\beta^2\varepsilon^2\right)t^2-\frac14\beta^2\varepsilon^2}-\frac{i}{2}\beta\varepsilon\right]^{-k}dt-\int_0^1(\alpha\varepsilon)^{n-1}t^{2n-3}\left[1-\sqrt{1-\alpha\varepsilon t^2-\frac14\beta^2\varepsilon^2}-\frac{i}{2}\beta\varepsilon\right]^{-k}dt\right\}$$

$$=\frac{2\pi^{n-1}}{k\Gamma(n-1)}(X_k-Y_k).$$

The absolute value of the integrand of the first integral is not greater than $\dfrac{t^{2n-3}}{(1-\sqrt{1-t^2})^k}$, it is an integrable function on the interval [0, 1] when $k<n-1$, so we have

$$\lim_{\varepsilon\to0}X_k=\int_0^1\frac{t^{2n-3}}{(1-\sqrt{1-t^2})^k}\,dt,$$

it is a real number. The absolute value of the integrand of the second integral is $O(\varepsilon^{n-k-1})$, so

$$\lim_{\varepsilon\to0}Y_k=0,$$

as $k<n-1$. It follows that when $k<n-1$

$$\lim_{\varepsilon\to0}\operatorname{Im}(H_k)=0.$$

Finally, we calculate $H_{n-1}$

$$H_{n-1}=\frac{2\pi^{n-1}}{\Gamma(n)}(X_{n-1}-Y_{n-1}),$$

where

$$X_{n-1}=\left(1-\frac14\beta^2\varepsilon^2\right)^{n-1}\int_0^1 t^{2n-3}\left[1-\sqrt{\left(1-\frac14\beta^2\varepsilon^2\right)(1-t^2)}-\frac{i}{2}\beta\varepsilon\right]^{-(n-1)}dt,$$

$$Y_{n-1}=(\alpha\varepsilon)^{n-1}\int_0^1 t^{2n-3}\left[1-\sqrt{1-\alpha\varepsilon t^3-\frac14\beta^2\varepsilon^2}-\frac{i}{2}\beta\varepsilon\right]^{-(n-1)}dt.$$

Let $1-\frac14\beta^2\varepsilon^2=\eta^2$, we have

$$X_{n-1}=\eta^{2n-2}\int_0^1 t^{2n-3}[1-\eta\sqrt{1-t^2}-i\sqrt{1-\eta^2}]^{-(n-1)}dt,$$

by [7], we obtain

$$\lim_{\varepsilon\to0}\operatorname{Im}(X_{n-1})=2^{n-3}\pi.$$

Since $$1-\sqrt{1-\alpha\varepsilon t^2-\frac{1}{4}\beta^2\varepsilon^2}-\frac{i}{2}\beta\varepsilon=\left(\frac{1}{2}\alpha t^2-\frac{i}{2}\beta\right)\varepsilon+O(\varepsilon^2),$$
we have
$$\lim_{\varepsilon\to0}Y_{n-1}=2^{n-1}\int_0^1\frac{t^{2n-3}\,dt}{\left(\frac{1}{2}\alpha t^2-\frac{i}{2}\beta\right)^{n-1}}=2^{n-1}\int_0^1\frac{t^{2n-3}\,dt}{(t^2-i\gamma)^{n-1}}=2^{n-2}\int_0^1\frac{x^{n-2}\,dx}{(x-i\gamma)^{n-1}}$$
$$=2^{n-2}\left\{\log\frac{1-i\gamma}{-i\gamma}-D_n+\sum_{k=1}^{n-2}C_k^{n-2}\frac{(-1)^k}{k}\right\},$$
where $\gamma=\frac{\beta}{\alpha}$, $D_n=\sum_{k=1}^{n-2}C_k^{n-2}\frac{1}{k}\left(\frac{i\gamma}{1-i\gamma}\right)^k$ $(n\geqslant3)$. It follows that
$$\lim_{\varepsilon\to0}\operatorname{Im}(Y_{n-1})=2^{n-2}\left[\operatorname{arc\,tg}\frac{1}{\gamma}-\operatorname{Im}(D_n)\right]\quad(n\geqslant3)\tag{3.2.4}$$
when $n=2$ $$\lim_{\varepsilon\to0}\operatorname{Im}(Y_1)=2\operatorname{Im}\left(\int_0^1\frac{t\,dt}{t^2-i\gamma}\right)-\operatorname{arc\,tg}\frac{1}{\gamma}.$$
If define $D_2=0$, then (3.2.4) holds too when $n\geqslant2$.

We now have
$$\lim_{\varepsilon\to0}\operatorname{Im}(H_{n-1})=\frac{2^{n-2}\pi^n}{\Gamma(n)}\left\{1-\frac{2}{\pi}\operatorname{arc\,tg}\frac{1}{\gamma}+\frac{2}{\pi}\operatorname{Im}(D_n)\right\},$$
here $n\geqslant2$. Consequently
$$\lim_{\varepsilon\to0}\frac{1}{\omega_{2n-1}}\int_{Q_\varepsilon}\frac{\dot u}{(1-\bar u_n)^n}=1-2^{n-2}\left(1-\frac{2}{\pi}\operatorname{arc\,tg}\frac{1}{\gamma}+\frac{2}{\pi}\operatorname{Im}(D_n)\right).\tag{3.2.5}$$
By (3.2.2), (3.2.3), (3.2.5), we obtain
$$\lim_{\varepsilon\to0}\frac{1}{\omega_{2n-1}}\int_{D_n(p_n,\varepsilon)}\frac{\dot u}{(1-\bar u_n)^n}=1-\frac{2^{n-1}}{\pi}\left(\operatorname{arc\,tg}\frac{1}{\gamma}-\operatorname{Im}(D_n)\right).\tag{3.2.6}$$

It is not hard to prove
$$\operatorname{Im}(D_n)=-\operatorname{arc\,tg}\frac{\beta}{\alpha}+h_n\left(\operatorname{arc\,tg}\frac{\beta}{\alpha}\right).$$
Substitute it into (3.2.6), this completes the proof of (3.1.1).

## References


[1] Muskhelishvili N. I., Singular integral equations, Translated from Russian, Groningen, 1953.

[2] Zygmund, A., Trigonometric series, *Cambridge University Press*, **2**(1959).

[3] Kung S. and Sun C. K., Integrals of Cauchy type in several complex variables I. *Acta Mathematica Sinica*, **15**(1965), 431—443 (In Chinese).

[4] Koranyi, A. and Vagi, S., Singular integrals in homogeneous spaces and some problems of classical analysis, *Ann. Scuola Normale Superiore Pisa*, **25** (1971) 575—648.

[5] Kung S. and Sun C. K., Integrals of Cauchy type in several complex variables III, *Acta Mathematica Sinica*, **15** (1965), 800—811 (In Chinese).

[6] Kung S. and Sun C. K., Integrals of Cauchy type in several complex variables II, *Acta Mathematica Sinica*, **15** (1965), 775—799 (*In Chinese*).

[7] Shi Ji-huai, On the Cauchy type integrals for the hypersphere, *Journal of the China University of the science and technology*, **2**(1980), 1—9 (In Chinese).

[8] Kung S. and Sun C. K., Singular integral equations on a complex hyperspher, *Acta Mathematica Sinica*, **16** (1966), 194—210 (*In Chinese*).

[9] Sun C. K., Regularization theorem of the singular integral equations on a complex hypersphere, *Acta Mathematica Sinica*, **20** (1977), 287—290 (In Chinse).

[10] Kung, S., A remark on integrals of Cauchy type in several complex variables, Proceeding of the Ist symposia of partial differential equation and differential geometry, Beijing, 1980.

[11] Ramirez De Arellano, Ein Divisionproblem und Randintegraldarstellungen in der Komplexen Analysis, *Math. Ann.*, **184** (1970), 172—187.

[12] Folland, G. B. and Stein, E. M., Estimates for the $\bar{\partial}$ complex and analysis on the Heisenberg group, *Comm. Pure. Appl. Math.*, **27** (1974), 429—522.

[13] Alt, W., Singuläre Mtegrale mit gemischten hologeneitäten auf Mannigfaltigkeiten und Anwendungen in der Funktionentheorie, *Math. Z.*, **137** (1974), 227—256.

[14] Kerzman, N. and Stein, E. M., The Szegö kernel in terms of Cauchy-Fantappiè kernels, *Duke Math* J., **45** (1978), 197—224.

[15] Henkin, G., Integral representations of functions holomorphic in strictly pseudoconvex domains and some applications, *Math. USSR, Sbornik*, **7** (1969), 597—616.

J. DIFFERENTIAL GEOMETRY
16 (1981) 147–151

# ON THE MINIMA OF YANG-MILLS FUNCTIONALS

CHAOHAO GU

*Dedicated to the author's teacher Professor Buchin Su*

In a previous paper [1] we found some lower bounds of the Yang-Mills functional on the tangential bundle over a 4-dimensional oriented manifold among all possible metrics with the Christoffel connections as the gauge potentials [2]. In this paper the results are generalized to vector bundles over a 4-dimensional oriented manifolds, provided the structure group (gauge group) $G$ is compact and its Lie algebra $g$ is nonsimple. Some lower bounds of the Yang-Mills functional are obtained and several cases for which the lower bounds are actually the absolute minimums are listed. In particular it is seen that for the Einstein manifolds or the conformally flat manifolds with zero scalar curvature the Yang-Mills functional attains its absolute minimum among all possible metrics on the manifold and all possible connections on the tangential bundles.

Let $G$ be a compact Lie group, and suppose that its Lie algebra $g$ be the direct sum of two Lie algebras

(1) $$g = g_1 + g_2.$$

An arbitrary element $\alpha$ of $g$ may be written in the form $\alpha = (\alpha_1, \alpha_2)$ with $\alpha_1 \in g_1$, $\alpha_2 \in g_2$. We define a linear mapping $*$ on the Lie algebra $g$ to itself by

$$*(\alpha_1, \alpha_2) = (\alpha_1, -\alpha_2).$$

Then the following relations evidently hold:

(2) (a) $$*^2 = I,$$

(3) (b) $$[\alpha, *\beta] = ([\alpha_1, \beta_1], [\alpha_2, -\beta_2]) = *([\alpha, \beta]),$$

(4) (c) $$\langle *\alpha, *\beta \rangle = \langle \alpha_1, \beta_1 \rangle + \langle -\alpha_2, -\beta_2 \rangle = \langle \alpha, \beta \rangle,$$

---

Received June 6, 1979. The author is grateful to Professor C. N. Yang for valuable discussions, and also to the State University of New York at Stony Brook for its hospitality.

where [ , ] and $\langle\ ,\ \rangle$ are the commutator and the invariant inner product respectively.

**Theorem 1.** *If g is a Lie algebra, and there exists a linear mapping* $*: g \to g$ *such that* (a) *and* (b) *hold, then g is the direct sum of the subalgebras, except for the trivial case* $* = \pm I$.

*Proof.* Let

$$g_1 = \{\alpha + *\alpha | \alpha \in g\}, \quad g_2 = \{\alpha - *\alpha | \alpha \in g\}. \tag{5}$$

Since $*$ is nontrivial, $g_1$ and $g_2$ are both nontrivial subspaces.

From (b) it is easily seen that

$$[\alpha + *\alpha, \beta + *\beta] = ([\alpha, \beta] + [*\alpha, \beta]) + *([\alpha, \beta] + [*\alpha, \beta]),$$
$$[\alpha - *\alpha, \beta - *\beta] = ([\alpha, \beta] - [*\alpha, \beta]) - *([\alpha, \beta] - [*\alpha, \beta]).$$

Hence $g_1$ and $g_2$ are both subalgebras. Moreover, from (a) and (b) we have

$$[\alpha + *\alpha, \beta - *\beta] = [\alpha, \beta] + [*\alpha, \beta] - [\alpha, *\beta] - [*\alpha, *\beta] = 0,$$

i.e., the elements of $g_1$ and those of $g_2$ are commutative. Finally, if $\gamma \in g_1 \cap g_2$, then $*\gamma = \pm\gamma$ which implies that $\gamma = 0$. Hence $g_1 \cap g_2 = \{0\}$, and the theorem is proved.

The mapping $*$ is called the generalized dual operator, since it contains the usual duality in $R^4$ as a particular case: Let $g$ be the Lie algebra so (4) formed by $4 \times 4$ skew-symmetric matrices $L = (l_{ab})$, $a, b = 1, \cdots, 4$. Then $g_1$ and $g_2$ are sets of self-dual and antiself-dual matrices respectively, i.e.,

$$L \in g_1 \quad \text{iff} \quad l_{ab} = \tfrac{1}{2}\varepsilon_{abcd} l_{cd},$$
$$L \in g_2 \quad \text{iff} \quad l_{ab} = -\tfrac{1}{2}\varepsilon_{abcd} l_{cd}.$$

The $*$ is the dual operator

$$*L = (l^*_{ab})$$

with

$$l^*_{ab} = \tfrac{1}{2}\varepsilon_{abcd} l_{cd}. \tag{6}$$

Now let $M$ be a 4-dimensional oriented Riemannian manifold, and $E \to M$ a vector bundle over $M$ with structure group $G$. Let $b$ be a connection or gauge potential on $E$. In a small patch $b$ is expressed as $g$-valued 1-form

$$b = b_\lambda(x)dx^\lambda, \tag{7}$$

and the field strength is the $g$-valued 2-form:

$$\begin{aligned} F &= db + \tfrac{1}{2}[b, b] = \tfrac{1}{2}F_{\lambda\mu}dx^\lambda \wedge dx^\mu \\ &= \tfrac{1}{2}(\partial_\lambda b_\mu - \partial_\mu b_\lambda + [b_\lambda, b_\mu])dx^\lambda \wedge dx^\mu. \end{aligned} \tag{8}$$

Define

$$I_W = \int_M \langle F_{pq}, F^{pq} \rangle dV = \int_M K_W dV, \tag{9}$$

$$I_E = \tfrac{1}{2} \int_M \langle F_{pq}, *F_{rs} \rangle \varepsilon^{pqrs} dV = \int_M K_E dV, \tag{10}$$

$$I_P = \tfrac{1}{2} \int_M \langle F_{pq}, F_{rs} \rangle \varepsilon^{pqrs} dV = \int_M K_P dV, \tag{11}$$

$$I_Q = \int_M \langle F_{pq}, *F^{pq} \rangle dV = \int_M K_Q dV, \tag{12}$$

where $I_W$ is the Yang-Mills functional, and $*$ the generalized dual operator. $I_P$ is the integral of the 2nd Chern class or 1st Pontrjagen class up to a constant factor. $I_E$ is the generalization of the Euler's characteristic number and is also a topological invariant, if the manifold is compact and free of boundary. This is a consequence of the Chern-Weil's theorem [3], since the bilinear form $f(\alpha, \beta) = \langle \alpha, *\beta \rangle$ on $g \times g$ is symmetric and invariant. In general $I_Q$ is not a topological invariant. However, for the tangential bundle $I_P = I_Q$ if the connection is the Christoffel connection of the Riemannian metric.

Define the inner product of two $g$-valued 2-form at each point of $M$

$$F \cdot \Phi = \langle F_{pq}, \Phi^{pq} \rangle. \tag{13}$$

It is easy to verify that

$$\begin{aligned} F \cdot F = F^{**} \cdot F^{**} &= F^{*\cdot} \cdot F^{*\cdot} = F^{\cdot *} \cdot F^{\cdot *} = K_W, \\ F \cdot F^{\cdot *} &= F^{*\cdot} \cdot F^{**} = K_P, \\ F \cdot F^{*\cdot} &= F^{\cdot *} \cdot F^{**} = K_Q, \\ F \cdot F^{**} &= F^{*\cdot} \cdot F^{\cdot *} = K_E. \end{aligned} \tag{14}$$

Here the 1st $*$ is the generalized dual operator, and the 2nd $*$ is the operator (6).

Let

$$\begin{aligned} F^{++} &= \tfrac{1}{4}(F^{\cdot\cdot} + F^{*\cdot} + F^{\cdot *} + F^{**}), \\ F^{-+} &= \tfrac{1}{4}(F^{\cdot\cdot} - F^{*\cdot} + F^{\cdot *} - F^{**}), \\ F^{+-} &= \tfrac{1}{4}(F^{\cdot\cdot} + F^{*\cdot} - F^{\cdot *} - F^{**}), \\ F^{--} &= \tfrac{1}{4}(F^{\cdot\cdot} - F^{*\cdot} - F^{\cdot *} + F^{**}). \end{aligned} \tag{15}$$

We have

$$
\begin{aligned}
F^{++}\cdot F^{++} &= \tfrac{1}{4}(K_W + K_Q + K_P + K_E),\\
F^{+-}\cdot F^{+-} &= \tfrac{1}{4}(K_W + K_Q - K_P - K_E),\\
F^{-+}\cdot F^{-+} &= \tfrac{1}{4}(K_W - K_Q + K_P - K_E),\\
F^{--}\cdot F^{--} &= \tfrac{1}{4}(K_W - K_Q - K_P + K_E),\\
F^{++}\cdot F^{+-} &= F^{++}\cdot F^{-+} = F^{++}\cdot F^{--} = F^{+-}\cdot F^{-+}\\
&= F^{+-}\cdot F^{--} = FF^{-+}\cdot F^{--} = 0.
\end{aligned}
\tag{16}
$$

Using these identities we can easily obtain

**Theorem 2.** *If $A$, $B$, $C$, $D$ are arbitrary constants and not all zero, then*

$$
\begin{aligned}
I_W = \frac{1}{A^2+B^2+C^2+D^2}\Big\{ &\int_M (AF^{++} + BF^{+-} + CF^{-+} + DF^{--})^2 dV\\
&+ \big[(-I_P - I_Q - I_E)A^2 + (I_P + I_E - I_Q)B^2\\
&+ (I_E - I_P + I_Q)C^2 + (I_Q + I_P - I_E)D^2\big]\Big\}.
\end{aligned}
\tag{17}
$$

Further, we have

**Theorem 3.**

$$
\begin{aligned}
\text{(a)}\quad I_W \geqslant \max\{&(-I_P - I_Q - I_E), (I_P + I_E - I_Q), (I_E - I_P + I_Q),\\
&(I_P + I_Q - I_E)\}.
\end{aligned}
\tag{18}
$$

(b) *The equality sign of the above inequality holds if and only if at least one of $F^{++}$, $F^{+-}$, $F^{-+}$, $F^{++}$ is zero.*

*Proof.* (a) follows directly from the identity (17).

(b) Suppose the equality sign holds, say $I_W = -I_P - I_Q - I_E$. Let $A = 1$. Then $B = C = D = 0$, and we have $F^{++} = 0$. Conversely, if $F^{++} = 0$, the same set of $A, B, C, D$ gives $I_W = -I_P - I_Q - I_E$. Moreover, from (18) it follows that

$$
\begin{aligned}
-I_P - I_Q - I_E &\geqslant I_P + I_E - I_Q,\\
-I_P - I_Q - I_E &\geqslant I_E - I_P + I_Q,\\
-I_P - I_Q - I_E &\geqslant I_P + I_Q - I_E,
\end{aligned}
$$

and hence (b) is proved.

Now we can list several cases in which the Yang-Mills functional attains its possible minimal value for given values of $I_P$, $I_E$ and $I_Q$:

| No | $F$ | $I_P, I_Q, I_E$ | $I_W$ |
|---|---|---|---|
| 1 | $F^{++}=0$ | $-I_Q-I_E>0, -I_P-I_E>0, -I_P-I_Q>0$ | $-I_P-I_Q-I_E$ |
| 2 | $F^{+-}=0$ | $I_P+I_E>0, -I_Q+I_P>0, -I_Q+I_E>0$ | $I_P+I_E-I_Q$ |
| 3 | $F^{-+}=0$ | $I_E+I_Q>0, I_Q-I_P>0, I_E-I_P>0$ | $I_E+I_Q-I_P$ |
| 4 | $F^{--}=0$ | $I_P+I_Q>0, I_P-I_E>0, I_Q-I_E>0$ | $I_P+I_Q-I_E$ |
| 1,2 | $F^{++}=F^{+-}=0$ | $-I_Q>I_E=-I_P>I_Q$ | $-I_Q$ |
| 1,3 | $F^{++}=F^{-+}=0$ | $-I_P>I_E=-I_Q>I_P$ | $-I_P$ |
| 1,4 | $F^{++}=F^{--}=0$ | $-I_E>I_P=-I_Q>I_E$ | $-I_E$ |
| 2,3 | $F^{+-}=F^{-+}=0$ | $I_E>I_P=I_Q>-I_E$ | $I_E$ |
| 2,4 | $F^{+-}=F^{--}=0$ | $I_P>I_E=I_Q>-I_P$ | $I_P$ |
| 3,4 | $F^{-+}=F^{--}=0$ | $I_Q>I_P=I_E>-I_Q$ | $I_Q$ |
| 2,3,4 | $F^{+-}=F^{-+}=F^{--}=0$ | $I_E=I_P=I_Q>0$ | $I_P$ |
| 1,3,4 | $F^{++}=F^{-+}=F^{--}=0$ | $-I_P=I_Q=-I_E>0$ | $-I_P$ |
| 1,2,4 | $F^{++}=F^{+-}=F^{--}=0$ | $-I_Q=-I_E=I_P>0$ | $I_P$ |
| 1,2,3 | $F^{++}=F^{+-}=F^{-+}=0$ | $-I_Q=-I_P=I_E>0$ | $-I_P$ |

In a general manifold the values of $I_E$, $I_P$, $I_Q$ place some constraints on both the Riemannian metric and the connection. For compact manifolds without boundary, $I_E$ and $I_P$ are topological invariants. So the cases (1,3), (1,4), (2,3) (2,4) (and hence (2,3,4) (1,3,4) (1,2,4), (1,2,3)) are more interesting than the other cases. In particular, consider the tangential bundle $E$ over an oriented compact manifold $M$ without boundary. If $M$ admits an Einstein metric or a conformally flat metric with zero scalar curvature, then we have cases (2,3) or (1,4) respectively. Hence the Yang-Mills function attains absolute minimum in both cases.

**Corollary.** *The Einstein metric or the conformally flat metric with zero scalar curvature on the compact manifold without boundary minimizes the Yang-Mills functional on the tangential bundle, provided the connection is the Christoffel connection of the metric.*

**Remark.** If the Lie algebra $g$ is decomposable to a direct sum of more than two components, then more detail results can be obtained in the same way.

## References

[1] C. H. Gu, H. S. Hu, D. Q. Li, C. L. Shen, Y. L. Xin & C. N. Yang, *Riemannian spaces with local duality and gravitational instantons*, Scientia Sinica **21** (1978) 475.

[2] C. N. Yang, *Integral formalism for gauge fields*, Lett. Phys. Rev. **33** (1974) 445–447.

[3] S. S. Chern, *Geometry of characteristic classes*, Proc. 13th Biennial Sem. Canadian Math. Congress, Vol. 1, 1972, 1–40.

STATE UNIVERSITY OF NEW YORK, STONY BROOK
FUDAN UNIVERSITY, SHANGHAI

*Chin. Ann. of Math.*
3 (4) 1982

# ON A CLASS OF MIXED PARTIAL DIFFERENTIAL EQUATIONS OF HIGHER ORDER

GU CHAOHAO (C. H. GU)
*(Institute of Mathematics, Fudan University)*

Dedicated to Professor Su Bu-chin on the Occasion of his 80th Birthday and his 50th Year of Educational Work

## I. Introduction

To the author's knowledge the existing results for mixed partial differential equations involve equations of 2nd order only. The theory of mixed equations of higher order is to be developed.

Busemann, A.[1], starting from the wave equation in three variables, derived a special mixed equation of second order in two variables. This equation has been applied to gasdynamics extensively[2,3]. Hua, L. K. also obtained the same equation from differential geometry and discussed various boundary value problrms[4]. Using the theory of positive symmetric systems[5,6], Gu, C. H. considered more general equations in $n$ variables and obtained a large class of well-posed boundary value problems[7]. The method can be used to treat much more general equations of second order, including some quasilinear equations[8,9,10]. Some new phenomena have been found. Hong, J. X. considered in detail the equations whose degenerate surface is characteristic[11]. On the basis of the appaoach in [8, 9], Sun, L. X. obtained some results on a class of equations with non-characteristic degenerate surface[12]. The results stated above mainly concerned the existence of $C^r$ solutions with $r\geqslant 2$, whereas many other papers on mixed equations in several variables considered only the existence of weak solutions or strong solutions[13].

The resluts in [7] can also be obtained through the properties of the wave equation without using the theory of symmetric positive systems. In the present paper we extend this approach to the equations of higher order, solve two kinds of boundary problems and consider the existence and uniqueness of $C^\infty$ solutions. The results in [7, 14] have been completely generalized to the cases of higher order. As a continuation of the present work, Hong, J. X. obtained some further results on mixed equations of higher order. However, his work cannot cover the results obtained here,

Manuscript received May 6, 1981.

since the related hyperbolic equations in the present paper may have multiple characteristics.

## II. A class of mixed equations of higher order

Let $(y, t)=(y_1, \cdots, y_n, t)$ be the coordinates of the point in $R^{n+1}$ and $(\xi, \tau)=(\xi_1, \cdots, \xi_n, \tau)$ be their dual variables. Thus

$$\xi\cdot y+\tau t=0 \tag{1}$$

is an $n$-dimensional subspace of $R^{n+1}$, provided $(\xi, \tau)\neq(0, \tau)$. Let

$$P(\xi, \tau)=\sum_{j=0}^{m} P_{m-j}(\xi)\tau^j \tag{2}$$

be a homogeneous hyperbolic polynomial of $m$-th degree. Here $P_{m-j}(\xi)$ are homogeneous polynomials of degree $m-j$ with real and constant coefficients. Moreover, (2) is hyperbolic with respect to $(\xi, \tau)=(0, 1)$, i. e., as an equation of $\tau$

$$P(\xi, \tau)=0 \tag{3}$$

admits real roots only. No loss of generality, we suppose that $P(0, 1)=1$. Corresponding to (2) we have the hyperbolic equation

$$P(\partial_y, \partial_t)u(y, t)=F(y, t). \tag{4}$$

In particular, if $u$ is a homogeneous function of degree $a+1$ defined on the half space $t>0$ of $R^{n+1}$

$$u=t^{a+1}\varphi\left(\frac{y}{t}\right), \ t>0, \tag{5}$$

then $F(y, t)$ must be a homogeneous function of degree $a+1-m$

$$F=t^{a+1-m}f\left(\frac{y}{t}\right). \tag{6}$$

Let $\left(\frac{y_1}{t}, \cdots, \frac{y_n}{t}\right)=(x_1, \cdots, x_n)$. From (4), (5) and (6) we obtain a partial differential equation

$$L(x, \partial_x, a)\varphi=f. \tag{7}$$

Here

$$\begin{aligned} L(x, \partial_x, a) &= \sum_{j=0}^{m} P_{m-j}(\partial_x)\prod_{k=1}^{j}(a+2-k-x\cdot\partial_x) \\ &= \sum_{j=0}^{m}\prod_{k=j}^{1}(a+1-m+k-x\cdot\partial_x)P_{m-j}(\partial_x) \end{aligned} \tag{8}$$

with

$$x\cdot\partial_x=\sum_{i=1}^{n} x_i\frac{\partial}{\partial x_i}.$$

In particular, for the case of

$$P(\xi, \tau)=\xi_1^2+\cdots+\xi_n^2-\tau^2,$$

we obtain the equation considered in [7, 8].

When (3) is satisfied by $(\xi, \tau)\neq(0, \tau)$, (1) is just the characteristic plane of (4). The envelop of the characteristic planes is the characteristic cone $\Lambda$, having the origin $O$ as the vertex. When (3) has multiple roots for some $\xi \neq 0$, the structure of $\Lambda$ may be very complicate[16].

Consider $R^n$ as the plane $t=1$ in $R^{n+1}$ and denote the intersection $\Lambda \cap R^n$ by $\Sigma$. Then $\Sigma$ is the envelop of the family of the planes

$$\xi \cdot x + \tau = 0 \quad (P(\xi, \tau) = 0). \tag{9}$$

Further $\Sigma$ is the degenerate surface of (7). The number of the real roots of the characteristic equation for the equation (7) is changed when the point $x$ moves across $\Sigma$ through a regular point of $\Sigma$.

For simplicity we denote $L(x, \partial_x, a)$ by $L(a)$.

**Lemma 1.** *The characteristic directions of the differential operator $L(a)$ at a point $x_0$ are the normal directions of planes (9), passing through $x_0$.*

*Proof* The vector $\xi \neq 0$ is a characteristic direction of $L(a)$ at $x_0$, if the characteristic equation for $L(a)$

$$Q(x_0, \xi) = \sum_{j=0}^{m} P_{m-j}(\xi)(-\xi \cdot x_0)^j = P(\xi, -\xi \cdot x_0) = 0 \tag{10}$$

is satisfied. The fact that the plane $\xi \cdot x + \tau = 0$ passes through $x_0$ is equivalent to $\tau = -\xi \cdot x_0$. The conclusion of Lemma 1 follows from (3) and (10) immediately.

**Lemma 2.** *The principal part of $L(a)$, $Q(x, \partial_x)$ is hyperbolic outside the convex hull of $\Sigma$.*

*Proof* The condition of hyperbolicity of $Q(x, \partial_x)$ at $x_0$ is that there exists a nonnull vector $-\eta$ such that

$$Q(x_0, \xi - \tau\eta) = P(\xi - \tau\eta, -\xi \cdot x + \tau(\eta \cdot x_0)) = 0, \tag{11}$$

as an equation with an unknown $\tau$, admits only real roots([17], Theorem 5.5.3). Let

$$\Gamma = \{(-\eta, \lambda) \mid P(-\eta, \sigma) > 0, \ \forall \sigma \geqslant \lambda\} \tag{12}$$

be a cone in the dual space of $R^{n+1}$. Evidently for each $-\eta$ we have $(-\eta, \lambda) \in \Gamma$ if $\lambda$ is suffeciently large. It is known that $\Gamma$ is a convex cone ([17], Theorem 5.5.6) and that

$$P(\xi - \tau\eta, \mu + \tau\lambda) = 0, \tag{13}$$

as an equation for $\tau$, admits only real roots for each non-null $(\xi, \mu)$. Suppose that $x_0 \in R^n$ and there is a vector $(-\eta, \lambda)$ such tha $(-\eta, \eta \cdot x_0) \in \Gamma$.

Comparing (11) with (13), we see that (11) admits only real roots and hence $Q(x, \partial_x)$ is hyperblic with respect to $\eta$. Here we require that there is a plane $-\eta \cdot x + \lambda = 0$ in $R^n$ or a plane $-\eta \cdot y + \lambda\tau = 0$ in $R^{n+1}$ such that $(-\eta, \lambda) \in \Gamma$ with $\lambda = \eta \cdot x_0$. This means that $(x_0, 1)$ lies outside the supporting planes of the dual cone of $\Gamma$ or $x_0$ lies outside the convex hull of $\Sigma$.

Q. E. D.

**Lemma 3.** *The formal adjoint of $L(a)$ is*

$$L^*(a) = (-1)^m L(-a - 3 + m - n). \tag{14}$$

*Proof* From (8) by diret calcuation we obtain

$$L^*(a)=\sum_{j=0}^{m}P^*_{m-j}(\partial_x)\prod_{k=1}^{j}(a+1-m+k+n+x\cdot\partial_x)$$

$$=(-1)^m\sum_{j=0}^{m}P_{m-j}(\partial_x)\prod_{k=1}^{j}(-a-3+m-n+2-k-x\cdot\partial_x)$$

$$=(-1)^mL(-a-3+m-n).$$

Let $\Omega$ be a bounded region, containing the convex hull of $\Sigma$. Suppose that $\partial\Omega$ is smooth and space-like, i. e., $(-n, n\cdot x_0)\in\Gamma$, where $n$ is the normal of $\partial\Omega$.

We consider the following two kinds of boundary value problems.

**Problem $T_1$.** To find the solution to equation (7) in $\Omega$ such that the boundary value conditions

$$\varphi|_{\partial\Omega}=n\cdot\partial_x\varphi|_{\partial\Omega}=\cdots=(n\cdot\partial_x)^{m-1}\varphi|_{\partial\Omega}=0 \tag{15}$$

are satisfied.

**Problem $T_2$.** To find the solution to equation (7) in $\Omega$ without any given boundary condition.

## III. Analytical Lemmas

Let $H_s(R^n)$ be the functional space with the norm

$$\|\varphi\|_s^2=\int(1+|\xi|^2)^s|\hat{\varphi}(\xi)|^2d\xi, \tag{16}$$

where $\hat{\varphi}(\xi)$ is the Fourier transform of $\varphi$. Moreover

$$\mathring{H}_s(\Omega)=\{u|u\in H_s(R^n),\ \mathrm{supp}\,u\in\overline{\Omega}\}. \tag{17}$$

$$\mathring{H}_s^{\mathrm{loc}}(\overline{R}_+^{u+1})=\{u|u\in H_s^{\mathrm{loc}}(R^{n+1}),\ \mathrm{supp}\,u\in\overline{R}_+^{n+1}\}. \tag{18}$$

Here $\overline{R}_+^{n+1}$ is the closed half space $t\geqslant0$ of $R^{n+1}$ and $u\in H_s^{\mathrm{loc}}$ means that $\varphi u\in H_s(R^{n+1})$ for each $\varphi\in C_0^\infty(R^{n+1})$.

Let $K_{\overline{\Omega}}$ be the conical region in $\overline{R}^{n+1}$

$$K_{\overline{\Omega}}=\left\{(y,\ t)\,|\,t>0,\ \ \frac{y}{t}\in\overline{\Omega}\quad\text{or } t=0,\ \ y=0\right\} \tag{19}$$

and

$$\Phi_p(y,\ t)=H(t)t^p\varphi\left(\frac{y}{t}\right). \tag{20}$$

Here $\varphi$ is a function defined on $\overline{\Omega}$ and $H(t)$ is the Heaviside function.

**Lemma 4.** (1) *Suppose that* $p>s-\frac{n}{2}-\frac{1}{2}$ $(s\geqslant0)$. $\varphi(x)\in\mathring{H}_s(\overline{\Omega})$, *iff* $\Phi_p(y,\ t)\in\mathring{H}_s^{\mathrm{loc}}(\overline{R}_+^{n+1})$ *and* $\mathrm{supp}\,\Phi_p\subset K_{\overline{\Omega}}$.

(2) *If* $p>-\frac{n}{2}-\frac{1}{2}$, $\varphi(x)\in\mathring{H}_{-s}(\Omega)$ $(s>0)$, *then*

$$\Phi_p(y,\ t)\in\mathring{H}_{-s}^{\mathrm{loc}}(\overline{R}_+^{n+1}) \tag{21}$$

and

$$\Phi_p(y,\ t)e^{-\varepsilon t}\in H_{-s}(\overline{R}_+^{n+1})\quad(\varepsilon>0). \tag{22}$$

*Proof* (1) Evidently supp $\Phi_p(y, t)\subset K_{\bar{\Omega}}$ is equivalent to supp $\varphi(x)\subset\bar{\Omega}$. From

$$\partial_t^j\partial_y^\alpha\left(\varphi\left(\frac{y}{t}\right)t^p\right)=t^{p-|\alpha|-j}\prod_{k=1}^{j}(p-|\alpha|+1-k-x\cdot\partial_x)\partial_x^\alpha\varphi\Big|_{x=\frac{y}{t}}$$

it is seen that

$$\int_0^{t_0}\int_{R^n}|\partial_t^j\partial_y^\alpha\Phi_p(y, t)|^2\,dy\,dt\leqslant Ct_0^{2(p-|\alpha|-j)+n+1}\int_{\Omega}\sum_{|\beta|\leqslant|\alpha|+j}|\partial_x^\beta\varphi|^2\,dx$$

for each $t_0>0$, if $p>s-\frac{n}{2}-\frac{1}{2}$ and $|\alpha|+j\leqslant s$. Here $C$ is a constant, independent of $t_0$. This proves the "only if" part.

Conversely, if $\Phi_p(y, t)\in\mathring{H}_s^{loc}(\bar{R}_+^{n+1})$ $(s\geqslant0)$, for each $t_0>0$ and $|\alpha|\leqslant s$, we have

$$\infty>\int_0^{t_0}\int_{R^n}|\partial_y^\alpha\Phi_p(y, t)|^2\,dy\,dt=\frac{t_0^{2(p-|\alpha|)+n+1}}{2(p-|\alpha|)+n+1}\int_{R^n}|\partial_x^\alpha\varphi|^2\,dx.$$

Moreover, supp $\varphi(x)\subset\bar{\Omega}$. Consequently, $\varphi(x)\in\mathring{H}_s(\Omega)$.

(2) From (20) it is seen that

$$\Phi_p(y, t)\in H_{-s}(R_y^n)$$

for each $t>0$. Then from

$$\int_{-\infty}^{t_0}\|\Phi_p(y, t)\|_{-s}^2\,dt=\int_0^{t_0}t^{2(p+n)}dt\int_{R^n}(1+|\xi|^2)^{-s}|\hat{\varphi}(\xi t)|^2d\xi$$

$$\leqslant\max(1, t_0^{2s})\int_0^{t_0}t^{2p+n}\,dt\cdot\|\varphi(x)\|_{-s}^2$$

we obtain (21). (22) can be obtained by the similar way. Lemma 4 is proved.

Suppose that $P(\xi, \tau)=0$, as an equation of $\tau$, has a root of multiplicity $l$ for some $\xi\neq0$ and has no root of multiplicity$\geqslant l+1$ for any $\xi\neq0$. As usual, let

$$\tilde{P}^2=\sum_{|\alpha|>0}|P^{(\alpha)}(\xi, \tau)|^2.$$

**Lemma 5.** (1) *For each polynomial $P$ mentioned above there is a constcnt $C$ such that*

$$\tilde{P}^2\geqslant C(1+|\xi|^2+|\tau|^2)^{m-l},\tag{23}$$

*where $(\xi, \tau)\in\mathbf{C}^{n+1}$ and $C>0$.*

(2) *For each polynimial $P$ mentioned above there is a constant $C'$ such that*

$$|P(\xi, \tau-i\varepsilon)|^2\geqslant C'(\varepsilon)(1+|\xi|^2+|\tau|^2)^{m-l},\tag{24}$$

*where $(\xi, \tau)\in R^{n+1}$, $\varepsilon>0$ and $C'(\varepsilon)>0$.*

*Proof* (1) For any $(\xi, \tau)\in\mathbf{C}^{n+1}$ with $|\xi|^2+|\tau|^2=1$, all $l$ th derivatives of $P(\xi, \tau)$ cannot vanish simultaneously. Otherwise, $P(\xi, \tau)$ would admit a root with multiplicity$\geqslant l+1$, since $P(\xi, \tau)$ is homogeneous. Hence there is a constant $C_1$ such that

$$\tilde{P}^2\geqslant\sum_{|\alpha|=l}|P^{(\alpha)}(\xi, \tau)|^2\geqslant C_1(|\xi|^2+|\tau|^2)^{m-l},$$

if $|\xi|^2+\tau^2\geqslant1$. Moreover, $\sum_{|\alpha|=m}|P^{(\alpha)}|^2$ is a positive constant. Hence we have (23).

(2) By the hyperbolicity of $P(\xi, \tau)$ we have $P(\xi, \tau-i\varepsilon)\neq0$ $(\varepsilon>0)$,Moreover

$$|\tilde{P}(\xi, \tau-i\varepsilon)|\leqslant C(\varepsilon)|P(\xi, \tau-i\varepsilon)|$$

(see Lemma 4.1.1[17]). Using inequality (23) we obtain (24).

The Cauchy problem for the hyperbolic operator $P(\partial_y, \partial_t)$ has a fundamental solution $E$ which satisfies

$$\widehat{e^{-\varepsilon t}E}=\frac{1}{P(\xi, \tau-i\varepsilon)}, \ \forall \varepsilon>0 \tag{25}$$

and

$$\text{supp } E\subset \Gamma_0=\{(y, t)\mid -\eta\cdot y+\tau t\geqslant 0, \ (-\eta, \tau)\in\Gamma\} \tag{26}$$

([17], Theorem 4.6.3). Moreover

$$E*F\in H^{\text{loc}}_{s+m-l}(\bar{R}^{n+1}_+), \ \forall F\in \mathring{H}^{\text{loc}}_s(\bar{R}^{n+1}_+) \tag{27}$$

is the unique distribution solution of (4), if the support of the solution belongs to $\bar{R}^{n+1}_+$. Here we have utilized Lemma 5(2).

**Lemma 6.** *Let $\{f_h\}$ be a sequence of distributions such that $f_h\in\mathscr{D}'(R^{n+1})$, supp $f_h\subset\bar{R}^{n+1}_+$. If there is $\varepsilon>0$ such that*

$$e^{-\varepsilon t}f_h\in S'(R^{n+1}) \text{ and } e^{-\varepsilon t}f_h\xrightarrow{S'}e^{-\varepsilon t}f \quad (h\to\infty), \tag{28}$$

*then*

$$u_h=E*f_h\xrightarrow{\mathscr{D}'}E*f=u \quad (h\to\infty). \tag{29}$$

*Here $S'$ is the space set of all temperate distributions*[17].

*Proof* From the properties of the convolution we have

$$e^{-\varepsilon t}u_h=e^{-\varepsilon t}E*e^{-\varepsilon t}f_h. \tag{30}$$

Considering the Fourier transforms of the both sides of (30) and using (25), we obtain

$$\widehat{e^{-\varepsilon t}u_h}=\frac{1}{P(\xi, \tau-i\varepsilon)}\widehat{e^{-\varepsilon t}f_h}. \tag{31}$$

From (24) it is seen that

$$\frac{1}{P(\xi, \tau-i\varepsilon)}\widehat{e^{-\varepsilon t}f_h}\xrightarrow{S'}\frac{1}{P(\xi, \tau-i\varepsilon)}\widehat{e^{-\varepsilon t}f}=e^{-\varepsilon t}\widehat{E*e^{-\varepsilon t}f} \quad (h\to\infty). \tag{32}$$

Since the Fourier transformation is a continuous map from $S'$ to itself, we have

$$e^{-\varepsilon t}u_h\xrightarrow{S'}e^{-\varepsilon t}u \quad (h\to\infty).$$

Consequently, for any $\psi\in C_0^\infty(R^{n+1})$

$$\langle u_h, \psi\rangle=\langle e^{-\varepsilon t}u_h, e^{\varepsilon t}\psi\rangle\to\langle e^{-\varepsilon t}u, e^{\varepsilon t}\psi\rangle=\langle u, \psi\rangle.$$

This proves Lemm 6.

## IV. Problems $T_1$ and $T_2$

For the problem $T_1$ we have

**Theorem 1.** *If $a>m+q-\frac{n}{2}-\frac{3}{2}$ $(q\geqslant 0)$ and $f\in\mathring{H}_s(\bar{\Omega})$ $(s\leqslant q)$, then in the space $\mathring{H}_{s+m-l}(\Omega)$ the problem $T_1$ has a unique solution which satisfies (7) in the sense of distribution. Moreover*

$$\|\varphi\|_{s+m-l}\leqslant C_s\|f\|_s \quad (C_s=\text{constant}). \tag{33}$$

*Proof* Firstly, suppose that $f(x)\in C^\infty(R^n)$ and supp $f(x)\subset\bar{\Omega}$. From Lemma 4 we have

$$F(y,\ t)=H(t)f\left(\frac{y}{t}\right)t^{a+1-m}\in \mathring{H}_q^{\mathrm{loc}}(\bar{R}_+^{n+1}). \tag{34}$$

Hence equation (4) admits a solution

$$u(y,\ t)=E*F(y,\ t)\in \mathring{H}_{q+m-l}^{\mathrm{loc}}(\bar{R}_+^{n+1}). \tag{35}$$

Since $F(y, t)$ is a homogeneous function of degree $a+1-m$, $u(\lambda y, \lambda t)/\lambda^{a+1}(\lambda>0)$ is a solution of (4) too. Moreover

$$u(\lambda y,\ \lambda t)=\lambda^{a+1}u(y,\ t).$$

Hence there is a function $\varphi(x)$ such that

$$u(y,\ t)=H(t)t^{a+1}\varphi\left(\frac{y}{t}\right).$$

From Lemma 4(1) we have

$$\varphi(x)\in H_{q+m-l}(R^n),$$

since $q+m-l\geqslant 0$. Further, from the properties of the convolution we have

$$\operatorname{supp} u\subset \operatorname{supp} E+\operatorname{supp} F\subset \Gamma_0+K_{\bar{\Omega}}\subset K_{\bar{\Omega}},$$

since $\bar{\Omega}$ contains the convex hull of $\Sigma$ and $\partial\Omega$ is spacelike. Hence $\varphi(x)\in H_{q+m-l}(\Omega)$

Now we prove that $\varphi(x)$ satisfies (7) in the sense of distribution. In fact, for any $\psi(t)\in C_0^\infty(R_+^1)$ and $V(x)\in C_0^\infty(R^n)$

$$\left\langle Pu,\ \psi(t)V\left(\frac{y}{t}\right)\right\rangle=\langle f(x),\ V(x)\rangle\langle t^{a+1-m+n},\ \psi(t)\rangle$$

holds. On the other hand

$$\left\langle Pu,\ \psi(t)V\left(\frac{y}{t}\right)\right\rangle=\left\langle t^{a+1-m}L(a)\varphi\left(\frac{y}{t}\right),\ \psi(t)V\left(\frac{y}{t}\right)\right\rangle$$
$$=\langle L(a)\varphi,\ V\rangle\langle t^{a+1-m+n},\ \psi(t)\rangle.$$

Choose $\psi(t)$ such that $\langle t^{a+1-m+n},\ \psi(t)\rangle\neq 0$, we have

$$\langle f(x),\ V(x)\rangle=\langle L(a)\varphi(x),\ V(x)\rangle \tag{36}$$

and hence (7) holds in the sense of distribution.

Let $f\in \mathring{H}_s(\Omega)$ $(s\leqslant q)$. There exists a sequence $\{f_h\}$ such that $f_h\in C_0^\infty(\Omega)$ and $f_h\to f$ in $\mathring{H}_s(\Omega)$ as $h\to\infty$. From Lemma 4 we have

$$e^{-\varepsilon t}F_h(y,\ t)\xrightarrow{H_s}e^{-\varepsilon t}F(y,\ t). \tag{37}$$

Since $f_h\in C_0^\infty(\Omega)$, there is $\varphi_h\in \mathring{H}_{q+m-l}(\Omega)$ such that $L(a)\varphi_h=f_h$. Let $u_h=E*F_h=\varphi_h\left(\frac{y}{t}\right)t^{a+1}$. From (36) and Lemma 6 it is seen that

$$u_h\xrightarrow{\mathscr{D}'(R^{n+1})}u=E*F(y,\ t). \tag{38}$$

Hence $\left\langle u_h,\ V\left(\frac{y}{t}\right)\psi(t)\right\rangle=\langle\varphi_h(x),\ V(x)\rangle\langle t^{a+1+n},\ \psi(t)\rangle\to\left\langle u,\ V\left(\frac{y}{t}\right)\psi(t)\right\rangle.$

Consequently $\langle\varphi_h(x),\ V(x)\rangle\to\left\langle u,\ \psi(t)V\left(\frac{y}{t}\right)\right\rangle\Big/\langle t^{a+1+n},\ \psi(t)\rangle$

holds for any $V\in C_0^\infty(R^n)$. Hence there is an element $\varphi(x)\in\mathscr{D}'(R^n)$ defined by

$$\langle\varphi(x),\ V(x)\rangle=\left\langle u,\ \psi(t)V\left(\frac{y}{t}\right)\right\rangle\Big/\langle t^{a+1+n},\ \psi(t)\rangle.$$

It is easily seen that

$$\varphi_h(x)\xrightarrow{\mathscr{D}'}\varphi(x),\ \text{supp}\ \varphi\subset\bar{\Omega}$$

and

$$L(a)\varphi=\lim_{h\to\infty}L(a)\varphi_h=f.$$

This proves the existence of the solution.

Suppose $\varphi\in\mathring{H}_{s+m-l}(\bar{\Omega})$ is a solution of (7). Then $u=H(t)\varphi\left(\frac{y}{t}\right)t^{a+1}\in\mathscr{D}'(R^{n+1})$ satisfies (4) in the sense of distribuion. From the uniqueness of the solution of Cauchy problem for (4) follows the uniqueness of the solution of problem $T_1$. Hence $L^{-1}(a)$ is a linear operator from $\mathring{H}_s(\Omega)$ to $\mathring{H}_{s+m-l}(\Omega)$.

Finally, we shall prove the boundedness of $L^{-1}(a)$ by using the closed graph theorem. Suppose that $\{f_h\}$ is a sequence in $\mathring{H}_s(\Omega)$ such that $f_h\to f$ in $\mathring{H}_s(\Omega)$ and $L^{-1}(a)f_h\to\varphi'$ in $\mathring{H}_{s+m-l}(\Omega)$ as $h\to\infty$. Let $\varphi=L^{-1}(a)f$. Repeating the above argument we see that

$$L^{-1}(a)f_h\xrightarrow{\mathscr{D}'}L^{-1}(a)f=\varphi. \tag{39}$$

Hence $\varphi'=\varphi$, i. e., $\{(f,\ L^{-1}(a)f\,|\,f\in\mathring{H}_s(\Omega)\}$ is a closed set in $\mathring{H}_s(\Omega)\times\mathring{H}_{s+m-l}(\Omega)$. From the closed graph theorem it follows that $L^{-1}(a)$ is bounded. The proof is completed.

**Theorem 2.** *If $a<-\frac{n}{2}-\frac{3}{2}-l$ and $f\in H_s(\Omega)$ with $s\geqslant l$, then the problem $T_2$ has unique solution in $H_{s+m-l}$.*

*Proof* From Lemma 3 we see that $L^*(a)=(-1)^m L(a^*)$ with

$$a^*=-a-3+m-n>m+l-\frac{n}{2}-\frac{3}{2}. \tag{40}$$

If $\psi\in H_m(\Omega)$, $L(a)\psi=0$, then the Green formula gives

$$(L(a^*)\varphi,\ \psi)=0$$

for all $\varphi\in\mathring{H}_m(\Omega)$. If $\sigma\in C_0^\infty(\Omega)$, then from Theorem 1 we see that there is a function $\varphi\in\mathring{H}_{2m-l}(\Omega)$ such that $L(a^*)\varphi=\sigma\psi$. Then

$$(\sigma\psi,\ \psi)=0$$

holds for all $\sigma\in C_0^\infty(\Omega)$. Hence $\psi=0$. This proves the uniqueness of the solution.

Let $f\in H_s(\Omega)$. For all $\psi\in C_0^\infty(\Omega)$ we have

$$|(f,\ \psi)|\leqslant\|f\|_s\|\psi\|_{-s}\leqslant c\|f\|_s\|L^*(a)\psi\|_{-s-m+l}. \tag{41}$$

Thus $(\psi,\ f)$ is a bounded linear functional defined on the subspace $\{L^*(a)\psi\,|\,\psi\in C_0^\infty(\Omega)\}$ of the space $\mathring{H}_{-s+l-m}$. From the Hahn-Banach Theorem it can be extended to a bounded linear functional on $\mathring{H}_{-s+l-m}$. According to the properties of $\mathring{H}_{-s+l-m}$, this functional can be expressed by an element $\varphi\in H_{s+m-l}$, i. e., for any $\psi\in C_0^\infty(\Omega)$

$$(f,\ \psi)=(\varphi,\ L^*(a)\psi)=(L(a)\varphi,\ \psi). \tag{42}$$

This means that $\varphi$ is the solution in $H_{s+m-l}(\Omega)$. The proof is completed.

Using Theorem 1 and 2, we obtain

**Corollary 1.** *If* $s \geqslant l+\left[\frac{n}{2}\right]+1$, $a>m+s-\frac{n}{2}-\frac{3}{2}$ *and* $f \in H_s$, *then the problem* $T_1$ *admits a* $C^m(\overline{\Omega})$ *solution uniquely.*

**Corollary 2.** *If* $s \geqslant l+\left[\frac{n}{2}\right]+1$, $a<-\frac{n}{2}-\frac{3}{2}-l$ *and* $f \in H_s$, *then the problem* $T_2$ *admits a* $C^m(\overline{\Omega})$ *solution uniquely.*

**Remark 1.** By the fundamental solution $E$ the solution to problem $T_1$ can be expressed as

$$\varphi(x)=\left(E * t^{a+1-m} f\left(\frac{y}{t}\right) H(t)\right)\Big|_{t=1,\, y=x}. \tag{43}$$

**Remark 2.** If we consider classical solutions, then the boundary conditions in problem $T_1$ may be non-homogeneous.

## V. $C^\infty$ Solutions

Let $f \in C^\infty(\overline{\Omega})$. we consider the $C^\infty(\overline{\Omega})$ solutions of (7). From Theorem 2 it follows that (7) admits only one $C^\infty(\overline{\Omega})$ solution, if $a<-\frac{n}{2}-\frac{3}{2}-l$. We next discuss the case $a \geqslant -\frac{n}{2}-\frac{3}{2}-l$. For convenience we use the comma as the notation for partial differentiation, e. g.

$$\varphi_{,i_1 \cdots i_s}=\frac{\partial^s \varphi}{\partial x_{i_1} \cdots \partial x_{i_s}} \tag{44}$$

$$\varphi_{i_1 \cdots i_{s-1}, i_s}=\frac{\partial}{\partial x_{i_s}} \varphi_{i_1 \cdots i_{s-1}}.$$

**Lemma 7.** *If $\varphi$ satisfies* (7), *then*

$$L(a-1) \varphi_{,i}=f_{,i}=(L(a) \varphi)_{,i}. \tag{45}$$

*Proof* Let $\Phi(y, t)=t^{a+1} \varphi\left(\frac{y}{t}\right)$ and let $\Phi$ satisfy

$$P(\partial_y, \partial_t) \Phi=t^{a+1-m} f\left(\frac{y}{t}\right).$$

Differentiating with respect to $y_i$, we obtain

$$P(\partial_y, \partial_t)\left[t^a \varphi_{,i}\left(\frac{y}{t}\right)\right]=t^{a-m} f_{,i}\left(\frac{y}{t}\right).$$

(45) follows from the definition of $L(a-1)$.

**Lemma 8.** *If we write $L(a)$ in the form*

$$L(a)=\sum_{|\alpha| \leqslant m} a_\alpha(x) \partial^\alpha, \tag{46}$$

then

$$a_0(a)=a_{0 \ldots 0}(a)=(a+1) \cdot a \cdots(a-m+2). \tag{47}$$

*Proof* (47) is a direct consequence of (8).

Suppose that $\varphi$ is a $C^\infty(\bar{\Omega})$ solution of (7). Then the s-th derivatives $\varphi_{,i_1\cdots i_s}$ satisfy

$$L(a-s)\varphi_{,i_1\cdots i_s}=f_{,i_1\cdots i_s}. \tag{48}$$

Conversely, choose $s$ such that $a-s<-\dfrac{n}{2}-\dfrac{3}{2}-l$. Then (48) admits one and only one solution $\varphi_{i_1\cdots i_s}\in C^\infty(\bar{\Omega})$. Since $f_{,i_1\cdots i_s}$ are symmetric with respect to the lower indices, the uniqueness of the solution gives that $\varphi_{i_1\cdots i_s}$ are also symmetric with respect to the lower indices. Differentiating (45), we obtain

$$L(a-s-1)\varphi_{i_1\cdots i_s,i_{s+1}}=f_{,i_1\cdots i_{s+1}}. \tag{49}$$

From the above argument, we have

$$\varphi_{i_1\cdots i_s,i_{s+1}}=\varphi_{i_1\cdots i_{s-1}i_{s+1},i_s}. \tag{50}$$

Because $\bar{\Omega}$ is simply connected, there exists a system of functions $\varphi_{i_1\cdots i_{s-1}}$ such that

$$\varphi_{i_1\cdots i_{s-1},i_s}=\varphi_{i_1\cdots i_s}. \tag{51}$$

$\varphi_{i_1\cdots i_{s-1}}$ are determined by (50) except for additional constants. We may choose a system of $\varphi_{i_1\cdots i_{s-1}}$ which are symmetric with respect to the lower indices. From (48) and (45) we obtain

$$(L(a-s+1)\varphi_{i_1\cdots i_{s-1}})_{,i_s}-L(a-s)\varphi_{i_1\cdots i_{s-1},i_s}-f_{,i_1\cdots i_s}$$

and hence

$$L(a-s+1)\varphi_{i_1\cdots i_{s-1}}-f_{,i_1\cdots i_{s-1}}=c_{i_1\cdots i_{s-1}}. \tag{52}$$

Here $c_{i_1\cdots i_{s-1}}$ are constants which are symmetric with respect to the indices. If $a_0(a-s+1)\neq 0$, we can change $\varphi_{i_1\cdots i_{s-1}}$ by adding a system of constants $a_{i_1\cdots i_{s-1}}$ which are determined uniquely such that $\varphi_{i_1\cdots i_{s-1}}$ satisfy the equations

$$L(a-s+1)\varphi_{i_1\cdots i_{s-1}}=f_{,i_1\cdots i_{s-1}}. \tag{53}$$

Using the same procedure successively, we see that if $a+1$ is not a non-negative integer the $C^\infty(\bar{\Omega})$ solution of (7) exists uniquely for any $f\in C^\infty(\bar{\Omega})$.

Now let $a+1$ be a non-negative integer. For example, let $a=-1$. The above procedure gives a system of $C^\infty(\bar{\Omega})$ functions $\varphi_{i_1}$, satisfying $L(a-1)\varphi_{i_1}=f_{,i_1}$ and $\varphi_{i_1,i_2}=\varphi_{i_2,i_1}$. Hence there is a $C^\infty(\bar{\Omega})$ function $\varphi$ such that $L(a)\varphi-f=C$. $\varphi$ is uniquely determined up to an additional constant. However, the additional constant does not effect the value of the constant $C$, for $a_0(a)=a_0(-1)=0$. Define the equivalent class of $C^\infty(\bar{\Omega})$ functions $[f(x)]_0$ as

$$[f(x)]_0=\{f(x)+c\,|\,c\in R^1\}.$$

In each class there is one and only one $f(x)$ such that (7) has a $C^\infty(\bar{\Omega})$ solution. If $\varphi(x)$ is a solution, then every function in the class $[\varphi]_0$ is also a solution.

More generally, let $a=-1+k$, where $k$ is an integer and $0\leqslant k\leqslant m-1$. Define an equivalent class of functions

$$[f(x)]_k=\{\tilde{f}(x)\,|\,\tilde{f}(x)\in C^\infty(\bar{\Omega}),\ \tilde{f}(x)-f(x) \text{ is a polynomial of degree } s\leqslant k\}.$$

From the above procedure and

$$a_0(-1)=a_0(0)=\cdots=a_0(-1+k)=0,$$

it is seen that in each class $[f(x)]_k$, there is one and only one function such that (7) admits a $C^\infty(\bar{\Omega})$ solution. Moreover, if $\varphi$ is a $C^\infty(\bar{\Omega})$ solution, then all functions in $[\varphi]_k$ are solutions.

Let $a=m-2+p$, where $p$ is a positive integer. Using the same argument, we obtain the following results: Let

$$[f(x)]_{p,m}=\{\tilde{f}(x)\,|\,\tilde{f}(x)\in C^\infty(\bar{\Omega}),$$
$$\tilde{f}_{,i_1\cdots i_p}-f_{,i_1\cdots i_p} \text{ are polynomials of degree } s\leqslant m-1\}.$$

In each class $[f(x)]_{p,m}$ there is one and only one function such that (7) admits a $C^\infty(\bar{\Omega})$ solution. Moreover, if $\varphi(x)$ is a $C^\infty(\bar{\Omega})$ solution of (7), then for each set of polynomials $k_{i_1\cdots i_p}$ which are symmetric with respect to $i_1\cdots i_p$ and of degree $<m$, there exists one and only one solution $\tilde{\varphi}(x)$ such that

$$\tilde{\varphi}_{,i_1\cdots i_p}=\varphi_{,i_1\cdots i_p}+k_{i_1\cdots i_p}.$$

The sets $[f(x)]_k$ and $[f(x)]_{p,m}$ have dimension

$$1+C_1^n+\cdots+C_k^{n+k-1}$$

and

$$C_p^{n+p-1}+C_{p+1}^{n+p}+\cdots+C_{p+m-1}^{n+p+m-2}$$

respectively. Thus we have obtained

**Theorem 3.** *Let $f(x)\in C^\infty(\bar{\Omega})$. If $a+2$ is not a positive integer, then (7) admits one and only one $C^\infty(\bar{\Omega})$ solution. If $a=-1+k$ $(k=0, 1, \cdots, m-1)$, then in the class $[f(x)]_k$ there is one and only one function such that (7) admits solutions. Further, if $\varphi(x)$ is a solution, then the whole set of solutions is $[\varphi(x)]_k$. If $a=m-2+p$, then in the class $[f(x)]_{p,m}$ there is one and only one function such that (7) admits $C^\infty(\bar{\Omega})$ solutions. The solutions are determined except for a set of polynomials $k_{i_1\cdots i_p}(x)$ which are symmetric with respect to their indices and of degree $s<m$.*

**Remark.** In IV we have seen that the problem $T_1$ admits a solution. However. even for the case $f\in C^\infty(\bar{\Omega})$, the problem $T_1$ does not admit a $C^\infty(\bar{\Omega})$ solution in general. As pointed in[14], for equation of mixed type the boundary conditions for solution in $C^\infty(\bar{\Omega})$ and for the solution in $C^r(\bar{\Omega})$ $(m\leqslant r<+\infty)$ may be quite different.

## References

[1] Bnsemann, A., Infinitesimale kegelige uberschallstromungenn, *Schriften Dtsch Akd. Lufo.*, **7B** (1943), 105—122.

[2] Calafoli, E., High speed aerodynamics, 1956, Bucarest.

[3] Lighthill, M. L., The diffraction of blast, *Proc. Roy. Soc.* **198** (1949), 454—470; *Proc. Roy. Soc.* **200** (1950), 554—565.

[4] Hua, L. K., Introduction to the mixed partial differential equation, *Journal of Chinese University of Science and Technology*, (1965), 1—27.

[5] Friedrichs, K. O., Symmetric positive linear differential equations, *Comm. pure Appl. Math.*, **11** (1958), 333—414.

[6] Gu, C. H., Differentiable solutions of symmetric positive partial differential equations, *Acta Math Sinica*, **14** (1964), 503—516, Chinese Math., 5 (1964), 541—444 (Eng).

[7] Gu, C. H., A class of mixed partial differential equations in $n$-dimensional space, *Scientia Sinica*, **14** (1965), 1574—1580.
[8] Gu, C. H., On boundary value problems for linear mixed equations with $n$-variables, *Bull. Sciences*, **67** (1978), 335—339.
[9] Gu, C. H., On the partial differential equations of mixed type in $n$-independent variables, *Comm. pure Appl. Math.*, **34**:3 (1981).
[10] Gu, C. H., Boundary value problems for quasilinear potitive symmetric systems and their applications to mixed equations, *Acta Math.* **21** (1978), 119—129.
[11] Hong, J. X., On boundary value problems for 2nd order mixed equations with characteristic degenerate surfaces, (to appear in *Chin. Ann. of Math*).
[12] Sun, L. X., On boundary value problems for 2nd order mixed equatons, (to appear).
[13] Karatopraklive, G. D., Theory of boundary value problems for mixed equations in multidimentional regions, *Differential Equations*, **13** (1977), 64—75 (Russ.), 43—51 (Eng.).
[14] Gu, C. H., On the $C^\infty$ solutions of a class of linear partial differential equations, *Comm. in PDE*, **5** (1980), 985—997.
[15] Hong, J. X., On mixed equation of higher order with characteristic degenerate sufaces, (to appear).
[16] Courant, R. and Hilbert, D., Methods of mathematical physics, vol. 2, 1962, Interscience.
[17] Hörmander, L., Linear partial differential operators, 1963, Springer-Verlag.

J. DIFFERENTIAL GEOMETRY
16 (1981) 161–177

# A GENERALIZATION OF A THEOREM OF DELAUNAY

WU-YI HSIANG & WEN-CI YU

*Dedicated to Professor Buchin Su on his 80th birthday*

## Introduction

In 1841 Delaunay [1] discovered the following beautiful way of constructing rotational symmetric hypersurfaces of constant mean curvature in euclidean 3-space $\mathbf{R}^3$, by rolling a given conic section on a line in a plane, and rotating about that line the trace of a focus, one obtains a hypersurface of constant mean curvature in $\mathbf{R}^3$. Conversely, all rotational symmetric hypersurfaces of constant mean curvature in $\mathbf{R}^3$, except those spheres, can be constructed in this way. Combining the above classical result with a theorem of Ruh and Vilms [6] which asserts that the Gauss map of a hypersurface of constant mean curvature in $\mathbf{R}^n$ is automatically harmonic, one can easily use the periodicity of such construction (for ellipse and hyperbola) to obtain harmonic maps of $T^2 \to S^2$. In this paper, we shall study generalized rotational hypersurfaces of constant mean curvature in higher dimensional euclidean spaces, and prove a generalization of Delaunay theorem, which again enables us to construct harmonic maps of $S^p \times S^1 \to S^{p+1}$.

From the viewpoint of equivariant differential geometry, a natural generalization of the $SO(2)$ rotation on $\mathbf{R}^3$ is simply an orthogonal transformation group $(G, \mathbf{R}^n)$ with codimension-two principal orbit type. Such orthogonal transformation groups are classified in [4]. Therefore it is quite natural to study solutions of hypersurfaces in $\mathbf{R}^n$ with constant mean curvature, which are invariant with respect to one of such orthogonal transformation groups with codimension-two principal orbit type.

In §1 we shall reduce the above equivariant geometric problem to a much simpler ordinary differential equation defined on a suitable two-dimensional manifold, namely, the orbit space $\mathbf{R}^n/G$. In the special case where $G = SO(n-1)$, the corresponding ordinary differential equation can be integrated explicitly in terms of hyperelliptic functions. Furthermore, it is

Received April 13, 1981.

possible to generalize Delaunay's rolling construction to this special case, and we shall exploit such geometric construction to discuss various properties such as periodicity of the solutions (cf. §2 and §3). In succeeding papers we shall further investigate the properties of hypersurfaces with constant mean curvature, which are invariant under other cohomogeneity-two orthogonal transformation groups.

## 1. Generalized rotational hypersurfaces of constant mean curvature in $\mathbf{R}^n$

Let $G$ be a compact connected Lie group, and $\Phi$ an orthogonal representation of $G$ acting on $\mathbf{R}^n$ with codimension-two principal orbit type. Such orthogonal transformation groups $(G, \Phi, \mathbf{R}^n)$ have been classified in [4], and the final result of such a classification can be neatly described as follows: Let $L/G$ be a symmetric space of rank 2, and $\Phi$ the isotropy representation of $G$ on the tangent space $\mathbf{R}^n$ of the base point. Then it follows from the definition of rank that $\mathbf{R}^n/G$ is of dimension two and hence it is an example of such orthogonal transformation groups. The final result of the classification of [4] simply asserts that every orthogonal transformation group $(G, \Phi, \mathbf{R}^n)$ with $\dim \mathbf{R}^n/G = 2$ is obtained from the isotropy representation of a symmetric space of rank 2.

As a generalization of rotational surfaces in $\mathbf{R}^3$, we shall call an invariant hypersurface in $(G, \Phi, \mathbf{R}^n)$ a *generalized rotational hypersurface* [with respect to the given type of cohomogeneity-two transformation $(G, \Phi, \mathbf{R}^n)$]. Hypersurfaces of constant mean curvature in $\mathbf{R}^n$ are locally given by a quasi-linear second order elliptic partial differential equation which is rather difficult to deal with. Therefore our first step in the study of rotational hypersurfaces of constant mean curvature will be to make effective use of the "symmetry" of such a geometric situation to reduce the above formidable analytic problem to a more manageable one. Let us recall here some basic generalities of equivariant differential geometry. Suppose $M$ is a Riemannian manifold with a given *isometric* transformation group $G$. Then it is rather natural to equip the orbit space $M/G$ with the following structures: (i) a generalized smooth structure on $M/G$ whose smooth functions are exactly the quotients of $G$-invariant smooth functions, namely, $C^\infty(M/G) \cong C^\infty(M)^G$, (ii) a generalized Riemannian metric $ds^2$ on $M/G$ which measures the distances between $G$-orbits, (we shall call it the orbital distance metric of $M/G$), (iii) a *smooth* function $f: M/G \to \mathbf{R}$ such that $f(\xi) = (\text{volume of } \xi)^2$ if $\xi$ is a *principal* $G$-orbit. In term of the above orbital geometric invariants of $(G, M)$, one has the following proposition of [3] which reduce the computation of mean curvature of an $G$-invariant submanifold $N \subset M$ to that of its image $N/G \subset M/G$.

**Proposition 1.** *Let $N$ be an invariant submanifold of $M$ with the same principal orbit type as that of $M$, $x$ be a point on a principal orbit $\xi \in N/G$, $\nu_x$ be an arbitrary unit normal vector of $N$ at $x$, and $\nu_\xi$ be its image vector at $\xi$. Then*

$$H(\nu_x) = H'(\nu_\xi) - \frac{1}{2}\frac{d}{d\nu_\xi} ln\, f, \tag{1}$$

*where $H(\nu_x)$, $H'(\nu_\xi)$ are the mean curvatures of $N$, $N/G$ in the direction of $\nu_x$, $\nu_\xi$ respectively, and $d/d\nu_\xi$ is the differentiation in the direction of $\nu_\xi$.*

Next, let us recall the result of some explicit computations of the orbital geometric invariants of those orthogonal transformation groups on $\mathbf{R}^n$ of cohomogeneity two [cf. 4]:

Suppose $(G, \Phi, \mathbf{R}^n)$ is an orthogonal transformation group of cohomogeneity two. Then, according to the classification of [4], $\Phi$ is the isotropy representation of a rank-2 symmetric space $L/G$. It follows from the maximal tori theorem of É. Cartan (for the case of symmetric space) that (i) there exists a 2-dimensional linear subspace $\mathbf{R}^2$ which interesects every $G$-orbit perpendicularly, (ii) the Weyl group of $L/G$, $W$, acts on $\mathbf{R}^2$ as a group generated by reflections and $\mathbf{R}^n/G \cong \mathbf{R}^2/W$. Therefore the orbit space $\mathbf{R}^n/G$ can be identified with a Weyl chamber of $(W, \mathbf{R}^2)$, and the orbital distance metric is *flat*. We shall choose an orthonormal coordinate system $(x, y)$ on $\mathbf{R}^2$ such that the Weyl chamber of $(W, \mathbf{R}^2)$ is given by $y \geqslant 0$ and $x \cdot \sin \pi/d - y \cos \pi/d \geqslant 0$, namely, it is a linear cone of angle $\pi/d$ as follows:

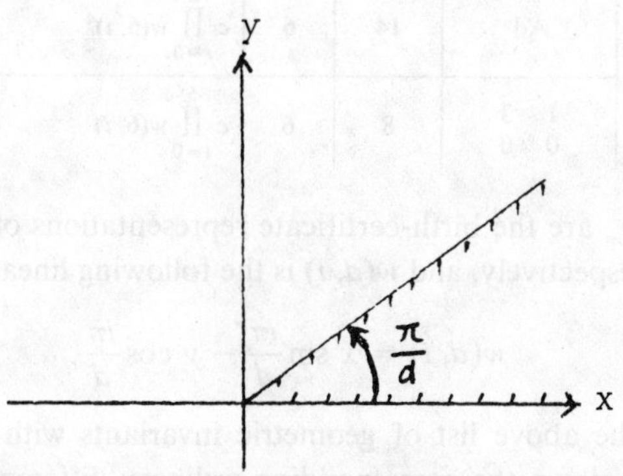


FIG. 1

We list the needed orbital geometric invariants of those cohomogeneity-two orthogonal transformation groups $(G, \Phi, \mathbf{R}^n)$ as follows:

| $G$ | $\Phi$ | $n = \dim \Phi$ | $d(W)$ | $f(\xi) = (\text{volume of } \xi)^2$ | *Asso. Sym. Space* $L/G$ |
|---|---|---|---|---|---|
| $SO(n-1)$ | $1 + \rho_{n-1}$ | $n$ | 1 | $cy^{2n-4}$ | $\mathbf{R}^1 \times S^{n-1}$ |
| $SO(l) \times SO(m)$ | $\rho_l + \rho_m$ | $l + m$ | 2 | $cx^{2m-2}y^{2l-2}$ | $S^l \times S^m$ |
| $SO(3)$ | $S^2\rho_3 - 1$ | 5 | 3 | $c \prod_{i=0}^{2} w(3, i)^2$ | $SU(3)/SO(3)$ |
| $SU(3)$ | Ad | 8 | 3 | $c \prod_{i=0}^{2} w(3, i)^4$ | $\frac{SU(3) \times SU(3)}{SU(3)}$ |
| $Sp(3)$ | $\Lambda^2\nu_3 - 1$ | 14 | 3 | $c \prod_{i=0}^{2} w(3, i)^8$ | $\frac{SU(6)}{Sp(3)}$ |
| $F_4$ | 1 | 26 | 3 | $c \prod_{i=0}^{2} w(3, i)^{16}$ | $\frac{E_6}{F_4}$ |
| $SO(5)$ | Ad | 10 | 4 | $cx^4y^4(x^2 - y^2)^4$ | $\frac{SO(5) \times SO(5)}{SO(5)}$ |
| $SO(2) \times SO(m)$ | $\rho_2 \otimes \rho_m$ | $2m$ | 4 | $c(x\,y)^{2m-4}(x^2 - y^2)^2$ | $\frac{SO(2+m)}{SO(2) \times SO(m)}$ |
| $S(U(2) \times U(m))$ | $[\mu_2 \otimes_{\mathbf{C}} \mu_m]_{\mathbf{R}}$ | $4m$ | 4 | $c(x\,y)^{4m-6}(x^2 - y^2)^4$ | $\frac{SU(2+m)}{S(U(2) \times U(m))}$ |
| $Sp(2) \times Sp(m)$ | $\nu_2 \otimes_{\mathbf{H}} \nu_m^*$ | $8m$ | 4 | $c(x\,y)^{8m-10}(x^2 - y^2)^8$ | $\frac{Sp(2+m)}{Sp(2) \times Sp(m)}$ |
| $U(5)$ | $[\Lambda^2\mu_5]_{\mathbf{R}}$ | 20 | 4 | $c(xy)^{10}(x^2 - y^2)^8$ | $\frac{SO(10)}{U(5)}$ |
| $U(1) \times Spin(10)$ | $[\mu_1 \otimes_{\mathbf{C}} \Delta_1^+]_{\mathbf{R}}$ | 32 | 4 | $c(xy)^{18}(x^2 - y^2)^{12}$ | $\frac{E_6}{U(1) \times Spin(10)}$ |
| $G_2$ | Ad | 14 | 6 | $c \prod_{i=0}^{5} w(6, i)^4$ | $\frac{G_2 \times G_2}{G_2}$ |
| $SO(4)$ | $\begin{smallmatrix} 1 & 3 \\ 0 & 0 \end{smallmatrix}$ | 8 | 6 | $c \prod_{i=0}^{5} w(6, i)$ | $G_2/SO(4)$ |

where $\rho_m$, $\mu_m$, $\nu_m$ are the birth-certificate representations of $SO(m)$, $SU(m)$ or $U(m)$, $Sp(m)$ respectively, and $w(d, i)$ is the following linear function:

$$w(d, i) = x \sin\frac{i\pi}{d} - y \cos\frac{i\pi}{d}.$$

Combining the above list of geometric invariants with Proposition 1, we can easily write down the corresponding ordinary differential equation satisfied by the image curve $N/G \subset \mathbf{R}^n/G \cong \mathbf{R}^2/W$ of a generalized rotational hypersurface $N$ of constant mean curvature in each of the above 14 cases. For example, in the first two cases, the corresponding equations are simply as follows:

(i) $(G, \Phi) = (SO(n-1), 1 + \rho_{n-1})$:

(2) $$\dot{x}\ddot{y} - \dot{y}\ddot{x} - (n-2)\frac{\dot{x}}{y} = h(constant),$$

(ii) $(G, \Phi) = (SO(l) \times SO(m), \rho_l + \rho_m)$:

(3) $$\dot{x}\ddot{y} - \dot{y}\ddot{x} - (l-1)\frac{\dot{x}}{y} + (m-1)\frac{\dot{y}}{x} = h\ (constant),$$

where $\dot{x}, \dot{y}, \ddot{x}, \ddot{y}$ are derivatives with respective to the euclidean arc length $ds^2 = dx^2 + dy^2$ of $\mathbf{R}^n/G \cong \mathbf{R}^2/W \subset \mathbf{R}^2$, and $(x(s), y(s))$ is the image curve $N/G$ of a generalized rotational hypersurface of constant mean curvature $h$.

In the first case, if one considers $y$ as a function of $x$, $y(x)$, then (2) is equivalent to the following:

(4) $$\left[\frac{y^{n-2}}{\sqrt{1+y'^2}} + \frac{h}{(n-1)}y^{n-1}\right]' = -y^{n-2}y'\left[\frac{y''}{(1+y'^2)^{3/2}} - \frac{(n-2)}{y(1+y'^2)^{1/2}} - h\right] = 0.$$

Therefore it is easy to integrate the above equation explicitly in terms of hyperelliptic functions, namely,

(5) $$\frac{y^{n-2}}{\sqrt{1+y'^2}} + \frac{h}{n-1}y^{n-1} = c,$$
$$x = \int\left[\left\{\frac{y^{n-2}}{c - \frac{h}{(n-1)}y^{n-1}}\right\}^2 - 1\right]^{-1/2} dy.$$

In the beginning case of $n = 3$, the above integral is an elliptical integral which is, by definition, closely related to the arc-length of an ellipse. The classical theorem of Delaunay neatly demonstrate the explicit relationship of the above geometrically arrived function $y(x)$ by a simple geometric construction, namely, by rolling an ellipse and plotting the trace of one of its focus. In the next section, we shall prove a generalization of the above theorem of Delaunay to construct higher dimensional rotational hypersurfaces of constant mean curvature by rolling.

## 2. Rolling construction and a generalization of Delaunay theorem

Suppose $\Gamma$ is a plane curve given by polar coordinate graph $r = r(\theta)$, e.g., an ellipse can be expressed as the polar coordinate graph of $r = ep/(1 - e\cos\theta)$ with origin at one of its focus. If one rolls $\Gamma$ on the $x$-axis, then the trace of the origin of the polar coordinate system attached to $\Gamma$ plots another curve $\Omega$. Let us first investigate the analytical relationship between $\Gamma$ and $\Omega$.

As indicated in Fig. 2, $\xi$ is the arc length of $\Gamma$ starting from $Q_0$. Let $s$ be arc length of $\Gamma$ starting from $P_0$, $\phi$ be the angle of the tangent vector of $\Omega$ at $P$, and $\bar{\phi}$ be the angle of $\overrightarrow{QP}$.

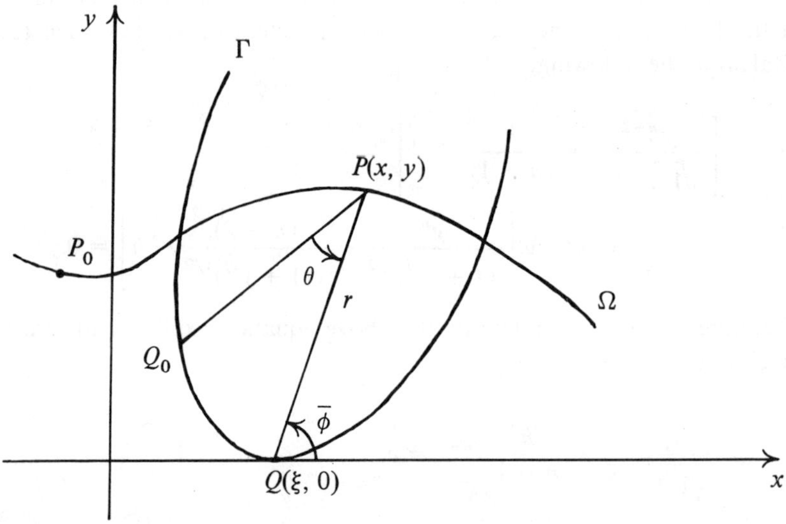


FIG. 2

Then it is easy to see the following relationships

$$x = \xi + r\cos\bar{\phi}, \quad y = r\sin\bar{\phi}, \quad dr = -\cos\bar{\phi}\cdot d\xi \tag{6}$$

It follows easily from (6) that

$$\cos\bar{\phi}\cdot dx + \sin\bar{\phi}\cdot dy = \cos\bar{\phi}\cdot d\xi + dr = 0. \tag{7}$$

Hence $PQ$ is normal to $\Omega$, and $\phi = \bar{\phi} - \pi/2$. By differentiating with respect to the arc length $s$ of $\Omega$, one obtains

$$\begin{aligned} r &= \frac{y}{\cos\phi}, \\ dr &= \frac{1}{\cos\phi}dy + \frac{y}{\cos^2\phi}\sin\phi\, d\phi = \tan\phi(1 + r\dot{\phi})ds; \end{aligned} \tag{8}$$

$$\xi = x + r \sin \phi, \tag{9}$$
$$d\xi = dx + \sin \phi \cdot dr + r \cos \phi \, d\phi = \sec \phi(1 + r\dot{\phi})ds.$$

Therefore it follows that

$$d\theta = \pm \frac{1}{r}\sqrt{d\xi^2 - dr^2} = \pm \frac{1}{r}(1 + r\dot{\phi})ds. \tag{10}$$

(By the choice of directions of the curves Γ and $U$ (cf. Fig. 2), we should take the + sign in (10).)

**Proposition 2.** *Suppose Ω is a $C^2$-curve given by $y = f(x) > 0$. If the center of curvature of Ω never lies on the x-axis, then there exists a unique polar coordinate graph Γ such that Ω is the trace of the origin by rolling Γ on the x-axis.*

*Proof.* Under the assumption of the proposition, one always has $1 + r\dot{\phi} \neq 0$. Therefore it never changes its sign. Choose a starting point $(x_0, y_0)$ and assign the corresponding values of $s = 0$, $\theta = 0$, $r = r_0 = y_0/\cos \phi_0$. Then

$$\theta = \theta(s) = \int_0^s \frac{1}{r}(1 + r\dot{\phi})\, ds \tag{11}$$

is clearly a strictly monotonic function of $s$. Hence one may solve for $s$ in terms of $\theta$ and substitute it into $r = y/\cos \phi = r(s)$ to obtain the polar coordinate equation $r = r(\theta)$. It is rather straightforward to verify that if one rolls the plane curve Γ (defined by the above polar coordinate graph $r = r(\theta)$) on the $x$-axis, then the trace of the origin of its attached polar coordinate system is exactly the given Ω. q.e.d.

Next let us assume that Ω is a *solution curve* of (2).

**Lemma 1.** *Suppose Ω is a solution curve of* (2). *If there exists a point P on Ω whose center of curvature lies on the x-axis or Ω intersect with the x-axis, then Ω is a circular arc with its center on the x-axis and* $(n - 1)/|h|$ *as its radius.*

*Proof.* The first case follows from the uniqueness of solution of (2) at a regular point, and the second case from the analyticity as well as the uniqueness of solutions of (2) at a boundary singular point (cf. 4). q.e.d.

In view of the above lemma, from now on we shall assume that Ω is a solution curve of (2) different from a circular arc. Therefore it follows from Proposition 2 that there exists a unique polar coordinate graph Γ given by $r = r(\theta)$.

**Theorem 1.** *Suppose Ω is not a circular arc. Then Ω satisfies* (2) *if and only if its corresponding Γ satisfies*

$$\frac{d^2}{d\theta^2}(\ln r) = \left[1 + \left(\frac{d}{d\theta}\ln r\right)^2\right]\frac{(n-2) + hr}{(n-1) + hr}. \tag{12}$$

*Proof.* It follows from (8) and (10) that

$$\frac{dr}{d\theta} = r \cdot \tan\phi, \quad \text{or} \quad \frac{d}{d\theta}(\ln r) = \tan\phi. \tag{13}$$

Observe that $\dot{x}\ddot{y} - \dot{y}\ddot{x} = \dot{\phi}$ and $\dot{x}/y = \cos\phi/y = 1/r$. Then it follows from (2) and (10) that

$$\frac{d\phi}{d\theta} = \frac{r\dot{\phi}}{1 + r\dot{\phi}} = \frac{(n-2) + rh}{(n-1) + rh}. \tag{14}$$

Differentiating (13), one has

$$\frac{d^2}{d\theta^2}(\ln r) = (1 + \tan^2\phi)\frac{d\phi}{d\theta},$$

which, together with (13) and (14), gives the differential equation (12) satisfied by $\Gamma$.

Conversely, suppose $\Gamma$ is a polar coordinate graph given by a solution $r = r(\theta)$ of (12), and $\Omega$ is the trace of the origin by rolling $\Gamma$ on the $x$-axis. We shall verify that $\Omega$ is the graph of $y = f(x)$ which satisfies (2). As shown in Fig. 2, $\phi = \bar{\phi} - \pi/2$, hence

$$\tan\phi = -\cot\bar{\phi} = \frac{1}{r}\frac{dr}{d\theta} = \frac{d}{d\theta}(\ln r). \tag{13}$$

Again it follows from (10) that $d\phi/d\theta = r\dot{\phi}/(1 + r\dot{\phi})$. Therefore by differentiating the above equation and making use of (12), one has

$$\frac{r\dot{\phi}}{1 + r\dot{\phi}} = \frac{(n-2) + hr}{(n-1) + hr} \quad \text{or} \quad r\dot{\phi} = (n-2) + hr. \tag{2}$$

This proves that $\Omega$ is a solution of $\dot{\phi} - (n-2)/r = h$ or $\dot{\phi} = (n-2)\dot{x}/y = h$. q.e.d.

Combining Proposition 2 with Theorem 1, one demonstrates that it is always possible to use the rolling construction to transform the problem of solving (2) to the problem of solving (12). As it was pointed out in §1, (2) can, in fact, already be solved explicitly in terms of hyperelliptic functions. Hence it is rather natural to seek explicit solutions of (12) and then try to analyze whether such a geometric transformation actually simplifies the original problem or at least improves our understanding of its solutions. In order to solve (12) explicitly, it is natural to introduce the following auxiliary variables, namely,

$$u = \ln r, \quad v = \frac{dr}{d\theta} = \frac{1}{r}\frac{dr}{d\theta}. \tag{14}$$

Then

$$\frac{d^2}{d\theta^2} \ln r = \frac{dv}{d\theta} = \frac{du}{d\theta}\frac{dv}{du} = v\frac{dv}{du},$$

and hence (12) becomes

$$v\frac{dv}{du} = (1 + v^2)\frac{(n-2) + he^u}{(n-1) + he^u},$$

or

$$\frac{d}{du} \ln(1 + v^2) = 2\frac{(n-2) + he^u}{(n-1) + he^u}. \tag{15}$$

Integrating both sides of (15), one has

$$\ln(1 + v^2) = \frac{2(n-2)}{(n-1)} u + \frac{2}{(n-1)} \ln|n - 1 + he^u| + \alpha_1. \tag{16}$$

That is,

$$(1 + v^2) = \alpha e^{2(n-2)/(n-1)} u|n - 1 + he^u|^{2/(n-1)}, \quad \alpha = e^{\alpha_1},$$

or

$$\frac{1}{r}\frac{dr}{d\theta} = v = \pm\left[\alpha r^{2(n-2)/(n-1)}|n - 1 + hr|^{2/(n-1)} - 1\right]^{-1/2}. \tag{17}$$

Let $w = 1/r$. Then

$$\frac{dw}{d\theta} = \mp \left\{\alpha|(n-1)w + h|^{2/(n-1)} - w^2\right\}^{1/2} = \mp b(w)^{1/2}, \tag{18}$$

where $b(w) = \alpha|(n-1)w + h|^{2/(n-1)} - w^2$. Hence

$$\theta = \mp \int b(w)^{-1/2} dw = \mp A(w) = c_1. \tag{19}$$

**Lemma 2.** *Let* $b_1(w) = \{\alpha[(n-1)w + h]^{2/(n-1)} - w^2\}$, $-h/(n-1) \leqslant w \leqslant \infty$, $n \geqslant 3$, $\alpha > 0$. *Then* $b_1(w)$ *satisfies the following properties*:

(i) *If* $h < 0$ *and* $\alpha \leqslant (-h/n-2)^{2(n-2)/(n-1)}$, *then* $b_1(w) \leqslant 0$.

(ii) *If* $h \geqslant 0$, *or* $h < 0$ *and* $\alpha > (-h/n-2)^{2(n-2)/(n-1)}$, *then* $b_1(w)$ *has exactly two simple roots* $w_1$ *and* $w_2$, *say* $w_1 < w_2$, *and* $b_1(w) \geqslant 0$ *for* $w \in [w_1, w_2]$.

(iii) $b_1(w) \geqslant -(w - w_1)(w - w_2)$, $w \in [w_1, w_2]$, *and*

$$w_1 = 0 \quad \text{if } h = 0,$$

$$\frac{-h}{n-1} < w_1 < 0 < w_2, \quad \text{if } h > 0,$$

$$\frac{-h}{(n-1)} < w_1 < \frac{-h}{(n-2)} < w_2, \quad \text{if } h < 0.$$

*Proof.* $b_1'(w) = 2\alpha[(n-1)w + h]^{(3-n)/(n-1)} - 2w,$

$$b_1''(w) = 2(3-n)\alpha[(n-1)\cdot w + h]^{(4-2n)/(n-1)} - 2, \quad n \geqslant 3.$$

Therefore $b_1''(w) \leqslant -2$, and hence $b_1'(w)$ has at most one simple root. From the above facts, it is easy to verify that $b_1(w)$ has all the above three properties.

**Lemma 2′.** *Let* $b_2(w) = \{\alpha[-(n-1)w - h]^{2/(n-1)} - w^2\}$, $h < 0$, $-\infty < w \leqslant -h/(n-1)$. *Then* $b_2(w)$ *satisfies the following properties*:

(i) $b_2(w)$ *has two simple roots* $w_1$, $w_2$ *and* $w_1 < 0 < w_2 < -h/(n-1)$.

(ii) $w_1 < -w_2$ *and* $b_2(w) \geqslant 0$ *if and only if* $w_1 \leqslant w \leqslant w_2$; *moreover*, $b_2(w) \geqslant -(w-w_1)(w-w_2)$ *for* $w \in [w_1, w_2]$.

*Proof.* Again one has $b_2''(w) \leqslant -2$, and hence $b_2'(w)$ has at most one simple root. Moreover, it is clear that $b_2(-h/(n-1)) < 0$ and $b_2(0) > 0$. Therefore it is straightforward to verify that $b_2(w)$ has properties (i) and (ii). q.e.d.

Based on the above properties of $b_i(w)$, $i = 1, 2$, one may define the antiderivatives $A_i(w)$ of $b_i(w)^{-1/2}$ more precisely as follows:

$$A_1(w) = \int_{\max(0, w_1)}^{w} \{\alpha[(n-1)t + h]^{2/(n-1)} - t^2\}^{-1/2} dt,$$

$$w \in [\max(0, w_1), w_2],$$

$$A_2(w) = \int_0^w \{\alpha[-(n-1)t - h]^{2/(n-1)} - t^2\}^{-1/2} dt, \quad w \in [0, w_2].$$

Summarizing the discussion of this section, we state the generalization of Delaunay theorem as follows.

**Theorem 2.** *Suppose* $N$ *is a generalized rotational surface of* $(SO(n-1), \rho_{n-1} + 1, \mathbf{R}^n)$*-type and of constant mean curvature* $h$, *and* $\Omega = N/G \subset \mathbf{R}^n/G \simeq \mathbf{R}^2/\mathbf{Z}_2$ *is its image curve which can be given as a graph of* $y = f(x) \geqslant 0$ *in the upper half plane. Then one has the following possibilities*:

I. *Case* $h < 0$.

(i) $\Omega$ *is a circular arc with its center on the x-axis and* $(n-1)/(-h)$ *as its radius*,

(ii) $\Omega$ *is given by* $y = (n-2)/(-h)$,

(iii) $y/\cos\phi < (n-1)/(-h)$, *and* $\Omega$ *is obtained by rolling a uniquely determined curve* $\Gamma$ *whose polar coordinate equation is given by* $\theta = \pm A_1(1/r) + c$.

(iv) $y/\cos\phi > (n-1)/(-h)$, *and* $\Omega$ *is obtained by rolling a uniquely determined curve* $\Gamma$ *whose polar coordinate equation is given by* $\theta = \pm A_2(1/r) + c$.

II. *Case* $h \geqslant 0$. $\Omega$ *is obtained by rolling a uniquely determined curve* $\Gamma$ *whose polar coordinate equation is given by* $\theta = \pm A_1(1/r) + c$.

We shall make use of the above theorem to investigate various properties of $\Omega$ via that of $\Gamma$ in §3.

## 3. Properties of generalized rotation hypersurfaces of constant mean curvature of $(SO(n-1), \mathbf{R}^n)$-type

In this section we shall apply the rolling construction of §2 to investigate some interesting properties of generalized rotational hypersurfaces of constant mean curvature of $(SO(n-1), \mathbf{R}^n)$-type. According to Theorems 1 and 2, the rolling construction enable us to transfer the equation of the generating curve $\Omega$, namely (2), to that of the polar coordinate function of the rolling curve $\Gamma$, namely (12), whose solutions can be explicitly expressed in terms of the following integral:

(19) $$\theta = \pm \int b_i(w)^{-1/2} dw, \quad i = 1, 2, \; w = \frac{1}{r}.$$

Therefore it is easy to use the above integral expression to study some interesting properties of $\Omega$ such as periodicity via the rolling construction. In the classical case of $n = 3$

$$b_1(w) = \alpha \cdot (2w + h) - w^2, \quad b_2(w) = -\left[\alpha(2w + h) + w^2\right]$$

are simply quadratic polynomials of $w$. Therefore it is straightforward to compute $\theta = \pm \int b_i(w) dw$ in terms of trigonometric functions. Explicit computation will show that

$$\frac{1}{r} = w = a + b\cos(\theta + c), \quad (a, b, c\text{: suitable constants}),$$

which is exactly the polar coordinate function of a conic section with a focus as its origin. This gives us the classical Delaunay theorem.

In the general case of $n \geqslant 4$, Lemmas 2 and 2′ provide those essential properties of the functions $b_i(w)$, $i = 1, 2$, for our investigation of the integral $\theta = \pm \int b_i(w)^{-1/2} dw$. We shall divide our discussion as follows.

### 3.1. Periodic solutions

In the classical case of $n = 3$, if $\Gamma$ is an ellipse then $\Omega$ is periodic, namely, $\Omega$ is given by a periodic function $y = f(x)$, $f(x + T) \equiv f(x)$. In the general case, $\Omega$ is periodic if and only if $\Gamma$ is periodic, and hence a necessary condition for $\Omega$ to be periodic will be that $r = r(\theta)$ is positive and bounded. In view of

Theorem 2 and Lemmas 2, 2′, the above necessary condition implies that $\theta = \pm \int b_1(w)^{-1/2}dw$, and $h < 0$, $\alpha > [-h/(n-2)]^{2(n-2)/(n-1)}$. In fact, it is not difficult to deduce the following theorem from Theorem 2 and Lemma 2 by straightforward computation.

**Theorem 3.** *$\Omega$ is periodic if and only if $\Gamma$ is given by the following polar coordinate function, namely, $\theta = \pm \int b_1(w)^{-1/2}dw$ with $w = 1/r$, $b_1(w) = \{\alpha[(n-1)w + h]^{2/(n-1)} - w^2\}$ and $h < 0$, $\alpha > [-h/(n-2)]^{2(n-2)/(n-1)}$. By Lemma 2, $b_1(w)$ has two simple roots $w_1, w_2$, $-h/(n-1) < w_1 < -h/(n-2) < w_2$. Then, up to a translation along the x-axis, $\Omega$ is given by $y = f(x)$ satisfying the following properties*:

(i) *$f(x)$ is a periodic function with its period $T < -2h(1+\pi)/(n-1)$,*

(ii) *$f(0) = 1/w_1 < -(n-1)/h$, $f(T/2) = 1/w_2$, and $f(x)$ is symmetric with respect to $x = T/2$,*

(iii) *$f(x)$ is strictly decreasing between 0 and $T/2$, and strictly increasing between $T/2$ and $T$.*

*Proof.* Theorem 2 reduces the verification of above theorem to the corresponding properties of $\Gamma$. Let $\theta = \int_{w_1}^{w} b_1(t)^{-1/2}dt$. Then the properties of $b_1(t)$ listed in Lemma 2 and simple estimates will show that

(i) $w$ is a strictly increasing function of $\theta$ between 0 and $\theta^* = \int_{w_1}^{w_2} b_1(t)^{-1/2}dt$,

(ii) $dw/d\theta = 0$ at both $\theta = 0$ and $\theta = \theta^*$.

Therefore it follows from the uniqueness of (12) that one may analytically continue the solution $w = w(\theta)$ of (12) simply by reflection with respect to $\theta = k\theta^*$, namely,

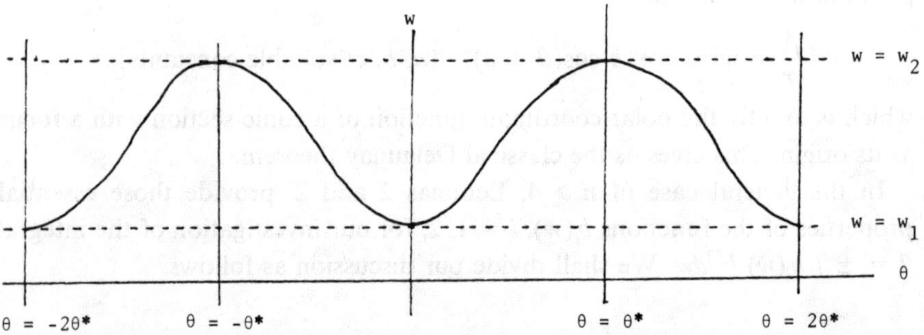


FIG. 3

Hence $\Gamma$ is a periodic curve of period $2\theta^*$, and consequently $\Omega$ is also periodic whose period $T$ is equal to the arc length of $\Gamma$ between $\theta = 0$ and $\theta = 2\theta^*$.

It is easy to use the properties $b_1(w) \geqslant -(w - w_1)(w - w_2)$ and $w_1 > -(n-1)/h$ to estimate that

$$T = 2\int_0^{\theta^*}\left[r^2 + \left(\frac{dr}{d\theta}\right)^2\right]^{1/2} d\theta < \frac{-2h}{(n-1)}(1+\pi).$$

COROLLARY. *As a corollary of Theorem 3 and a theorem of Ruh and Vilms* [6], *the Gauss map of the generalized rotational hypersurface of $\Omega$ in $\mathbf{R}^n$ is a periodic harmonic map of $\mathbf{R}^1 \times S^{n-2} \to S^{n-1}$, and hence its quotient is a harmonic map of degree zero of $S^1 \times S^{n-2} \to S^{n-1}$.*

### 3.2. Generalized rotational minimal hypersurfaces in $\mathbf{R}^n$

In the classical case of $n = 3$, the constant mean curvature $h = 0$ if and only if $\Gamma$ is a parabola. The generating curve $\Omega$ is given by $y = \frac{1}{c}\cosh(c(x - x_0))$. In the case $n \geqslant 4$, it is to specialize the explicit formula of (5) to obtain the following:

$$x = \frac{1}{c}g(c \cdot y) + x_0, \quad g(u) = \int_1^u \frac{dt}{\sqrt{t^{2n-4} - 1}}. \tag{12'}$$

Therefore the properties of $\Omega$ can easily be investigated directly from the above formula (without going through the following construction).

### 3.3. The hyperbolic type

Suppose $n = 3$, and $\Gamma$ is a hyperbola. Then we have the following two different possibilities, namely,

(i) if one rolls a branch of hyperbola on the upper side of the $x$-axis, then the focus inside of the branch traces a curve $\Omega$ whose rational surface has constant mean curvature $h > 0$.

(ii) if one rolls a branch of hyperbola on the lower side of the $x$-axis, then the focus outside of the branch traces a curve $\Omega$ whose rotational surface has constant mean curvature $h < 0$.

Correspondingly, one also has the following two possibilities for the general case of $n \geqslant 4$, namely,

(i) $h > 0$. In this case $b_1(w) = \{\alpha[(n-1)w + h]^{2/(n-1)} - w^2\}$ has exactly two simple roots $w_1, w_2$ and $(-h)/(n-1) < w_1 < 0 < w_2$. The curve $\Gamma$ is given by the polar coordinate relationship:

$$\theta = \pm\int_0^w b_1(t)^{-1/2}dt + \theta_0.$$

Let $\theta^* = \int_0^{w_2} b_1(t)^{-1/2}$. Then it is not difficult to show that $\Gamma$ is a curve of the following shape:

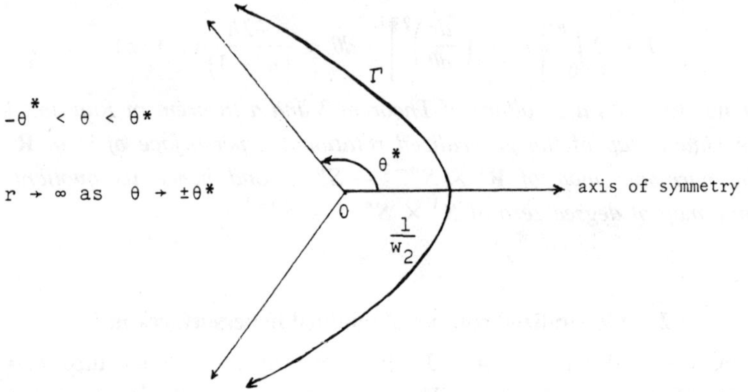


FIG. 4

$\Omega$ is obtained by rolling the above $\Gamma$ on the upper side of the $x$-axis.

(ii) $h < 0$ and $\Gamma$ is given by the following polar coordinate relationship:

$$\theta = \pm \int_0^w b_2(t)^{-1/2} dt + \theta_0,$$

$$b_2(t) = \left\{ \alpha\left[-(n-1)t - h\right]^{2/(n-2)} - t^2 \right\}.$$

By Lemma 2′, $b_2(w)$ has two simple roots $w_1$, $w_2$, $w_1 < 0 < w_2 < (-h)/(n-1)$. Let $\theta^* = \int_0^{w_2} b_2(t)^{-1/2} dt$. Again it is not difficult to see that $\Gamma$ is then a curve of the following shape:

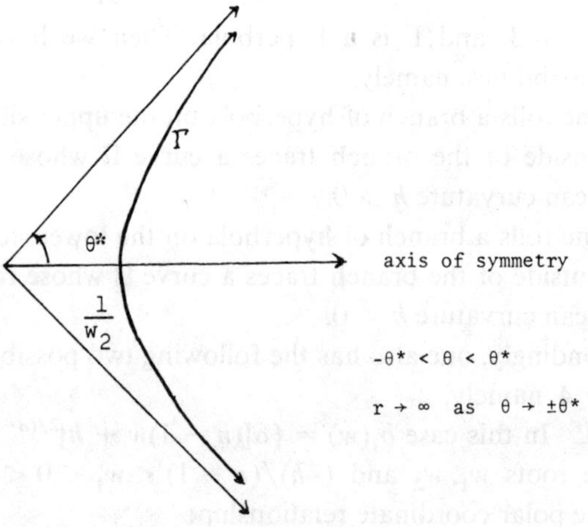


FIG. 5

$\Omega$ is obtained by rolling $\Gamma$ on the lower side of the $x$-axis.

It is rather straightforward to use the rolling construction and the integral expression of Γ to investigate further properties of Ω in the above two cases of hyperbolic type. We state some of their basis properties as follows without proof.

**Theorem 3′.** *The generating curve Ω of hyperbolic type has the following properties*:

(a) *If $h > 0$, then, up to a translation along the x-axis, Ω is given by $y = f(x)$ satisfying the following properties*:

(i) *$f(x)$ is a convex function symmetric with respect to $x = 0$,*

(ii) *Ω is a finite length open arc defined over $(-\beta, \beta)$ with $\beta < 1/h$, $\lim_{x\to\beta} f'(x) = +\infty$ and $\lim_{x\to\beta} f(x) = \gamma < +\infty$.*

(b) *If $h' < 0$, then, up to a translation along the x-axis, Ω is given by $y = f(x)$ satisfying the following properties*:

(i) *$f(x)$ is defined over $[-\beta', \beta']$, $\beta' < (n-1)/|h'|$ and symmetric with respect to $x = 0$,*

(ii) *$f(x)$ is strictly increasing over $(-\beta', 0)$ and strictly decreasing over $(0, \beta')$, $f''(x) > 0$, $\lim_{x\to\beta'} f'(x) = -\infty$, $\lim_{x\to\beta'} f(x) = \gamma' < +\infty$.*

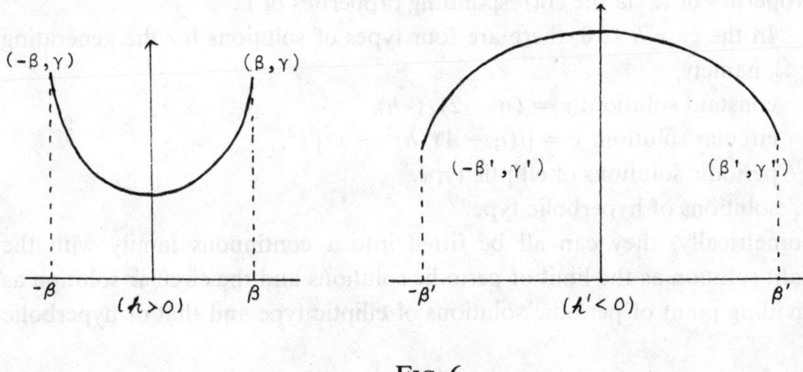


FIG. 6

(c) *By suitably adjusting the value of α, one may realize an arbitrary positive value for γ or γ′. Therefore it is easy to piece the above two kinds of generating curves with $h' = -h$ and $\gamma' = \gamma$ by translation and uniqueness. In this way, one obtains periodic solutions of the following type whose Gauss map provides harmonic maps of $S^1 \times S^{n-2} \to S^{n-1}$, which covers $S^{n-1}$ twice generically, and hence also those examples which cover $S^{n-1}$ $2k$ times generically by taking longer period.*

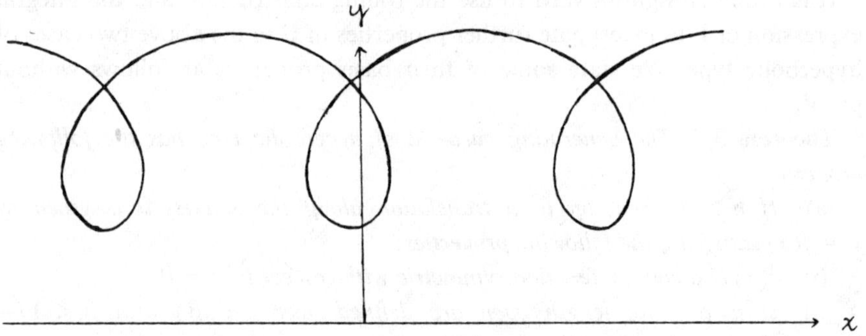


FIG. 7

## 3.4. Concluding remarks

(i) The rolling construction is a simple geometric construction which establishes a simple, rigid, analytic relationship between the cartesian expression of $\Omega$ and the polar expression of $\Gamma$. In the case that $\Omega$ is a generating curve of a generalized rotational hypersurface of constant mean curvature in $\mathbf{R}^n$, the polar coordinate expression of $\Gamma$ is given by a rather simple integral, namely, $\theta = \pm \int b_i(w)\, dw$. Therefore it is usually advantageous to investigate the properties of $\Omega$ via the corresponding properties of $\Gamma$.

(ii) In the case $h < 0$, there are four types of solutions for the generating curve $\Omega$, namely,

(1) constant solution: $y = (n-2)/(-h)$,

(2) circular solution: $y = [((n-1)/h)^2 - x^2]^{1/2}$,

(3) periodic solutions of elliptic type,

(4) solutions of hyperbolic type.

Geometrically, they can all be fitted into a continuous family with the constant solution as the limit of periodic solutions and the circular solution as the dividing point of periodic solutions of elliptic type and that of hyperbolic type.

(iii) In the periodic case, one constructs a family of harmonic maps of $S^k \times T^l \to S^{k+1} \times T^{-1}$. It is not difficult to check that the composition $S^k \times T^l \to S^{k+1} \times T^{l-1} \to S^{k+2} \times T^{l-2}$ is no longer harmonic. However, it is rather tempting to see if some suitable modification of the above composition will yield a harmonic map of $S^k \times T^l \to S^{k+2} \times T^{l-2}$.

## References

[1] C. Delaunay, *Sur la surface de revolution dont la courbure mayenne est constante*, J. Math. Pures. Appl. Sér. 1, **6** (1841) 309–320.

[2] J. Eells, *On the surfaces of Delaunay and their Gauss maps*, to appear.

[3] W. Y. Hsiang & W. T. Hsiang, *On the construction of exotic and/or knotted spheres of constant mean curvature in the standard sphere*, to appear.

[4] W. Y. Hsiang & B. Lawson, *Minimal submanifolds of low cohomogeneity*, J. Differential Geometry **5** (1971) 1–38.

[5] K. Kenmotsu, *Surface of revolution with prescribed mean curvature*, Tôhoku Math. J. **32** (1980) 147–153.

[6] E. A. Ruh & J. Vilms, *The tension field of the Gauss map*, Trans. Amer. Math. Soc. **149** (1970) 569–573.

FUDAN UNIVERSITY, SHANGHAI
UNIVERSITY OF CALIFORNIA, BERKELEY

# SOME UNIQUENESS THEOREMS ON TWO-DIMENSIONAL RIEMANNIAN MANIFOLDS IMMERSED IN A GENERAL EUCLIDEAN SPACE

*(Dedicated to Professor Buchin Su on his 80th birthday)*

## 1. Introduction

Let $M_n$ be a differentiable manifold of dimension $n \geqslant 2$, and $X: M_n \to E_{n+m}$ a differentiable map of $M_n$ into a Euclidean space $E_{n+m}$ of dimension $n+m$ with $m > 0$. The map $X$ is called an immersion, and $M_n$, or rather $M_n$ together with map $X$, is called an immersed submanifold of $E_{n+m}$ if the functional matrix of $X$ is of rank $n$ everywhere. When $m = 1$, an immersed submanifold $M_n$ of the space $E_{n+m}$ is called an immersed hypersurface.

Let us consider now an oriented immersed submanifold $M_n$. For each point $P \in M_n$ there is a unique linear space $N$ of dimension $m$ normal to $X(M_n)$ at the point $X(P)$. For any unit normal vector $e_r(P)$ in the normal space $N$ at the point $X(P)$, we define the first fundamental form I, which is positive definite, and the second fundamental form $\mathrm{II}_r$ associated with $e_r(P)$ to be

$$(1.1) \qquad \mathrm{I} = \mathrm{d}X \cdot \mathrm{d}X, \qquad \mathrm{II}_r = \mathrm{d}X \cdot \mathrm{d}e_r,$$

where $\mathrm{d}X$ and $\mathrm{d}e_r$ are vector-valued linear differential forms on $M_n$, and the dot denotes the scalar product of two vectors in $E_{n+m}$. The eigenvalues $h_1(e_r), \ldots, h_n(e_r)$ of $\mathrm{II}_r$ relative to I are called the principal curvatures of the manifold $M_n$ associated with $e_r(P)$, and the $k$th mean curvature $H^{(k)}(e_r)$ of $M_n$ at $X(P)$ with respect to $e_r(P)$ is defined to be the $k$th elementary symmetric function of $h_1(e_r), \ldots, h_n(e_r)$ divided by the number of terms:

$$(1.2) \qquad \binom{n}{k} H^{(k)}(e_r) = \sum_{i_1 < \cdots < i_k} h_{i_1}(e_r) \ldots h_{i_k}(e_r), \quad (k = 1, \ldots, n),$$

where $\binom{n}{k}$ is a binomial coefficient. In particular, we have the first and $n$th mean curvatures $H^{(1)}(e_r) = (1/n)\Sigma_{i=1}^{n} h_i(e_r)$ and $H^{(n)}(e_r) = h_1(e_r) \ldots h_n(e_r)$; the latter is also called the Gauss–Kronecker curvature of $M_n$ with respect to $e_r(P)$, which is simply the Gaussian curvature when $m = 1$.

It is well known that for a manifold $M_n$ immersed in an $E_{n+m}$ for any $m > 0$, the normal vector

$$(1.3) \qquad \sum_{r=n+1}^{n+m} H^{(1)}(e_r) e_r$$

*Geometriae Dedicata* **12** (1982) 35–51. 0046–5755/82/0121–0035$02.55.

is invariant under a Euclidean motion in the normal space $N$ of $M_n$ at the point $X(P)$. The invariant normal vector (1.3) is called the *mean curvature vector* of $M_n$ at the point $X$, and its magnitude

$$(1.4) \qquad \sum_{r=n+1}^{n+m} [(H^{(1)}(e_r))^2]^{1/2}$$

is called the *mean curvature* of $M_n$ at the point $X$. Recently, C. C. Hsiung and L. R. Mugridge [6] extended the Gaussian curvature of a $M_2$ immersed in $E_3$ by proving that the scalar

$$(1.5) \qquad \sum_{r=n+1}^{n+m} H^{(2)}(e_r)$$

is invariant under a Euclidean motion in $N$ at the point $X$. The new invariant (1.5) will be called the *generalized Gaussian curvature* of $M_n$ at $X$.

A normal vector $e_r$ is said to be parallel in the normal bundle of $M_n$ if $de_r$ is tangent to $M_n$ everywhere. For simplicity we shall write $H_r$ and $K_r$ for the first mean curvature $H^{(1)}(e_r)$ and the Gauss–Kronecker curvature $H^{(n)}(e_r)$ with respect to $e_r$, respectively.

Let $M_N$ be a compact oriented Riemannian manifold immersed in a Euclidean space $E_{n+m}$. By a normal frame $Xe_{n+1} \ldots e_{n+m}$ on the manifold $M_n$ we mean a point $X$ of the manifold $M_n$ and an ordered set of mutually orthogonal unit vectors $e_{n+1}, \ldots, e_{n+m}$ normal to the manifold $M_n$ at the point $X$. The manifold $M_n$ is called a star manifold if there exists a point 0, called a pole, in $E_{n+m}$ and class $C^2$ field of normal frames $Xe_{n+1} \ldots e_{n+m}$ over the manifold $M_n$ such that the Gauss–Kronecker curvature $H^{(n)}(e_r)$ of $M_n$ and the support function $X \cdot e_r$ with respect to the pole 0 are positive for every normal vector $e_r, n+1 \leqslant r \leqslant n+m$, at every point of the manifold $M_n$. This normal frame $Xe_{n+1} \ldots e_{n+m}$ is called a fundamental normal frame of the star manifold $M_n$ at the point $X$. An $n$-dimensional star manifold with boundary is an $n$-dimensional compact subset of an $n$-dimensional star manifold.

In 1903, Minkowski [7] established his well-known uniqueness theorem on compact convex surfaces in $E_3$, which states that a diffeomorphism between two compact convex surfaces in $E_3$ such that the two surfaces have parallel inner normal vectors and the same Gaussian curvature at every pair of corresponding points is a translation in $E_3$. Since then several authors have given different proofs of this uniqueness theorem (for example, see [4] for a brief history). In 1957, S. S. Chern [1] proved the theorem by deriving some integral formulas. In 1958 C. C. Hsiung [4] extended the uniqueness theorem to convex surfaces with boundary. In 1959, S. S. Chern [2] gave a proof of Minkowski's uniqueness theorem for convex hypersurfaces in $E_{n+1}$ for any $n \geqslant 2$.

The purpose of the present paper is to extend Minkowski's uniqueness theorem to two-dimensional Riemannian manifolds $M_2$ immersed in $E_{2+m}$ for any $m > 0$ and to obtain a new uniqueness theorem for the immersed submanifolds $M_2$ as follows (for generalizations of Christoffel's uniqueness theorem which is similar to Minkowski's theorem see [5]):

THEOREM I. *Let $M_2$ and $M_2^*$ be compact oriented two-dimensional Riemannian manifolds immersed in a Euclidean space $E_{2+m}$ of dimension $2+m$ for any $m > 0$. Suppose that*

(i) *$M_2$ is a star manifold with respect to a fundamental normal frame $e_3, \ldots, e_{2+m}$,*
(ii) *the $m-1$ normal vector fields $e_3, \ldots, e_{2+m-1}$ are parallel in the normal bundle of $M_2$ and*
(iii) *the mean curvatures of $M_2$ with respect to the normal vectors $e_3, \ldots, e_{2+m}$ are all equal, that is, $H_r = H_s$ $(r, s = 3, \ldots, 2+m)$.*

*If there exists an orientation-preserving diffeomorphism $f$ of $M_2$ onto $M_2^*$ such that at each pair of corresponding points*

(iv) *$M_2$ and $M_2^*$ have the same normal frame $e_3, \ldots, e_{2+m}$,*
(v) *$M_2$ and $M_2^*$ have the same generalized Gaussian curvature, that is, $\Sigma_{r=3}^{2+m} K_r = \Sigma_{r=3}^{2+m} K_r^*$, where $K_r^*$ denotes the Gauss–Kronecker curvature of $M_2^*$ with respect to $e_r$,*
(vi) *the mean curvatures of $M_2^*$ with respect to the normal vectors $e_3, \ldots, e_{2+m}$ are all equal, that is, $H_r^* = H_s^*$ $(r, s = 3, \ldots, 2+m)$, and*
(vii) *the support function $X^* \cdot e_r$ of $M_2^*$ with respect to the same pole 0 is positive for $r = 3, \ldots, 2+m$,*

*then $f$ is a translation in $E_{2+m}$.*

It is clear that Theorem I with $m = 1$ is Minkowski's uniqueness theorem for surfaces in $E_3$.

THEOREM II. *Let $M_2$ and $M_2^*$ be compact two-dimensional oriented Riemannian manifolds immersed in a Euclidean space $E_{2+m}$ for any $m > 0$. If there exists an orientation-preserving diffeomorphism $f$ of $M_2$ onto $M_2^*$ such that at each pair of corresponding points $M_2$ and $M_2^*$ have*

(i) *a common normal frame $e_3, \ldots, e_{2+m}$,*
(ii) *the same mean curvature vector, that is, $\Sigma_{r=3}^{2+m} H_r e_r = \Sigma_{r=3}^{2+m} H_r^* e_r$, and*
(iii) *the same nonzero generalized Gaussian curvature, that is, $\Sigma_{r=3}^{2+m} K_r = \Sigma_{r=3}^{2+m} K_r^* \neq 0$, and*
(iv) *there exists a positive integer $r$ $(3 \leqslant r \leqslant 2+m)$ for which $K_r > 0$ throughout $M_2$,*

*then $f$ is a translation in $E_{2+m}$.*

It should be remarked that the manifold $M_2$ in Theorem II need not be a star manifold with respect to the normal frame $e_3, \dots, e_{2+m}$.

## 2. Immersed submanifolds in Euclidean space

Suppose that a Euclidean space $E_{2+m}$ is oriented. By a frame $Xe_1 \dots e_{2+m}$ in the space $E_{2+m}$ we mean a point $X \in E_{2+m}$ and an ordered set of mutually orthogonal unit vectors $e_1, \dots, e_{2+m}$ with an orientation coherent with that of the space $E_{2+m}$ so that the determinant $|e_1, \dots, e_{2+m}|$ is equal to $+1$. To avoid confusion we shall use the following ranges of indices hereafter:

$$(2.1) \qquad 1 \leqslant \alpha, \beta \leqslant 2, \qquad 3 \leqslant r, s, t \leqslant 2+m, \qquad 1 \leqslant i, j, k \leqslant 2+m.$$

Then we have

$$(2.2) \qquad e_i \cdot e_j = \delta_{ij},$$

where $\delta_{ij}$ are the Kronecker deltas. Let $F(2, m)$ be the space of all frames in the space $E_{2+m}$, so that $\dim F(2, m) = \frac{1}{2}(2+m)(2+m+1)$. In $F(2, m)$ we introduce the linear differential forms $\omega_i'$, $\omega_{ij}'$ by the equations

$$(2.3) \qquad dX = \sum_i \omega_i' e_i, \qquad de_i = \sum_j \omega_{ij}' e_j,$$

where

$$(2.4) \qquad \omega_{ij}' + \omega_{ji}' = 0.$$

Since $d(dX) = 0$ and $d(de_i) = 0$, from (2.3) we have

$$(2.5) \qquad d\omega_i' = \sum_j \omega_j' \wedge \omega_{ji}', \qquad d\omega_{ij}' = \sum_k \omega_{ik}' \wedge \omega_{kj}',$$

where $\wedge$ denotes the exterior product.

As explained in Section 1, an immersed two-dimensional submanifold in the space $E_{2+m}$ is an abstract manifold $M_2$ and a differentiable map $X: M_2 \to E_{2+m}$ such that the induced map $X_*$ on the tangent space is injective everywhere. Analytically, our map can be defined by a vector-valued function $X(P)$, $P \in M_2$. Our assumption implies that the differential $dX(P)$ of $X(P)$, which is a linear differential form on $M_2$ with value in $E_{2+m}$, has as values a linear combination of exactly two vectors $t_1, t_2$. Since $X_*$ is injective, we identify the tangent space of $M_2$ at the point $P$ with the plane formed by $t_1, t_2$. A linear combination of the vectors $t_1, t_2$ is called a *tangent vector*, and a vector orthogonal to $t_1$ and $t_2$ is called a *normal vector*. The immersion of $M_2$ in $E_{2+m}$ gives rise to a bundle $B$, whose bundle space is the subset of $M_2 \times F(2, m)$ consisting of

$$(P, X(P)e_1 e_2 e_3 \dots e_{2+m}) \in M_2 \times F(2, m)$$

such that $e_1, e_2$ are tangent vectors, and $e_3, \dots, e_{2+m}$ are normal vectors at the point $X(P)$.

Consider the inclusion map $\phi$ and the projection $p$:

$$B \xrightarrow{\phi} M_2 \times F(2, m) \xrightarrow{p} F(2, m).$$

By putting

$$(2.6) \qquad \omega_i = (p\phi)^* \omega_i', \qquad \omega_{ij} = (p\phi)^* \omega_{ij}',$$

from (2.4) and (2.5) we have

$$(2.7) \qquad \omega_{ij} + \omega_{ji} = 0,$$

$$(2.8) \qquad d\omega_i = \sum_j \omega_j \wedge \omega_{ji}, \qquad d\omega_{ij} = \sum_k \omega_{ik} \wedge \omega_{kj}.$$

Here $\omega_i$ and $\omega_{ij}$ are linear differential forms in the bundle $B$. From the definition of the bundle $B$ it follows that

$$(2.9) \qquad \omega_r = 0 \quad (r = 3, \dots, 2 + m)$$

and that $\omega_1$ and $\omega_2$ are linearly independent. Thus for the immersion $X: M_2 \to \to E_{2+m}$ we have

$$(2.10) \qquad dX = \sum_{\alpha=1}^{2} \omega_\alpha e_\alpha, \qquad de_i = \sum_{j=1}^{2+m} \omega_{ij} e_j.$$

From (2.8) and (2.9) it follows that

$$\sum_\alpha \omega_\alpha \wedge \omega_{r\alpha} = 0 \quad (r = 3, \dots, 2 + m),$$

from which by Cartan's lemma on exterior algebra we have

$$(2.11) \qquad \omega_{r\alpha} = \sum_\beta A_{r\alpha\beta} \omega_\beta, \qquad A_{r\alpha\beta} = A_{r\beta\alpha}, \quad (r = 3, \dots, 2 + m).$$

If the determinant det $(A_{r\alpha\beta}) \neq 0$ for some $r$, by introducing the matrix $(\lambda_{r\alpha\beta})$ inverse to the matrix $(A_{r\alpha\beta})$ we have

$$(2.12) \qquad \omega_\alpha = \sum_\beta \lambda_{r\alpha\beta} \omega_{r\beta}, \qquad \lambda_{r\alpha\beta} = \lambda_{r\beta\alpha} \quad (r = 3, \dots, 2 + m).$$

By means of (2.2), (2.10), (2.11) and (2.12), Equations (1.1) can be written as

$$(2.13) \qquad \begin{aligned} \mathrm{I} &= \sum_\alpha \omega_\alpha^2, \\ \mathrm{II}_r &= \sum_\alpha \omega_{r\alpha} \omega_\alpha = \sum_{\alpha,\beta} A_{r\alpha\beta} \omega_\alpha \omega_\beta = \sum_{\alpha,\beta} \lambda_{r\alpha\beta} \omega_{r\alpha} \omega_{r\beta}. \end{aligned}$$

Thus by the definitions of $H_r$ and $K_r$ we obtain

$$(2.14) \qquad K_r = A_{r11} A_{r22} - A_{r12} A_{r21} = \det(A_{r\alpha\beta}) = 1/\det(\lambda_{\alpha\beta}),$$

$$(2.15) \qquad H_r = \tfrac{1}{2}(A_{r11} + A_{r22}) = \tfrac{1}{2}\,\mathrm{trace}(A_{r\alpha\beta}).$$

Through a point in a Euclidean space $E_{2+m}$ let $A_1, \dots, A_{2+m-1}$ be $2 + m - 1$ differentiable vector functions of 2 variables $u^1, u^2$, and let $J$ be any vector. The scalar product of the vector $J$ and the vector product $A_1 \times \ldots \times A_{2+m-1}$ of the vectors $A_1, \ldots, A_{2+m-1}$ is given by

$$J \cdot (A_1 \times \ldots \times A_{2+m-1}) = (-1)^{m-1} |J, A_1, \ldots, A_{2+m-1}|, \tag{2.16}$$

where $|J, A_1, \ldots, A_{2+m-1}|$ is a determinant, the elements of each of whose columns are the contravariant components of the vector indicated. It is obvious that the vector $A_1 \times \ldots \times A_{2+m-1}$ is orthogonal to each of the vectors $A_1, \ldots, A_{2+m-1}$ and that an interchange of any two vectors in the vector product changes the vector product only by a sign. Thus we have

$$e_1 \times \ldots \times \hat{e}_r \times \ldots \times e_{2+m} = (-1)^{m+r} e_r, \tag{2.17}$$

where the circumflex over $e_r$ indicates that the vector $e_r$ is to be deleted.

The vector product of vectors and the exterior product of differential forms can be combined to define a convenient operation $\hat{\times}$ as follows (for example, see [3] or [5]):

$$\begin{aligned} &A_1 \mathbin{\hat{\times}} \ldots \mathbin{\hat{\times}} A_{i-1} \mathbin{\hat{\times}} dA_i \mathbin{\hat{\times}} A_{i+1} \mathbin{\hat{\times}} \ldots \mathbin{\hat{\times}} A_{j-1} \mathbin{\hat{\times}} dA_j \mathbin{\hat{\times}} A_{j+1} \mathbin{\hat{\times}} \\ &\mathbin{\hat{\times}} \ldots \mathbin{\hat{\times}} A_{2+m-1} \\ &= (A_1 \times \ldots \times A_{i-1} \times A_{i,\alpha_i} \times A_{i+1} \times \ldots \times A_{j-1} \times A_{j,\alpha_j} \times \\ &\times A_{j+1} \times \ldots \times A_{2+m-1}) du^{\alpha_i} \wedge du^{\alpha_j}, \end{aligned} \tag{2.18}$$

where a repetition of an index means summation over the range of the index, $\alpha_i, \alpha_j = 1, 2$, and $A_{i,\alpha_i} = \partial A_i / \partial u^{\alpha_i}$. It is readily seen that the vector (2.18) is independent of the order of the vectors $dA_i$, $dA_j$.

Let $dA$ be the area element of an immersed submanifold $M_2$ in the space $E_{2+m}$. Then by the means of the operation $\hat{\times}$ we obtain

$$dX \mathbin{\hat{\times}} dX \mathbin{\hat{\times}} e_3 \mathbin{\hat{\times}} \ldots \mathbin{\hat{\times}} \hat{e}_r \mathbin{\hat{\times}} \ldots \mathbin{\hat{\times}} e_{2+m} = (-1)^{m+r} 2 e_r \, dA. \tag{2.19}$$

On the other hand, by (2.10), (2.17) and (2.18) we have

$$dX \mathbin{\hat{\times}} dX \mathbin{\hat{\times}} e_3 \mathbin{\hat{\times}} \ldots \mathbin{\hat{\times}} \hat{e}_r \mathbin{\hat{\times}} \ldots \mathbin{\hat{\times}} e_{2+m} = (-1)^{m+r} 2 e_r \omega_1 \wedge \omega_2.$$

Comparison of this equation with (2.19) yields

$$dA = \omega_1 \wedge \omega_2. \tag{2.20}$$

From (2.11), (2.14), (2.15) and (2.20) it follows that

$$K_r \, dA = \omega_{r1} \wedge \omega_{r2}, \tag{2.21}$$

$$2H_r \, dA = \omega_{r1} \wedge \omega_2 - \omega_{r2} \wedge \omega_1 \quad (r = 3, \ldots, 2+m). \tag{2.22}$$

## 3. Integral formulas for a pair of immersed manifolds

Let $M$ be a compact differentiable manifold of dimension 2, and let $M_2$ and $M_2^*$ be compact oriented immersed manifolds in $E_{2+m}$ given by $X: M \to E_{2+m}$ and $X^*: M \to E_{2+m}$, respectively. Then Section 2 can be applied to the manifold $M_2$, and for the corresponding quantities and equations for the manifold $M_2^*$ we shall use the same symbols and numbers with a star respectively.

Suppose that there is an orientation-preserving diffeomorphism $f$ of $M_2$ onto $M_2^*$ such that at each pair of corresponding points $M_2$ and $M_2^*$ have a common normal frame and hence parallel tangent spaces. Without loss of generality we may assume that

$$(3.1)\qquad e_i^* = e_i \quad (i = 1, \ldots, 2+m).$$

From (2.10), (2.10)* and (3.1) it follows that

$$(3.2)\qquad \omega_{r\alpha}^* = \omega_{r\alpha} \quad (r = 3, \ldots, 2+m; \alpha = 1, 2).$$

Suppose also that $M_2$ and $M_2^*$ have the same nonzero generalized Gaussian curvature at each pair of corresponding points, so that

$$(3.3)\qquad \sum_r K_r^* = \sum_r K_r \neq 0.$$

From (3.2), (2.21) and (2.21)* we obtain

$$(3.4)\qquad K_r^*\, \mathrm{d}A^* = K_r\, \mathrm{d}A \quad (r = 3, \ldots, 2+m),$$

and hence

$$(3.5)\qquad \sum_r K_r^*\, \mathrm{d}A^* = \sum_r K_r\, \mathrm{d}A.$$

By (3.3) and (3.5) we thus have

$$(3.6)\qquad \mathrm{d}A^* = \mathrm{d}A,$$

which together with (3.4) gives

$$(3.7)\qquad K_r^* = K_r \quad (r = 3, \ldots, 2+m).$$

For the pair of immersed manifolds $M_2$ and $M_2^*$ under the assumption that the $m-1$ normal vector fields $e_3, \ldots, e_{2+m-1}$ are parallel in the normal bundle of $M_2$ and there exists a fixed integer $t\,(3 \leqslant t \leqslant 2+m)$ for which $K_t > 0$, we introduce a linear differential form

$$(3.8)\qquad \begin{aligned} C_{2+m} &= \sum_{r=3}^{2+m} (-1)^{r+1} |X^*, X, \mathrm{d}X, e_3, \ldots, \hat{e}_r, \ldots, e_{2+m}| \\ &= \sum_{r=3}^{2+m} (-1)^{r+1} D_r. \end{aligned}$$

To calculate $\mathrm{d}C_{2+m}$ we first calculate $\mathrm{d}D_r$ as follows:

$$(3.9)\qquad \begin{aligned} \mathrm{d}D_r &= \mathrm{d}|X^*, X, \mathrm{d}X, e_3, \ldots, \hat{e}_r, \ldots, e_{2+m}| \\ &= (\mathrm{I}) + (\mathrm{II}) - \sum_{s=3}^{2+m} (\mathrm{III})_s, \end{aligned}$$

where

(3.10) $(\mathrm{I}) = |\mathrm{d}X^*, X, \mathrm{d}X, e_3, \ldots, \hat{e}_r, \ldots, e_{2+m}|$,

(3.11) $(\mathrm{II}) = |X^*, \mathrm{d}X, \mathrm{d}X, e_3, \ldots, \hat{e}_r, \ldots, e_{2+m}|$,

(3.12) $(\mathrm{III})_s = |X^*, X, \mathrm{d}X, e_3, \ldots, \mathrm{d}e_s, \ldots, \hat{e}_r, \ldots, e_{2+m}|$.

By means of (2.10), (2.10)*, (3.1), (2.16), (2.17), (2.18) and (3.10) we have

$$\begin{aligned}(\mathrm{I}) &= -|X, \mathrm{d}X^*, \mathrm{d}X, e_3, \ldots, \hat{e}_r, \ldots, e_{2+m}| \\ &= (-1)^m X \cdot (\mathrm{d}X^* \hat{\times} \mathrm{d}X \hat{\times} e_3 \hat{\times} \ldots \hat{\times} \hat{e}_r \hat{\times} \ldots \hat{\times} e_{2+m}) \\ &= (-1)^m X \cdot (e_1 \times \ldots \times \hat{e}_r \times \ldots \times e_{2+m})(\omega_1^* \wedge \omega_2 - \omega_2^* \wedge \omega_1) \\ &= (-1)^r (X \cdot e_r)(\omega_1^* \wedge \omega_2 - \omega_2^* \wedge \omega_1).\end{aligned} \tag{3.13}$$

From (2.12), (2.12)*, (3.2), (2.21), (2.14), (2.14)* and (3.7) it follows that for $3 \leqslant t \leqslant 2+m$

$$\begin{aligned}&\omega_1^* \wedge \omega_2 - \omega_2^* \wedge \omega_1 \\ &= (\lambda_{t11}^* \omega_{t1}^* + \lambda_{t12}^* \omega_{t2}^*) \wedge (\lambda_{t21}\omega_{t1} + \lambda_{t22}\omega_{t2}) - \\ &\quad - (\lambda_{t21}^* \omega_{t1}^* + \lambda_{t22}^* \omega_{t2}^*) \wedge (\lambda_{t11}\omega_{t1} + \lambda_{t12}\omega_{t2}) \\ &= (\lambda_{t11}^*\lambda_{t22} - \lambda_{t12}^*\lambda_{t21} - \lambda_{t21}^*\lambda_{t12} + \lambda_{t22}^*\lambda_{t11})(\omega_{t1} \wedge \omega_{t2}) \\ &= [-(\lambda_{t11}^* - \lambda_{t11})(\lambda_{t22}^* - \lambda_{t22}) + (\lambda_{t12}^* - \lambda_{t12})(\lambda_{t21}^* - \lambda_{t21}) + \\ &\quad + (\lambda_{t11}^*\lambda_{t22}^* - \lambda_{t12}^*\lambda_{t21}^*) + (\lambda_{t11}\lambda_{t22} - \lambda_{t12}\lambda_{t21})] \times \\ &\quad \times (\omega_{t1} \wedge \omega_{t2}) = [-\det(\lambda_{t\alpha\beta}^* - \lambda_{t\alpha\beta}) + \det(\lambda_{t\alpha\beta}^*) + \\ &\quad + \det(\lambda_{\alpha\beta})] K_t \mathrm{d}A = -\det(\lambda_{t\alpha\beta}^* - \lambda_{t\alpha\beta}) K_t \mathrm{d}A + 2\mathrm{d}A.\end{aligned} \tag{3.14}$$

Substituting (3.14) in (3.13) gives

(3.15) $(\mathrm{I}) = (-1)^{r+1} p_r \det(\lambda_{t\alpha\beta}^* - \lambda_{t\alpha\beta}) K_t \, \mathrm{d}A + (-1)^r 2 p_r \mathrm{d}A$,

where $p_r = X \cdot e_r$, and $3 \leqslant r, t \leqslant 2+m$.

By means of (2.10), (2.16), (2.19) and (3.11) we obtain

(3.16) $(\mathrm{II}) = (-1)^m X^* \cdot (-1)^{m+r} 2 e_r \, \mathrm{d}A = (-1)^{r+1} 2 p_r^* \, \mathrm{d}A$,

where $p_r^* = X^* \cdot e_r$.

The vector-valued functions $X$ and $X^*$ can be written

(3.17) $X = \sum_{i=1}^{2+m} X_i e_i, \qquad X^* = \sum_{i=1}^{2+m} X_i^* e_i.$

If $e_3, \ldots, e_{2+m-1}$ are parallel in the normal bundle of $M_2$, then from the definition it follows immediately that

(3.18) $\omega_{rs} = 0 \quad (s, r = 3, \ldots, 2+m)$,

so that

$$(3.19)\qquad de_s = \sum_{\alpha=1}^{2} \omega_{s\alpha} e_\alpha \quad (s = 3,\ldots,2+m).$$

Using the above method together with (2.22), (3.12), (3.17) and (3.19), we can calculate $(\mathrm{III})_s$ for $s < r$ as follows:

$$(3.20)\qquad \begin{aligned}(\mathrm{III})_s &= (-1)^{m+s} X^* \cdot (X \hat{\times} dX \hat{\times} de_s \hat{\times} e_3 \hat{\times} \ldots \hat{\times} \hat{e}_s \hat{\times} \ldots \hat{\times} \hat{e}_r \hat{\times} \\ &\qquad \hat{\times} \ldots \hat{\times} e_{2+m}) = \\ &= (-1)^{m+s} X^* \cdot [X_s(\omega_1 \wedge \omega_{s2} - \omega_2 \wedge \omega_{s1})(-1)^{s-1} e_1 \times \\ &\qquad \times \ldots \times \hat{e}_r \times \ldots \times e_{2+m} + X_r(\omega_1 \wedge \omega_{s2} - \\ &\qquad - \omega_2 \wedge \omega_{s1})(-1)^{r-2} e_1 \times \ldots \times \hat{e}_s \times \ldots \times e_{2+m}] = \\ &= (-1)^{r+1} (\sum_i X_i^* e_i) \cdot [(X_s e_r - X_r e_s)(\omega_1 \wedge \omega_{s2} - \\ &\qquad - \omega_2 \wedge \omega_{s1})] = \\ &= (-1)^r 2(X_s^* X_r - X_r^* X_s) H_s \, dA.\end{aligned}$$

A similar calculation yields the same expression of $(\mathrm{III})_s$ for $s > r$. Therefore for $s \neq r$ we obtain

$$(3.21)\qquad (\mathrm{III})_s = (-1)^r 2(X_s^* X_r - X_r^* X_s) H_s \, dA \quad (s = 3,\ldots,2+m).$$

Substituting (3.15), (3.16) and (3.21) in (3.9) gives

$$(3.22)\qquad \begin{aligned}dD_r &= (-1)^{r+1} p_r \det(\lambda^*_{t\alpha\beta} - \lambda_{t\alpha\beta}) K_t \, dA + \\ &\quad + (-1)^r 2p_r \, dA + (-1)^{r+1} 2p_r^* \, dA + \\ &\quad + 2 \sum_{s=3}^{2+m} (-1)^{r+1} (X_s^* X_r - X_r^* X_s) H_s \, dA,\end{aligned}$$

which together with (3.8) implies

$$(3.23)\qquad \begin{aligned}dC_{2+m} &= \sum_{r=3}^{2+m} p_r \det(\lambda^*_{t\alpha\beta} - \lambda_{t\alpha\beta}) K_t \, dA - \\ &\quad - 2 \sum_{r=3}^{2+m} p_r \, dA + 2 \sum_{r=3}^{2+m} p_r^* \, dA + \\ &\quad + 2 \sum_{r=3}^{2+m} \sum_{s=3}^{2+m} (X_s^* X_r - X_r^* X_s) H_s \, dA.\end{aligned}$$

Since

$$\begin{aligned}&\sum_{r=3}^{2+m} \sum_{s=3}^{2+m} (X_s^* X_r - X_r^* X_s) H_s \, dA \\ &= \sum_{\substack{r,s=3 \\ s>r}}^{2+m} (X_s^* X_r - X_r^* X_s)(H_s - H_r) \, dA,\end{aligned}$$

we can rewrite (3.23) as

$$dC_{2+m} = \sum_{r=3}^{2+m} p_r \det(\lambda^*_{t\alpha\beta} - \lambda_{t\alpha\beta}) K_t \, dA - \tag{3.24}$$
$$- 2\sum_{r=3}^{2+m} p_r \, dA + 2\sum_{r=3}^{2+m} p^*_r dA +$$
$$+ 2\sum_{\substack{r,s=3 \\ s>r}}^{2+m} (X^*_s X_r - X^*_r X_s)(H_s - H_r)\, dA.$$

By interchanging the roles of $M_2$ and $M^*_2$ in (3.24) we obtain

$$dC^*_{2+m} = \sum_{r=3}^{2+m} p^*_r \det(\lambda^*_{t\alpha\beta} - \lambda_{t\alpha\beta}) K_t \, dA - \tag{3.25}$$
$$- 2\sum_{r=3}^{2+m} p^*_r dA + 2\sum_{r=3}^{2+m} p_r \, dA +$$
$$+ 2\sum_{\substack{r,s=3 \\ s>r}}^{2+m} (X_s X^*_r - X_r X^*_s)(H^*_s - H^*_r)\, dA,$$

where

$$C^*_{2+m} = \sum_{r=3}^{2+m} (-1)^{r+1} |X, X^*, dX^*, e_3, \ldots, \hat{e}_r, \ldots, e_{2+m}|. \tag{3.26}$$

Addition of (3.24) and (3.25) yields

$$d(C_{2+m} + C^*_{2+m}) = \sum_{r=3}^{2+m} (p_r + p^*_r) \det(\lambda^*_{t\alpha\beta} - \lambda_{t\alpha\beta}) K_t \, dA + \tag{3.27}$$
$$+ 2\sum_{\substack{r,s=3 \\ s>r}}^{2+m} (X^*_s X_r - X^*_r X_s)((H_s - H_r) -$$
$$- (H^*_s - H^*_r))\, dA.$$

Integrating the differential form (3.27) over the compact manifold $M_2$ and using Green's theorem, we obtain the integral formula

$$\int_{M_2} \sum_{r=3}^{2+m} (p_r + p^*_r) \det(\lambda^*_{t\alpha\beta} - \lambda_{t\alpha\beta}) K_t \, dA + \tag{3.28}$$
$$+ 2\int_{M_2} \sum_{\substack{r,s=3 \\ s>r}}^{2+m} (X^*_s X_r - X^*_r X_s)((H_s - H_r) - (H^*_s - H^*_r))\, dA = 0.$$

In order to derive an integral formula for a pair of manifolds $M_2$ and $M^*_2$ satisfying the conditions of Theorem II, we introduce a linear differential form

$$A_{2+m} = |e_3, \ldots, e_{2+m}, W, \mathrm{d}X|, \tag{3.29}$$

where

$$W = X^* - X. \tag{3.30}$$

Then we have

$$\mathrm{d}A_{2+m} = \sum_{r=3}^{2+m} (i)_r + (\mathrm{ii}), \tag{3.31}$$

where

$$(\mathrm{i})_r = |e_3, \ldots, \mathrm{d}e_r, \ldots, e_{2+m}, W, \mathrm{d}X|, \tag{3.32}$$

$$(\mathrm{ii}) = |e_3, \ldots, e_{2+m}, \mathrm{d}W, \mathrm{d}X|. \tag{3.33}$$

From (2.10), (2.16), (2.17), (2.18) and (2.22) we can easily obtain

$$(\mathrm{i})_r = -2(W \cdot e_r)H_r \,\mathrm{d}A. \tag{3.34}$$

By substituting (3.30) in (3.33) we can write

$$(\mathrm{ii}) = (\mathrm{ii})_a - (\mathrm{ii})_b, \tag{3.35}$$

where

$$\begin{aligned} (\mathrm{ii})_a &= |e_3, \ldots, e_{2+m}, \mathrm{d}X^*, \mathrm{d}X|, \\ (\mathrm{ii})_b &= |e_3, \ldots, e_{2+m}, \mathrm{d}X, \mathrm{d}X|. \end{aligned} \tag{3.36}$$

Using (2.10), (2.10)* and (3.14) we can readily have

$$\begin{aligned} (\mathrm{ii})_a &= (-1)^{2+m-1} e_3 \cdot [e_4 \,\hat{\times}\, \ldots \,\hat{\times}\, e_{2+m} \,\hat{\times}\, \mathrm{d}X^* \,\hat{\times}\, \mathrm{d}X] = \\ &= \omega_1^* \wedge \omega_2 - \omega_2^* \wedge \omega_1 = \\ &= \det(\lambda^*_{\iota\alpha\beta} - \lambda_{\iota\alpha\beta})K_\iota \,\mathrm{d}A + 2\,\mathrm{d}A, \end{aligned} \tag{3.37}$$

$$(\mathrm{ii})_b = 2\,\mathrm{d}A. \tag{3.38}$$

A combination of (3.31) with (3.34), (3.35), (3.37), (3.38) gives

$$\mathrm{d}A_{2+m} = -2 \sum_{r=3}^{2+m} (W \cdot e_r)H_r \,\mathrm{d}A - \det(\lambda^*_{\iota\alpha\beta} - \lambda_{\iota\alpha\beta})K_\iota \,\mathrm{d}A. \tag{3.39}$$

By integrating the differential form (3.39) over the compact manifold $M_2$ we obtain the integral formula

$$2 \int_{M_2} \sum_{r=3}^{2+m} (W \cdot e_r)H_r \,\mathrm{d}A + \int_{M_2} \det(\lambda^*_{\iota\alpha\beta} - \lambda_{\iota\alpha\beta})K_\iota \,\mathrm{d}A = 0. \tag{3.40}$$

Similarly, we introduce the linear differential form

$$A^*_{2+m} = |e_3, \ldots, e_{2+m}, W, \mathrm{d}X^*|, \tag{3.41}$$

and obtain its differential

$$\mathrm{d}A^*_{2+m} = -2\sum_{r=3}^{2+m}(W\cdot e_r)H^*_r\,\mathrm{d}A + \det(\lambda^*_{t\alpha\beta} - \lambda_{t\alpha\beta})K_t\,\mathrm{d}A, \tag{3.42}$$

and therefore the integral formula

$$\int_{M_2} 2\sum_{r=3}^{2+m}(W\cdot e_r)H^*_r\,\mathrm{d}A - \int_{M_2}\det(\lambda^*_{t\alpha\beta} - \lambda_{t\alpha\beta})K_t\,\mathrm{d}A = 0. \tag{3.43}$$

Subtracting (3.40) from (3.43), we thus arrive at the integral formula

$$\int_{M_2}\sum_{r=3}^{2+m}(W\cdot e_r)(H^*_r - H_r)\mathrm{d}A - \int_{M_2}\det(\lambda^*_{t\alpha\beta} - \lambda_{t\alpha\beta})K_t\,\mathrm{d}A = 0. \tag{3.44}$$

## 4. Proofs of Theorems I and II

We need the following elementary lemma in the proofs of Theorems I and II:

LEMMA 4.1. *Let*

$$ax + 2bxy + cy^2, \qquad a^*x + 2b^*xy + c^*y^2 \tag{4.1}$$

*be two positive definite quadratic forms such that*

$$ac - b^2 = a^*c^* - b^{*2}. \tag{4.2}$$

*Then*

$$\begin{vmatrix} a^* - a & b^* - b \\ b^* - b & c^* - c \end{vmatrix} \leqslant 0, \tag{4.3}$$

*where the equality holds if and only if the two forms are identical.*

Lemma 4.1 is well known, so its proof is omitted here.

*Proof of Theorem I.* Because of assumptions (i), (ii), (iv), (v) in Theorem I we have the integral formula (3.28), whose second term vanishes due to assumptions (iii) and (v) so that (3.28) is reduced to

$$\int_{M_2}\sum_{r=3}^{2+m}(p_r + p^*_r)\det(\lambda^*_{t\alpha\beta} - \lambda_{t\alpha\beta})K_t\,\mathrm{d}A = 0. \tag{4.4}$$

On the other hand, assumptions (i) and (vii) immediately imply that over the whole manifold $M_2$, $\Sigma_{r=3}^{2+m}(p_r + p^*_r) > 0$ and $K_t > 0$, which together with (4.3) show that the integrand of the integral (4.4) is nonpositive, and therefore that (4.4) holds if and only if

$$\det(\lambda^*_{t\alpha\beta} - \lambda_{t\alpha\beta}) = 0. \tag{4.5}$$

Thus, by Lemma 4.1 again we obtain

$$\lambda^*_{t\alpha\beta} = \lambda_{t\alpha\beta} \quad (\alpha, \beta = 1, 2). \tag{4.6}$$

From (4.6), (3.2), (2.12), (2.12)*, (2.10), (2.10)* it follows readily that

$$d(X^* - X) = 0. \tag{4.7}$$

Hence $X^* - X$ is a constant vector, that is, $f$ is a translation.

*Proof of Theorem II.* Because of assumptions (i), (iii) and (iv) in Theorem II we have the integral formula (3.44), whose first term vanishes due to assumption (ii) so that (3.44) is reduced to

$$\int_{M_2} \det(\lambda^*_{t\alpha\beta} - \lambda_{t\alpha\beta}) K_t \, dA = 0. \tag{4.8}$$

Hence a repetition of the arguments used in the proof of Theorem I shows that $f$ is a translation. q.e.d.

Theorems I and II can be easily extended to two-dimensional Riemannian manifolds with boundary immersed in a Euclidean space $E_{2+m}$ as follows:

THEOREM I′. *Let $M_2$ and $M_2^*$ be compact oriented two-dimensional Riemannian manifolds, with boundaries $B_2$ and $B_2^*$ respectively, immersed in a Euclidean space $E_{2+m}$ for any $m > 0$. Suppose that there exists an orientation-preserving diffeomorphism $f$ of $M_2$ onto $M_2^*$, and that $M_2$, $M_2^*$ and $f$ all satisfy the conditions of Theorem I. If $f$ restricted to $B_2$ is a translation in $E_{2+m}$ carrying $B_2$ onto $B_2^*$, then $f$ is a translation carrying the whole manifold $M_2$ onto the whole manifold $M_2^*$.*

THEOREM II′. *Let $M_2$ and $M_2^*$ be compact oriented two-dimensional Riemannian manifolds, with boundaries $B_2$ and $B_2^*$ respectively, immersed in a Euclidean space $E_{2+m}$ for any $m > 0$. Suppose that there exists an orientation-preserving diffeomorphism $f$ of $M_2$ onto $M_2^*$, and that $M_2$, $M_2^*$ and $f$ all satisfy the conditions of Theorem II. If the diffeomorphism $f$ restricted to the boundary $B_2$ is a translation in $E_{2+m}$ carrying $B_2$ onto $B_2^*$, then $f$ is a translation carrying the whole manifold $M_2$ onto the whole manifold $M_2^*$.*

*Proof of Theorem I′.* Since $M_2$ has a boundary $B_2$, the integral formula (3.28) now becomes

$$\begin{aligned} \int_{B_2} (C_{2+m} + C^*_{2+m}) &= \int_{M_2} \sum_{r=3}^{2+m} (p_r + p_r^*) \det\cdot(\lambda^*_{t\alpha\beta} - \lambda_{t\alpha\beta}) K_t \, dA + \\ &\quad + 2 \int_{M_2} \sum_{\substack{r,s=3 \\ s>r}}^{2+m} (X_s^* X_r - X_r^* X_s)((H_s - H_r) - (H_s^* - H_r^*)) dA. \end{aligned} \tag{4.9}$$

From (3.8) and (3.26) it follows that

$$(4.10)\qquad \begin{aligned} &C_{2+m}+C^*_{2+m}=\\ &=\sum_{r=3}^{2+m}(-1)^{r+1}|X^*, X, \mathrm{d}X-\mathrm{d}X^*, e_3,\ldots,e_r,\ldots,e_{2+m}|.\end{aligned}$$

On $B_2$, since $f$ is a translation, we obtain $\mathrm{d}X-\mathrm{d}X^*=0$ and therefore $C_{2+m}+C^*_{2+m}=0$. Thus the integral formula (4.9) is reduced to (3.28) again, and Theorem I′ is proved by the arguments in the remainder of the proof of Theorem I.

*Proof of Theorem II′*. Since $M_2$ has a boundary $B_2$, by using (3.39) and (3.42) we can easily see that the integral formula (3.44) now becomes

$$(4.11)\qquad \begin{aligned} &\int_{B_2}(A^*{}_{2+m}-A_{2+m})=\\ &=2\int_{M_2}\sum_{r=3}^{2+m}(W\cdot e_r)(H_r-H^*_r)\,\mathrm{d}A+\\ &\quad+2\int_{M_2}\det(\lambda^*_{t\alpha\beta}-\lambda_{t\alpha\beta})K_t\,\mathrm{d}A.\end{aligned}$$

From (3.29) and (3.41) it follows that

$$(4.12)\qquad A^*_{2+m}-A_{2+m}=|e_3,\ldots,e_{2+m}, W, \mathrm{d}X^*-\mathrm{d}X|.$$

On $B_2$, since $f$ is a translation, we obtain $\mathrm{d}X^*-\mathrm{d}X=0$ and therefore $A^*_{2+m}-A_{2+m}=0$. Thus the integral formula (4.11) is reduced to (3.44) again, and Theorem II′ is proved by the arguments in the remainder of the proof of Theorem II.

## 5. Example

As an example of a Riemannian manifold $M_2$ satisfying conditions (i), (ii) and (iii) of Theorem I we consider a two-sphere immersed in a Euclidean space $E_4$.

Consider a unit sphere

$$(5.1)\qquad S_2=\{(x,y,z)\,|\,x^2+y^2+z^2=1\}$$

parametrized by

$$(5.2)\qquad \phi(u,v)=(\sin u\cos v, \sin u\sin v, \cos u),\qquad 0<u<\pi, 0<v<2\pi.$$

Suppose that $S_2$ is immersed in $E_4$ by an immersion $X\colon S_2\to E_4$ which is defined by

$$X(x, y, z) = \left(\frac{1}{\sqrt{2}}(x+y), \frac{1}{\sqrt{2}}(x-y), \frac{1}{\sqrt{2}}z, \frac{1}{\sqrt{2}}z\right). \tag{5.3}$$

Then we have

$$\begin{aligned}(X \circ \phi)(u, v) = \Big(&\frac{1}{\sqrt{2}}\sin u\,(\cos v + \sin v),\\ &\frac{1}{\sqrt{2}}\sin u\,(\cos v - \sin v), \frac{1}{\sqrt{2}}\cos u, \frac{1}{\sqrt{2}}\cos u\Big)\end{aligned} \tag{5.4}$$

with partial derivatives:

$$\begin{aligned}(X \circ \phi)_u = \Big(&\frac{1}{\sqrt{2}}\cos u\,(\cos v + \sin v),\\ &\frac{1}{\sqrt{2}}\cos u(\cos v - \sin v), -\frac{1}{\sqrt{2}}\sin u, -\frac{1}{\sqrt{2}}\sin u\Big),\\ (X \circ \phi)_v = \Big(&\frac{1}{\sqrt{2}}\sin u(\cos v - \sin v),\\ &-\frac{1}{\sqrt{2}}\sin u(\cos v + \sin v), 0, 0\Big).\end{aligned} \tag{5.5}$$

The vectors $(X \circ \phi)_u$ and $(X \circ \phi)_v$ are orthogonal to each other, and their lengths are given by

$$\|(X \circ \phi)_u\| = 1, \qquad \|(X \circ \phi)_v\| = \sin u. \tag{5.6}$$

Now we choose the unit tangent vectors $e_1, e_2$ and normal vectors $e_3, e_4$ of the immersed surface $X \circ \phi$ as follows:

$$e_1 = (X \circ \phi)_u, \qquad e_2 = \frac{1}{\sin u}(X \circ \phi)_v. \tag{5.7}$$

$$\begin{aligned}e_3 = \tfrac{1}{2}(&\sin u(\cos v + \sin v),\\ &\sin u(\cos v - \sin v), \cos u - 1, \cos u + 1),\\ e_4 = \tfrac{1}{2}(&\sin u(\cos v + \sin v),\\ &\sin u(\cos v - \sin v), \cos u + 1, \cos u - 1).\end{aligned} \tag{5.8}$$

It is easy to see that the unit vectors $e_1, e_2, e_3, e_4$ are mutually orthogonal and that

$$|e_1, e_2, e_3, e_4| = 1. \tag{5.9}$$

Furthermore, the partial derivatives of $e_3$ and $e_4$ are:

$$(5.10)\qquad \begin{aligned} e_{3u} &= e_1/\sqrt{2}, \qquad e_{3v} = e_2 \sin u/\sqrt{2},\\ e_{4u} &= e_1/\sqrt{2}, \qquad e_{4v} = e_2 \sin u/\sqrt{2}. \end{aligned}$$

Thus we have

$$(5.11)\qquad \mathrm{d}e_3 = \frac{1}{\sqrt{2}} e_1\, \mathrm{d}u + \frac{\sin u}{\sqrt{2}} e_2\, \mathrm{d}v, \qquad \mathrm{d}e_4 = \frac{1}{\sqrt{2}} e_1\, \mathrm{d}u + \frac{\sin u}{\sqrt{2}} e_2\, \mathrm{d}v.$$

On the other hand, from (5.3), (5.4) and (5.8) it follows that

$$(5.12)\qquad \mathrm{d}(X \circ \phi) = (X \circ \phi)_u\, \mathrm{d}u + (X \circ \phi)_v\, \mathrm{d}v = e_1\, \mathrm{d}u + e_2 \sin u\, \mathrm{d}v.$$

By (1.1), (5.11) and (5.12), for the immersed surface $X \circ \phi$ we obtain the first and second fundamental forms

$$(5.13)\qquad \begin{aligned} I &= \mathrm{d}(X \circ \phi) \cdot \mathrm{d}(X \circ \phi) = (\mathrm{d}u)^2 + \sin^2 u (\mathrm{d}v)^2,\\ II_r &= \mathrm{d}(X \circ \phi) \cdot \mathrm{d}e_r = (\mathrm{d}u)^2/\sqrt{2} + \sin^2 u (\mathrm{d}v)^2/\sqrt{2} \quad (r = 3, 4), \end{aligned}$$

and therefore the principal curvatures $h_1(e_r)$ and $h_2(e_r)$ associated with the normal vector $e_r$ are

$$(5.14)\qquad h_1(e_r) = h_2(e_r) = 1/\sqrt{2} \quad (r = 3, 4),$$

which together with (1.2) gives

$$(5.15)\qquad K_r = 1/2, \qquad H_r = 1/\sqrt{2} \qquad (r = 3, 4).$$

From (5.4) and (5.8) follow the support functions $X \cdot e_r$:

$$(5.16)\qquad (X \circ \phi) \cdot e_3 = 1/\sqrt{2}, \qquad (X \circ \phi) \cdot e_4 = 1/\sqrt{2}.$$

Hence the two-sphere $S_2$ immersed in $E_4$ satisfies condition (i) of Theorem I by (5.15) and (5.16), condition (ii) by (5.11), and condition (iii) by (5.15).

## BIBLIOGRAPHY


1. Chern, S. S.: 'A Proof of the Uniqueness of Minkowski's Problem for Convex Surfaces', *Amer. J. Math.* **79** (1957), 949–950.
2. Chern, S. S.: 'Integral Formulas for Hypersurfaces in Euclidean Space and Their Applications to Uniqueness Theorems', *J. Math. Mech.* **8** (1959), 947–955.
3. Hsiung, C. C.: 'Curvature and Betti Numbers of Compact Riemannian Manifolds with Boundary', *Rend. Sem. Mat. Univ. e Politec. Torino* **17** (1957–1958), 95–131.
4. Hsiung, C. C.: 'A Uniqueness Theorem for Minkowski's Problem for Convex Surfaces with Boundary', *Illinois J. Math.* **2** (1958), 71–75.
5. Hsiung, C. C.: 'Some Uniqueness Theorems on Riemannian Manifolds with Boundary', *Illinois J. Math.* **4** (1960), 526–540.

6. Hsiung, C. C. and Mugridge, L. R.: 'Euclidean and Conformal Invariants of Submanifolds', *Geom. Dedicata* **8** (1979), 31–38.
7. Minkowski, H.: 'Volumen und Oberfläche', *Math. Ann.* **57** (1903), 447–495.

*Authors' Addresses*

Chuan-Chih Hsiung,
Lehigh University,
Bethlehem,
PA 18015,
U.S.A.

Kyu Sam Park,
Kutztown State College,
*Kutztown*,
PA 19530,
U.S.A.

(Received September 8, 1980)

*Chin. Ann. of Math.*
3 (4) 1982

# ON THE STATIC SOLUTIONS OF MASSIVE YANG-MILLS EQUATIONS

HU HESHENG (H. S. HU)
*(Institute of Mathematics, Fudan University)*

Dedicated to Professor Su Bu-chin on the Occasion of his 80th Birthday and his 50th Year of Educational Work

## §1. Introduction

Usually, a pure Yang-Mills field over Minkowski spacetime $R^{1,n-1}$ is considered as a field of massless particles. Its action integral is[1]

$$L=\int -\frac{1}{4}(f_{\lambda\mu},\ f^{\lambda\mu})d^nx. \tag{1}$$

Here $f_{\lambda\mu}$ is the strength of gauge field with a compact gauge group $G$ and ( , ) denotes the cartan's inner product of the Lie algebra $g$ of $G$. However, many particles in nature are not massless and hence it is a problem of general interest to consider the massive Yang-Mills fields. It has been proved in [2] and [3], from different points of view, that the following gauge invariant functional

$$L_m=\int\left[-\frac{1}{4}(f_{\lambda\mu},\ f^{\lambda\mu})-\frac{m^2}{2}(b_\lambda-\omega_\lambda,\ b^\lambda-\omega^\lambda)\right]d^nx \tag{2}$$

may be considered as the action integral of the massive Yang-Mills field. Here $b_\lambda$ is the gauge potential, $\omega_\lambda$ is defined by

$$\omega_\lambda=U^{-1}\partial_\lambda U \tag{3}$$

and $U$ is a $G$-valued function which is a section of the product bundle $R^{n-1}\wedge G$.

One may think that the choice of $U$ is a choice of gauge and that the gauge is a reference system of measuring the generalized phase of a gauge field. Let $U$ be the variational variables as well as $b_\lambda$, the Euler equations of the action integral (1) and (2) are the massless Yang-Mills equations and massive Yang-Mills equations respectively.

We mentioned that the massive Yang-Mills field is also attractive for its relationship with harmonic mapping. The functional (2) is nothing else than the coupling of the pureYang-Mills functional and the following action integral of the harmonic maps from $R^{1,n-1}$ to the gauge group $G$

$$S(U)=\int(\omega_\lambda,\ \omega^\lambda)d^nx. \tag{4}$$

Manuscript received May 12, 1981.

There are guite a lot of papers devoted to the solutions to Yang-Mills equations. One problem of considerable interest is whether there exists any static solution to the Yang-Mills equations such that it has finite energy and no singularities. Recently the following facts concerning the nonexistence of the global solutions are discovered.

(a) If $n\neq 5$, the pure Yang-Mills equations on an $n$-dimensional spacetime $R^{1,n-1}$ do not admit any static solution which has (i) finite energy (ii) no singularities and (iii) the field strength approaching to zero sufficiently fast at infinity. (Deser, S.[5])

Thus, for $n=4$, i. e. on the real spacetime, there does not exist such solution. For $n=5$, solutions do exist[6,7], since the instantons in 4-dimensional Euclidean space may be regarded as static solutions in 5-dimensional spacetime.

(b) In an $n$-dimensional spacetime with $n\neq 4$, the massive Yang-Mills equations with real mass do not admit any static solution which has (i) finite energy (ii) no singularities and (iii) the field strength and potential approaching to zero sufficiently fast at infinity (Hu, H. S. [8])

Comparing these two results, we discovered that there is a "discontinuity" as $m\to 0$ in 5-dimensional spacetime, i. e. for $n=5$ and $m\neq 0$, no such solution, but when $m=0$ such solutions do exist. Deser, S. and Isham, C. J. in a recent paper[9] wrote that this is the first explicit example which make us recognize that there exists a classical "discontinuity". In their paper, the results are extended to the gauge field with "soft" mass, i. e. Yang-Mills-Higgs-Kibble field. For $n=5$, the "discontinuity" holds in general.

In the present paper, we will show that in the results (a) and (b) not only condition (iii) can be removed, but also the finite energy condition (ii) can be weakened. In other words, when the total energy within the sphere of radius $r$ approaches to infinity quite slowly as $r\to\infty$, the above nonexistence theorem holds true also. In the proof we use a certain technique used in[10] with some improvement. Since finite energy and infinite energy is essentially different in physics, this new discovery may be of interest in physics.

The method of proving the main theorem of the present paper is utilizable for more general case. For example, in the case of the Yang-Mills-Higgs-Kibble field, the results for "soft" mass is improved similarly.

## § 2. Massive Yang-Mills fields

By choosing the Lorentz gauge, the gauge invariant functional becomes

$$L=-\int\left\{\frac{1}{4}(f_{\lambda\mu},\ f^{\lambda\mu})+\frac{m^2}{2}(b_\lambda,\ b^\lambda)\right\}d^{n-1}x \quad (\lambda,\ \mu=0,\ 1,\ \cdots,\ n-1). \tag{5}$$

Here the field strength is

$$f_{\lambda\mu}=b_{\lambda,\mu}-b_{\mu,\lambda}-[b_\lambda,\ b_\mu]\quad \left(b_{\lambda,\mu}=\frac{\partial b_\lambda}{\partial x^\mu}\right) \tag{6}$$

and the metric of spacetime is

$$ds^3=\eta_{\lambda\mu}\,dx^\lambda dx^\mu=-dx^{0^2}+dx^{1^2}+\cdots+dx^{n-1^2}. \tag{7}$$

The massive Yang-Mills equation become

$$J_\alpha-m^2 b_\alpha=0, \tag{8}$$

$$\eta^{\lambda\mu}b_{\lambda,\mu}=0. \tag{9}$$

Here

$$J_\alpha=\eta^{\lambda\mu}f_{\alpha\lambda|\mu}=\eta^{\lambda\mu}(f_{\alpha\lambda,\mu}+[b_\mu,\ f_{\alpha\lambda}]). \tag{10}$$

moreover, it is interesting to note that (9) is a consequence of (8), so we only need to consider equation (8).

We always assume the gauge group $G$ is a compact group. Under the Lorentz gauge the energy monentum tensor

$$T_{\alpha\beta}=(f_{\alpha\nu},\ f^\nu_\beta)-\frac{1}{4}\eta_{\alpha\beta}(f_{\mu\nu},\ f^{\mu\nu})+m^2(b_\alpha,\ b^\alpha)-\frac{m^2}{2}\eta_{\alpha\beta}(b_\lambda,\ b^\lambda) \tag{11}$$

and we have the conservation law

$$\frac{\partial T^\beta_\alpha}{\partial x^\alpha}=0. \tag{12}$$

Especially, the energy density is

$$T_{00}=\frac{1}{2}\left[(f_{0i},\ f_{0i})+\frac{1}{2}(f_{ij},\ f_{ij})\right]+\frac{m^2}{2}(b_0,\ b_0)+\frac{m^2}{2}(b_i,\ b_i)$$
$$(i,\ j=1,\ 2,\ \cdots,\ n-1) \tag{13}$$

and

$$T_{ii}=T^\alpha_\alpha-T^0_0=-\frac{1}{2}(n-3)(f_{0i},\ f^{0i})+\frac{1}{4}(5-n)(f_{ij},\ f^{ij})$$
$$+\frac{m^2}{2}(3-n)(b_\lambda,\ b^\lambda)+m^2(b_0,\ b_0). \tag{14}$$

For a static gauge field, $b_\lambda$ is independent of $x^0$.

The total energy of the field is

$$\int_{R^{n-1}} T_{00}d^{n-1}x, \tag{15}$$

where $R^{n-1}$ is $x^0=$const. In the previous works, one often assume that the total energy is finite, now the condition is weakened to

$$\int\frac{T_{00}}{\psi(r)}d^{n-1}x<\infty, \tag{16}$$

where $\psi(r)$ is a positive, unbounded, continuous function of $r$ satisfying

$$\int_R^\infty\frac{dr}{r\psi(r)}=\infty\quad (R>0). \tag{17}$$

If $\psi(r)=1$, the energy is finite. But the energy may be infinite, for example, in the case $\psi(r)=O(\log r)$ (as $r\to\infty$). Hence when (16) holds true, the total enegy may be either finite or infinite. In this paper the energy is called "slowly divergent energy" if $\int T_{00}d^{n-1}x=\infty$ and (16) holds.

## § 3. Nonexistence of the static solution

In the following we give the precise statement and the proof of the main theorem concerning the massive Yang-Mills equations.

**Theorem.** *In an n-dimensional spacetime $R^{1,n-1}$ with $n\neq 4$, the compact group Yang-Mills field with real mass does not possess any non-trivial static solution which is free of singularities and has finite or "slowly divergent" energy.*

*Proof* From the expression (10) for $J_0$ and the static condition $b_{a,0}=0$, we have

$$\begin{aligned}(b_0, J_0) &= (b_0, f_{0i|i}) = (b_0, f_{0i})_{,i} - (b_{0,i}, f_{0i}) + (b_0, [b_i, f_{0i}]) \\ &= (b_0, f_{0i})_{,i} - (b_{0,i}, f_{0i}) - ([b_i, b_0], f_{0i}) \\ &= (b_0, f_{0i})_{,i} + (f_{0i}, f_{0i}). \end{aligned} \tag{18}$$

Consider the integral

$$0=\int_0^\infty \omega(r)dr\int_{|x|<r}(J_0-m^2b_0, b_0)d^{n-1}x, \tag{19}$$

where $|x|=\{(x^1)^2+\cdots+(x^{n-1})^2\}^{\frac{1}{2}}$ and $\omega(r)$ will be defined later. Using (18), we have

$$0=\int_0^\infty \omega(r)dr\int_{|x|<r}K\,d^{n-1}x+\int_0^\infty \omega(r)dr\int_{|x|=r}(b_0, f_{0i})\frac{x^i}{r}dS, \tag{20}$$

where

$$K=-(f_{0i}, f_{0i})-m^2(b_0, b_0)\leqslant 0, \tag{21}$$

The equality in (21) holds if and only if $f_{0i}=b_0=0$. If $K$ does not equal zero identically, then there exists a constant $R>0$ and a positive constant $\varepsilon$ such that

$$\int_{|x|<R_1}Kd^{n-1}x<-\varepsilon \quad (R_1\leqslant R). \tag{22}$$

Choose

$$\omega(r)=\begin{cases}0, & r<R,\\ \dfrac{1}{r\psi(r)}, & R\leqslant r\leqslant R_1,\\ 0, & r>R_1,\end{cases} \tag{23}$$

where $\psi(r)$ is positive unbounded, continuous functon of $r$ satisfying

$$\int_R^\infty \frac{dr}{r\psi(r)}=\infty. \tag{24}$$

Then, from (21), we have

$$\begin{aligned}0&<-\varepsilon\int_R^{R_1}\frac{dr}{r\psi(r)}+\int_R^{R_1}\frac{dr}{r\psi(r)}\int_{|x|=r}\{(b_0, b_0)+(f_{0i}, f_{0i})\}dS\\ &<-\varepsilon\int_R^{R_1}\frac{dr}{r\psi(r)}+\frac{1}{R}\int_{|x|<R_1}\frac{(b_0, b_0)+(f_{0i}, f_{0i})}{\psi(r)}d^{n-1}x.\end{aligned} \tag{25}$$

Choose $R_1$ sufficiently large, it is easily seen that the right side should be negative. This is a contradiction. Consequently, we should have $K=0$ identically, i. e.

$$b_0=0, \quad f_{0i}=0. \tag{26}$$

Thus, we have

$$T_{ii}=\frac{1}{4}(5-n)(f_{ij}, f_{ij})+\frac{m^2}{2}(3-n)(b_i, b_i) \tag{27}$$

and (12) is reduced to

$$T_{ij,i}=0. \tag{28}$$

Consider the integral

$$0=\int_0^\infty \omega(r)dr\int x^j T_{ij,i}d^{n-1}x=\int_0^\infty \omega(r)dr\int_{|x|<r}\{(x^j T_{ij})_{,i}-T_{ii}\}d^{n-1}x$$

$$=\int_0^\infty \omega(r)dr\int_{|x|=r}(x^j T_{ij})\frac{x^i}{r}-\int_0^\infty \omega(r)dr\int_{|x|<r}T_{ii}d^{n-1}x. \tag{29}$$

It is easily seen that there exists a constant $A$ such that

$$|T_{ij}|\leqslant AT_{00}. \tag{30}$$

Moreover, from (27), we have

(a) If $n\geqslant 5$, then $T_{ii}\leqslant 0$

and the equality holds only when $b_i=0$.

(b) If $n\leqslant 3$, then $T_{ii}\geqslant 0$

and the equality holds only when $b_i=0$.

In either case, if $T_{ii}\neq 0$, we have $T_{ii}<0$ (or $T_{ii}>0$) in some region. Hence there exist two constant $R>0$ and $\varepsilon>0$ such that

$$\int_{|x|<R_1}T_{ii}d^{n-1}x<-\varepsilon \quad (\text{or} >\varepsilon) \quad (R_1\geqslant R). \tag{31}$$

Choosing the same $\omega(r)$ as in (23), for the case (a), we have

$$0<-\varepsilon\int_R^{R_1}\frac{dr}{r\psi(r)}+A\int_0^\infty\int\frac{T_{00}}{\psi(r)}\,dS\,dr. \tag{32}$$

By the assumption that the energy is finite or "slowly divergent", we can choose $R_1$ sufficiently large, and it is easily seen that the right side of equation (32) should be negative. This gives a contradiction again. Consequently, we should have

$$T_{ii}=0.$$

For the case (b), the situation is quite similar. Consequently

( i ) when $n\neq 3, 4, 5$, we have $f_{ij}=0$, $b_i=0$;

( ii ) when $n=5$, we have $b_i=0$, hence $f_{ij}=0$;

(iii) when $n=3$, from the field equation (8) and $f_{ij}=0$ we have $b_i=0$.

In other words, when $n\neq 4$, the solution should be a trivial one. Thus the Theorem is proved completely.

**Remarks.**

**1.** For the massless case $m=0$. Deser's Theorem is also improved similarly.

**2.** Consider the Yang-Mills-Higgs-Kibble field (the gauge field with "soft" mass)

$$I=\int\left(-\frac{1}{4}F_{\mu\nu}F^{\mu\nu}-\frac{m^2}{2}b_\mu b^\mu-\frac{1}{4}\nabla_\mu\phi\nabla^\mu\phi-V(\phi)\right), \tag{33}$$

where $\phi$ is a scalar invariant and $V(\phi)$ is the potential. By using the same method, the result of [9] can be improved and extended to the case of "slowly divergent" energy and the classical "discontinuity" holds also for $n=5$.

In the following section, we shall specialize that the condition for the energy in our theorem cannot be omitted, because for any dimensional spacetime in massive and massless Yang-Mills field we can find static regular solutions with energy diverges sufficiently fast. In the meanwhile we obtain all the static solutions of strictly spherically symmetric gauge field.

## § 4. Static solutions of strictly spherically symmetric gauge field

Suppose the gauge potential of stricty spherically symmetric static gauge fields are in the canonical form[11]

$$b_i(x)=\phi(r)x_i, \quad b_0=\sigma(r), \tag{34}$$

where $\phi(r)$, $\sigma(r)$ are $g$-valued functions, depending only on $r$. In order to solve the massive Yang-Mills equations, we substitute (34) in (9) and we have

$$r\phi'(r)+(n-1)\phi(r)=0. \tag{35}$$

Hence we obtain

$$\phi(r)=\frac{\text{const}}{r^{n-1}}. \tag{36}$$

The requirement of regularity at the origin implies that $\phi(r)=0$. From (8) we obtain

$$\Delta\sigma(r)-m^2\sigma(r)=0 \tag{37}$$

or

$$\frac{d^2\sigma}{dr^2}+\frac{(n-2)}{r}\frac{d\sigma}{dr}-m^2\sigma=0. \tag{38}$$

Let

$$mr=R, \quad \sigma=R^{-p}q(R) \quad \left(p=\frac{n-3}{2}\right). \tag{39}$$

We obtain

$$q''+\frac{q'}{R}-\left(1+\frac{p^2}{R^2}\right)q=0, \tag{40}$$

this is the modified Bessel equation[12]. It is known that this equation admits the following solutions which are everywhere regular.

$$q=q_0I_p(R), \tag{41}$$

where $q_0$ is an element of $g$, $I_p(R)$ is the Bessel function with purely imaginary argument. Hence the equation (8) posses the following everywhere regular solutions

$$b_i=0, \quad b_0=q_0(mr)^{-\frac{n-3}{2}}I_p(mr). \tag{42}$$

Since when $r\to\infty$

$$I_p(mr)\sim\frac{e^{n+r}}{(2\pi mr)^{\frac{1}{2}}}, \tag{43}$$

the energy of such solutions is infinite and is not "slowly divergent".

From the above discussion we conclude that for any $n$, the condition of energy in the theorem cannot be omitted. That is to say, the massive Yang-Mills equation admits infinite many static regular solution whose energy is infinite and divergent sufficienty fast.

At the conclusion of the present paper we give two open problems.

1. In the case $n=4$, does there exist a static regular solution of massive (or massless i. e. pure) Yang-Mills equation with finite energy or "slowly divergent" energy?

2. Are there static solutions of the massless Yang-Mills equations in $R^{4+1}$ with $b_0 \neq 0$? This problem arises due to the fact that the instantons of $R^4$ are static solutions in $R^{4+1}$ with $b_0=0$ and the solutions in $R^{4+1}$ with $b_0 \neq 0$ may be consider as some solutions of a certain coupled Yang-Mills equations in 4-dimensional Euclidean space.

## References

[1] Yang, C. N., Integral formulism of Gauge fields, *Phys. Rev. Letters*, **33** (1974), 445—447.

[2] Shizuya, K., *Nucl. Phys.* **B94** (1975), 260.

[3] Gu Chaohao, On the mass of gauge particles, a report at the seminar of modern physics of Fudan University, 1978.

[4] Gu Chaohao, The loop phase-factor approach to gauge field, *Physica Energiae Fortis et Physica Nuclearis*, **2** (1978), 97—108.

[5] Deser, S., Absence of static solutions in source-free Yang-Mills theory, Ref. TH. 2214 CERN, 1976.

[6] Jackiw, R., Nohl, C., Rebbi, C., *Phys. Rev.*, **D15** (1977), 1642—1645.

[7] Gu Chaohao, Hu Hesheng & Shen Chunli, A geometrical interpretation of instanton solutions in Euclidean space, *Scientia Sinica*, **21** (1978), 767—772.

[8] Hu Hesheng, On equations of Yang-Mills gauge fields with mass, *Research Reports of Mathematics of Fudan Univ.* (1979, Spring), *Kexue Tongbao* **25** (1980), 191—195.

[9] Deser, S., Isham, C. J,. Static solution of Yang-Mills-Higgs-Kibble system, *Kexue Tongbao*, **25** (1980), 773—776.

[10] Weder, R., *Phys. Lett.*, **85B** (1979), 249.

[11] Gu Chaohao & Hu Hesheng, On spherically symmetric $SU_2$ gauge fields and monopoles, *Acta Physica Sinica*, **26** (1977), 155—168.

[12] Whittaker, E. T. & Watson, G. N., *A course of modern analysis*, (1955), 360—361.

Math. Z. 176, 319–325 (1981)



# On the Existence of Closed Geodesics on Spherical Manifolds

Wilhelm Klingenberg

Mathematisches Institut der Universität Bonn,
Wegelerstraße 10, D-5300 Bonn 1, Federal Republic of Germany

Dedicated to Professor Buchin Su on his 80th birthday

In this note we take up anew the old problem of the existence of closed geodesics on an $n$-dimensional spherical manifold $M$, i.e., a riemannian manifold which as underlying manifold has the $n$-sphere $S^n$.

Our goal is to construct closed geodesics on $M$ with the help of the space of circles on $S^n$. This has been done before. After Lyusternik [6] had proved the existence of $n$ closed geodesics in this manner, Alber [1] claimed to have proved the existence of at least $g(n)$ closed geodesics obtainable from the space of circles. Here, $g(n)=2n-s-1$ with $s$ determined by $0 \leqq s = n-2^k < 2^k$. As was pointed out recently by W. Ballmann, Alber's proof is based on an incorrect statement on the cohomology of the space of non-constant unparameterized closed curves, cf. also [4] where we repeated Alber's mistake.

In this paper we will show how the situation can be remedied by working with the cohomology of the space of circles instead with the cohomology of the bigger space of all closed curves on the sphere.

We will use extensively the Hilbert manifold $\Lambda M$ of closed $H^1$-curves on $M$, together with its riemannian metric and the gradient flow stemming from the energy integral $E\colon \Lambda M \to \mathbb{R}$. To keep this note short we refer for this theory to our monography [4] or our forthcoming book [5]. Here we only recall the definitions.

**1.** By $\Lambda M$ we denote the Hilbert manifold of $H^1$-mappings $c\colon S = [0,1]/\{0,1\} \to M$. $\Lambda M$ carries a canonical riemannian metric. On $\Lambda M$ we have the differentiable function $E\colon \Lambda M \to \mathbb{R}$, the so-called energy integral. $E \geqq 0$ and $\{E=0\}$ consists of the point curves = constant mappings $c$. The critical points of $E$ are, besides the point curves, the closed geodesics. For every $\kappa \geqq 0$ we put $\{E \leqq \kappa\} = \Lambda^\kappa M$.

The integral curve $\phi_s c$ of the -grad $E$ vector field, starting at $c$, is defined for all $s \in \mathbb{R}$. We thus have an $\mathbb{R}$-action $\phi_s\colon \Lambda M \to \Lambda M$.

On $\Lambda M$ there is given the canonical $S^1$-action

$$(z = e^{2\pi i r}, c(t)) \in S^1 \times \Lambda M \mapsto (z \cdot c(t) = c(t+r)) \in \Lambda M.$$

The quotient mapping w.r.t. this $S^1$-action we denote by

$$\tilde{\pi}\colon \Lambda M \to \tilde{\Pi} M = \Lambda M/S^1$$

We also have the $\mathbb{Z}_2$-action generated by the involution

$$\theta\colon \Lambda M \to \Lambda M; \quad (c(t) \mapsto (c(1-t)).$$

Let

$$^{-}\colon \Lambda M \mapsto \bar{\Lambda} M = \Lambda M/_{\theta}\mathbb{Z}_2$$

be the quotient mapping w.r.t. this $\mathbb{Z}_2$-action.

If $z \in S^1$, $c \in \Lambda M$, then $\theta z \cdot c = \bar{z} \cdot \theta c$. Hence, $\theta$ induces an involution on $\tilde{\Pi} M$ which we again denote by $\theta$.

$$^{-}\colon \tilde{\Pi} M \to \Pi M = \tilde{\Pi} M/_{\theta}\mathbb{Z}_2$$

shall be the corresponding quotient mapping. For $^{-} \circ \tilde{\pi}\colon \Lambda M \to \Pi M$ we also simply write $\pi$.

Since $\pi(c) = \pi(\theta c)$, we get a mapping $\tilde{\pi}\colon \bar{\Lambda} M \to \Pi M$ such that $\tilde{\pi} \circ ^{-} = ^{-} \circ \tilde{\pi} = \pi$.

The operations $z\colon \Lambda M \to \Lambda M$; $c \mapsto z \cdot c$, and $\theta\colon \Lambda M \to \Lambda M$; $c \mapsto \theta c$, are isometries and leave $E$ invariant. Hence, $\phi_s\colon \Lambda M \to \Lambda M$ commutes with the $S^1$-action and the $\mathbb{Z}_2$-action, i.e., $\phi_s$ is equivariant w.r.t. these actions.

Of particular interest to us is the $\mathbb{Z}_2$-action generated by the involution

$$e^{i\pi}\colon \Lambda M \to \Lambda M; \quad c(t) \mapsto c(t+1/2).$$

Together with the $\mathbb{Z}_2$-action of $\theta$, we get what we are going to call the $(\mathbb{Z}_2 \times \mathbb{Z}_2)$-action on $\Lambda M$. We introduce the following notations for the quotient mappings.

$$^{*}\colon \Lambda M \to \Lambda^{*} M = \Lambda M/_{e^{i\pi}}\mathbb{Z}_2$$

$$^{-*}\colon \Lambda M \to \bar{\Lambda}^{*} M = \Lambda M/(\mathbb{Z}_2 \times \mathbb{Z}_2).$$

Let $\Lambda_{\theta} M$ be the fixed point set of $\theta\colon \Lambda M \to \Lambda M$. $\Lambda_{\theta} M$ is a totally geodesic submanifold of $\Lambda M$ which is carried into itself by $\phi_s$. Moreover, the closed geodesics are bounded away from $\Lambda_{\theta} M$, cf. [5].

The fixed point set $\Lambda_{e^{i\pi}} M$ of $e^{i\pi}\colon \Lambda M \to \Lambda M$ again is a totally geodesic submanifold, invariant under $\phi_s$. This time, however, we do have closed geodesics in $\Lambda_{e^{i\pi}} M$, i.e., all closed geodesics $c$ which have as isotropy group $\tilde{I}(c) = \{z \in S^1,\ z \cdot c = c\}$ a group of even order.

**2.** We consider the sphere $S^n = \left\{ \sum_0^n x_i^2 = 1 \right\}$ with the standard metric. On $S^n$ we have the space $\mathrm{A}S^n$ of (parameterized) circles. The carrier of a circle is the intersection of $S^n$ with a 2-plane having distance $\leqq 1$ from $0 \in \mathbb{R}^{n+1}$. $\mathrm{A}^0 S^n$ denotes the subset of constant circles, thus, $\mathrm{A}^0 S^n \cong S^n$. To every circle $c_0 \in \mathrm{A}S^n - \mathrm{A}^0 S^n$ we associate the corresponding (parameterized) great circle which lies on the intersection of $S^n$ with the 2-plane through $0 \in \mathbb{R}^{n+1}$, parallel to the carrier of $c_0$. This yields the $D^{n-1}$-bundle

$$\alpha\colon \mathrm{A}S^n - \mathrm{A}^0 S^n \to \mathrm{B}S^n$$

over the space $\mathrm{B}S^n$ of (parameterized) great circles on $S^n$.

$\mathrm{B}S^n$ is a homogeneous space $\cong \mathbb{O}(n+1)/\mathbb{O}(n-1)$. We view $\mathrm{A}S^n$ and $\mathrm{B}S^n$ as subspace of $\Lambda S^n$. These subspaces are carried into themselves by the $S^1$-action and $\mathbb{Z}_2$-action on $\Lambda S^n$. In particular, we have the quotient mapping by the $(\mathbb{Z}_2\times\mathbb{Z}_2)$-action,

$$\bar{\ }^*: (\mathrm{A}S^n, \mathrm{B}S^n) \to (\bar{\mathrm{A}}^* S^n, \bar{\mathrm{B}}^* S^n)$$

which commutes with the projection $\alpha$. Thus, we get the $D^{n-1}$-bundle

$$\bar{\alpha}^*: \bar{\mathrm{A}}^* S^n - \bar{\mathrm{A}}^{*0} S^n \to \bar{\mathrm{B}}^* S^n.$$

$\bar{\mathrm{B}}^* S^n$ can be viewed as the total space of a $P^{n-1}$-bundle over $P^n$, where $P^j$ denotes the $j$-dimensional (real) projective space. As a homogeneous space, this reads,

$$\mathbb{O}(n+1)/\mathbb{O}(1)\times\mathbb{O}(1)\times\mathbb{O}(n-1) \to \mathbb{O}(n+1)/\mathbb{O}(1)\times\mathbb{O}(n)$$

with fibre $\mathbb{O}(n)/\mathbb{O}(1)\times\mathbb{O}(n-1)$.

According to Borel [2], the $\mathbb{Z}_2$-cohomology ring of $\bar{\mathrm{B}}^* S^n$ can be written in the form

$$S(u_1)\otimes S(u_2)\otimes S(u_3,\dots,u_{n+1})/S^+(u_1,u_2,u_3,\dots,u_{n+1}).$$

Here, $S(u_1,\dots,u_p)$ is the algebra of symmetric polynomials in $u_1,\dots,u_p$ and $S^+(u_1,\dots,u_p)$ is the ideal generated by the non-constant polynomials. The two generators $u_1, u_2$ correspond to the generators $\theta$ and $\theta e^{i\pi}$ of the $(\mathbb{Z}_2\times\mathbb{Z}_2)$-action. The cohomology ring $H^*(\Delta S^n)$ of the space $\Delta S^n = \pi \mathrm{B}S^n = G(2, n-1)$ of the unparameterized great circles can be written in the form

$$S(v_1, v_2)\otimes S(v_3,\dots,v_{n+1})/S^+(v_1,v_2,v_3,\dots,v_{n+1}).$$

The projection mapping

$$\bar{\pi}^*: \bar{\mathrm{B}}^* S^n \to \Delta S^n$$

induces for the cohomology the canonical inclusion of the last mentioned ring into the first one, with $S(v_1, v_2)$ going into $S(u_1,u_2)\subset S(u_1)\otimes S(u_2)$ and $S(v_3,\dots,v_{n+1})$ going into $S(u_3,\dots,u_{n+1})$.

The Thom isomorphism yields

$$H^k(\bar{\mathrm{A}}^* S^n, \bar{\mathrm{A}}^{*0} S^n) = H^{k-(n-1)}(\bar{\mathrm{B}}^* S^n); \quad n-1 \leqq k \leqq 3n-2$$
$$H^k(\Gamma S^n, \Gamma^0 S^n) = H^{k-(n-1)}(\Delta S^n); \quad n-1 \leqq k \leqq 3n-3.$$

For a later application we introduce still another representation $(\tilde{\mathrm{A}}S^n, \tilde{\mathrm{A}}^0 S^n)$ of the subcomplex $(\mathrm{A}S^n, \mathrm{A}^0 S^n)$ of $(\Lambda S^n, \Lambda^0 S^n)$ which is $(\mathbb{Z}_2\times\mathbb{Z}_2)$-equivariantly homotopic to $(\mathrm{A}S^n, \mathrm{A}^0 S^n)$. For that purpose we consider the fibre $\alpha^{-1}(c_0)$ over the great circle $c_0$ in the bundle $\alpha$: $\mathrm{A}S^n - \mathrm{A}^0 S^n \to \mathrm{B}S^n$. The initial point $p_0 = c_0(0)$ of $c_0$ determines an equator $E(p_0, \bar{p}_0)$ on $S^n$, i.e., the set of points having distance $\pi/2$ from $p_0$ and the antipodal point $\bar{p}_0$ of $p_0$. Each circle $c \neq$ const with $\alpha(c) = c_0$ meets the equator in exactly two points $r_c, r'_c$. We replace $c$ by the closed curve formed by the two half great circles from $p_0$ to $\bar{p}_0$, passing through $r_c$ and $r'_c$ respectively. Every point circle $r$ on $E(p_0, \bar{p}_0)$ with distance $\pi/2$ from the great circle $c_0$ we replace by the half great circle from $p_0$ to $\bar{p}_0$ through $r$, passed through back and forth. A curve of the latter type clearly can be retracted into $r$ using parts of the half great circle from $p_0$ to $\bar{p}_0$ through $r$.

Consider now the space $\tilde{\mathsf{A}}S^n \subset \Lambda S^n$ of parameterized closed curves which we obtain in this manner for each great circle $c_0$. $\tilde{\mathsf{A}}S^n \bmod \tilde{\mathsf{A}}^0 S^n$ clearly is $\mathbb{Z}_2 \times \mathbb{Z}_2$-equivariantly homotopic to $\mathsf{A}S^n \bmod \mathsf{A}^0 S^n$. The image of $\tilde{\mathsf{A}}S^n$ under $\pi$ we denote by $\tilde{\Gamma}S^n$. Then $\Gamma S^n \bmod \Gamma^0 S^n$ is homologous in $\Pi S^n$ to $\tilde{\Gamma}S^n \bmod \tilde{\Gamma}^0 S^n$.

**3.** We now describe a family of homology classes in $\bar{\mathsf{A}}^* S^n \bmod \bar{\mathsf{A}}^{*0} S^n$ which, under the projection mapping $\pi$, yield the cycles of $\Gamma S^n \bmod \Gamma^0 S^n$ which one uses in the literature, cf. [1, 4].

For $(j, k)$, $0 \leqq j \leqq k \leqq n-1$, let $\mathsf{B}_{j,k} \subset \mathsf{B}S^n$ be the set of those parameterized great circles which start on the sphere $S^j = \left\{\sum_0^j x_1^2 = 1\right\}$ and lie on the sphere $S^{k+1} = \left\{\sum_0^j x_1^2 = 1\right\}$. Put $\alpha^{-1}(\mathsf{B}_{j,k}) = \mathsf{A}_{j,k} - \mathsf{A}^0_{j,k}$. $\mathsf{A}^0_{j,k}$ shall denote the set of point circles in the closure of $\mathsf{A}_{j,k} - \mathsf{A}^0_{j,k}$. Since $\mathsf{B}_{j,k}$ is transformed into itself by the $(\mathbb{Z}_2 \times \mathbb{Z}_2)$-action on $\mathsf{B}S^n$, we also can form the quotient spaces $\bar{\mathsf{B}}^*_{j,k}$, $\bar{\mathsf{A}}^*_{j,k} - \bar{\mathsf{A}}^*_{j,k}$, $\bar{\mathsf{A}}^{*0}_{j,k}$. $\alpha | \mathsf{A}_{j,k} - \mathsf{A}^0_{j,k}$ determines the $D^{n-1}$-bundle

$$\bar{\alpha}^*_{j,k}\colon\ \bar{\mathsf{A}}^*_{j,k} - \bar{\mathsf{A}}^{*0}_{j,k} \to \bar{\mathsf{B}}^*_{j,k}$$

We denote by $\langle j, k\rangle$ the $(j+k+n-1)$-dimensional homology class determined by the cycle $\bar{\mathsf{A}}^*_{j,k} \bmod \bar{\mathsf{A}}^{*0}_{j,k}$. The image of this class in $\Gamma S^n \bmod \Gamma^0 S^n$ is the class which in [4] was denoted by $\{j, k\}$.

Following Alber [1], we have in $\Gamma S^n \bmod \Gamma^0 S^n$ a chain of $g(n) = 2n - s - 1$ pairwise subordinated homology classes. The classes of this chain are obtained by forming the cap product of the maximal class $\{n-1, n-1\}$ with the sequence of cocycles of $\Gamma S^n - \Gamma^0 S^n \cong \mathsf{B}S^n$ given by

$$(n-1, n-1) = (0,1)^{2n-2s-2}(1,1)^s, \ldots, (0,1)^{2n-2s-2}, \ldots, (0,1), (0,0).$$

Here, $(0,0) \in H^0(\Lambda S^n)$ while $(0,1) \in H^1(\Lambda S^n)$ and $(1,1) \in H^2(\Lambda S^n)$ are the first and second Stiefel-Whitney class of the canonical $\mathbb{R}^2$-bundle over $\Lambda S^n \cong G(2, n-1)$. In the Borel representation given above, $(0,1)$ and $(1,1)$ are the elements $v_1 + v_2$ and $v_1 v_2$, while the generator $(n-1, n-1)$ of $H^{2n-2}(\Lambda S^n)$ is the element $v_1^{n-2} v_2^{n-2}$. Cf. [4] for the notations and computations in $H^*(\Lambda S^n)$.

We use this chain of subordinated homology classes in $\Gamma S^n \bmod \Gamma^0 S^n$ to define the corresponding chain of homology classes in $\bar{\mathsf{A}}^* S^n \bmod \bar{\mathsf{A}}^{*0} S^n$. Take successively the cap product of $\langle n-1, n-1\rangle$ with the sequence of cocycles

$$u_1^{n-1} u_2^{n-1} = (u_1 + u_2)^{2n-2s-2}(u_1, u_2)^s, \ldots, (u_1 + u_2)^{2n-2s-2}, \ldots, (u_1 + u_2), 1$$

of $\bar{\mathsf{A}}^* S^n - \bar{\mathsf{A}}^{*0} S^n \cong \bar{\mathsf{B}}^* S^n$.

Since $\pi^*(n-1, n-1) = u_1^{n-1} u_2^{n-1}$, i.e.,

$$\pi_*(\langle n-1, n-1\rangle \cap u_1^{n-1} u_2^{n-1}) = \{n-1, n-1\} \cap (n-1, n-1) = \{0, 0\},$$

we indeed get a chain of $g(n)$ subordinated classes in $\bar{\mathsf{A}}^* S^n \bmod \bar{\mathsf{A}}^{*0} S^n$ which projects under $\pi^*$ onto Alber's chain of subordinated homology classes of $\Gamma S^n \bmod \Gamma^0 S^n$.

Replacing $\mathsf{A}S^n$ and $\Gamma S^n$ by $\tilde{\mathsf{A}}S^n$ and $\tilde{\Gamma}S^n$, we get by the same procedure chains of $g(n)$ subordinated homology classes of $\bar{\tilde{\mathsf{A}}}^* S^n \bmod \bar{\tilde{\mathsf{A}}}^{*0} S^n$ and $\tilde{\Gamma} S^n \bmod \tilde{\Gamma}^0 S^n$.

**4.** We now consider a riemannian manifold $M$, given by $S^n$ with an arbitrary riemannian metric. The inclusion $(\mathsf{A}S^n, \mathsf{A}^0 S^n) \hookrightarrow (\Lambda M, \Lambda^0 M)$ is a $\theta$- and $S^1$-equivariant mapping. In particular, the restriction of this inclusion to the various subspaces $(A_{j,k}, A^0_{j,k})$ determines a mapping of the homology classes $\langle j, k\rangle$ into $(\bar{\Lambda}^* M, \bar{\Lambda}^{*0} M)$ and $\{j, k\}$ into $(\Pi M, \Pi^0 M)$ which commutes with the mapping $\pi_*$.

We define the $\phi$-family $\mathscr{A}_{j,k}$ in the sense of [4] as consisting of the set of images under the inclusion of relative chains, homologous to the $(j+k+n-1)$-chain $\mathsf{A}_{j,k} \bmod \mathsf{A}^0_{j,k}$. Moreover, $\mathscr{A}_{j,k}$ shall comprise the image of a $(\mathbb{Z}_2 \times \mathbb{Z}_2)$-equivariant homotopy of such an inclusion. Finally, we include into $\mathscr{A}_{j,k}$ all restrictions of the previously defined chains of $\Lambda M \bmod \Lambda^0 M$ to subcycles which are homologous to the given one.

The critical value of the $\phi$-family $\mathscr{A}_{j,k}$ is given by

$$\kappa_{j,k} = \inf_{A \in \mathscr{A}_{j,k}} \sup E|A$$

cf. [4]. Our main result now is the following.

**Theorem.** *Let $\kappa_1, \ldots, \kappa_{g(n)}$ be the critical values of the chain of $\phi$-families, obtained from the chain of $g(n)$ subordinated homology classes of $\bar{\mathsf{A}}^* S^n \bmod \bar{\mathsf{A}}^{*0} S^n$, as defined in* (**3**). *Then*

$$0 < \kappa_1 \leqq \kappa_2 \leqq \ldots \leqq \kappa_{g(n)}. \tag{*}$$

*If in $(*)$ there occurs equality at one place, say $\kappa_l = \kappa_{l+1} =$ (briefly) $\kappa$, then the number of $S^1$-orbits at E-level $\kappa$ in $\Lambda M$ is infinite. Hence, there exists an infinite number of unparameterized non-oriented closed geodesics of E-value $\kappa$ on $M$.*

*Remarks.* The first $n$ values of the sequence $(*)$ are the critical values $\kappa_{0,0}, \ldots, \kappa_{0,n-1}$ of the families $\mathscr{A}_{0,0}, \ldots, \mathscr{A}_{0,n-1}$. Here, the Theorem is due to Lyusternik [6]. For this case it would have been sufficient to work with the $\mathbb{Z}_2$-action generated by $\theta$. Instead of the space $\mathsf{A}_{0,k}$ we could have taken the space $\mathsf{A}_k$ of circles which lie over the space $\mathsf{B}_k$ of great circles on $S^{k+1}$ which start at the point $(1, 0, \ldots, 0) \in S^n$. Then $\bar{\mathsf{A}}_k \bmod \bar{\mathsf{A}}^0_k$ is a $(k+n-1)$-cycle of $\bar{\mathsf{A}}S^n \bmod \bar{\mathsf{A}}^0 S^n$. Under the inclusion $(\bar{\mathsf{A}}_k, \bar{\mathsf{A}}^0_k) \hookrightarrow (\bar{\Lambda} M, \bar{\Lambda}^0 M)$, this cycle remains non-homologous to zero as one sees e.g. by taking the cap product with the $k$-th power of the 1-dimensional cocycle of $\bar{\Lambda} M - \bar{\Lambda}_\theta M$.

*Proof.* The relation $\kappa_l \leqq \kappa_{l+1}$ follows from the fact that we consider a chain of $\phi$-families where each element in the $(l+1)$-th family yields an element of the $l$-th family when forming for the domain of the cycle the cap product with a cocycle of $\bar{\mathsf{A}}^* S^n - \bar{\mathsf{A}}^* S^n \cong \bar{\mathsf{B}}^* S^n$. The relation $0 < \kappa_1$ holds since the elements in the first family are images of cycles homologous to the $(n-1)$-cycle

$(\mathsf{A}_{0,0}, \mathsf{A}^0_{0,0}) \to (\Lambda M, \Lambda^0 M)$ which represents a generator of the group $H_{n-1}(\Lambda S^n, \Lambda^0 S^n) \cong \mathbb{Z}_2$, cf. also [4].

It remains to discuss the case $\kappa_l = \kappa_{l+1} =$ (briefly) $\kappa$, for some $l$, $1 \leqq l < g(n)$. We derive a contradiction from the assumption that the set $Cr\kappa$ of critical points of $E$-value $\kappa$ in $\Lambda M$ consists of only finitely many $S^1$-orbits $(S^1 \cdot c_\iota, S^1 \cdot \theta c_\iota)$ where $\iota$ runs through a finite set $\mathsf{I}$. For each $\iota \in \mathsf{I}$ we choose an open tubular $S^1$-invariant neighbourhood $\mathcal{U}_\iota$ of $S^1 \cdot c_\iota$ such that $\mathcal{U}_\iota \cap \theta \mathcal{U}_\kappa = \varnothing$, all $\iota, \kappa$ in $I$, and $\mathcal{U}_\iota \cap \mathcal{U}_\kappa = \varnothing$, all $\iota, \kappa$ with $\iota \neq \kappa$ in $\mathsf{I}$, and $\mathcal{U}_\iota \subset \Lambda M - \Lambda_\theta M$. The union of the $\mathcal{U}_\iota$, $\theta \mathcal{U}_\iota$, $\iota \in I$, is an open neighbourhood $\mathcal{U}$ of $Cr\kappa$. $\mathcal{U}$ contains no curve from some $c_0 \in \mathcal{U}$ to $\theta c_0 \in \mathcal{U}$.

A standard result of Lyusternik-Schnirelman theory is now the existence of an element in the $(l+1)$-th $\phi$-family belonging to $\mathcal{U} \cup \Lambda^{\kappa-} M$. Here we make essential use of the fact that the deformation $\phi_s$ is $(\mathbb{Z}_2 \times \mathbb{Z}_2)$-equivariant. Let $u_{l+1}\colon \mathsf{A}_{l+1} \to \Lambda M$ be such an element. Then $\mathcal{O} = u_{l+1}^{-1}(\mathcal{U}) \subset \mathsf{A}_{l+1} - (\mathsf{A}_{l+1} \cap \mathsf{A}^0 S^n)$. $\mathcal{O}$ is transformed into itself by the $(\mathbb{Z}_2 \times \mathbb{Z}_2)$-action. No $c_0 \in \mathcal{O}$ can be joined inside $\mathcal{O}$ to $\theta c_0$.

On the other hand, we obtain an element of the $l$-th $\phi$-family by restricting $\bar{u}^*_{l+1}\colon \bar{\mathsf{A}}^*_{l+1} \to \bar{\Lambda}^* M$ to the cap product of $\bar{\mathsf{A}}^*_{l+1}$ with a cocycle in the cohomology class $(u_1 + u_2)$ or $u_1 u_2$, as the case may be, having its carrier in the open set $\bar{u}^{*-1}_{l+1}(\bar{\Lambda}^{*\kappa-})$. Consider now the $(\mathbb{Z}_2 \times \mathbb{Z}_2)$-invariant counter image in $\mathsf{A} S^n$. Thus, $\kappa_l < \kappa_{l+1}$.

**5.** As for the applications of the Theorem, it should be noted that it does not yield the existence of $g(n)$ geometrically distinct closed geodesics on $M$. Here we call two geodesics geometrically distinct if they are not only different, when considered as elements of the space $\Pi M$, but also, that they are not different coverings of the same underlying prime closed geodesics. The problem of the existence of geometrically distinct closed geodesics has been treated by much more delicate methods in [4].

However, under suitable restrictions of the range of sectional curvature $K$ of $M$, we get the following results which yield the existence of relatively short simple closed geodesics. Here, a geodesic is called simple if it has no self-intersections.

**Corollary 1.** *Assume that the sectional curvature of the simply connected, compact manifold $M$ satisfies $0 < \max K/4 < K \leqq \max K$. Then there exist on $M$ $n$ simple closed geodesics $c$ with length bounded by $2\pi/\sqrt{\max K} \leqq L(c) < 4\pi/\sqrt{\max K}$.*

**Corollary 2.** *Assume that the sectional curvature of the simply connected compact manifold $M$ satisfies $0 < 4 \max K/9 < K \leqq \max K$. Then there exist $g(u)$ simple closed geodesics on $M$ with length bounded by $2\pi/\sqrt{\max K} \leqq L(c) < 3\pi/\sqrt{\max K}$.*

*Note.* Corollary 1 is proved with the methods of [3] while the proof of Corollary 2 uses the methods of Thorbergsson [7].

*Proof.* By renormalization of the metric on $M$, we can assume $\max K = 1$. Under the hypothesis of Corollary 1, there exists a homeomorphism $h\colon S^n \to M$, carrying $p_0 = (1, 0, \ldots, 0)$, $\bar{p}_0 = (-1, 0, \ldots, 0)$ into points $(p, \bar{p})$ of maximal distance on $M$. $h$

transforms $\Lambda S^n$ into $\Lambda M$ and, in particular, the half great circles from $p_0$ to $\bar{p}_0$ into once broken geodesics from $p$ to $\bar{p}$ with length in the interval $[\pi, 2\pi[$. The break occurs at the intersection with the 'equator' $E(p,\bar{p})=\{r\in M,\ d(p,r)=d(p,r)\}$. For more details see [3–5].

Replace now in the definition of the families $\mathscr{A}_{0,k}$, $0\leqq k\leqq n-1$, the cycles $\mathrm{A}_{0,k} \bmod \mathrm{A}^0_{0,k}$ of $\Lambda S^n \bmod \Lambda^0 S^n$ by the cycles $\tilde{\mathrm{A}}_{0,k} \bmod \tilde{\mathrm{A}}^0_{0,k}$ in $\tilde{\Lambda} S^n \bmod \tilde{\Lambda}^0 S^n$, cf. **3.** Then one sees that the latter can be mapped into sets of closed curves all of which have length $<4\pi$. Thus, the critical values of these families all are $<8\pi^2$, i.e., there exist at least $n$ different unparameterized closed geodesics of length $<4\pi$. Under our hypothesis, all such geodesics are simple and of length $\geqq 2\pi$, cf. the references given above.

Similarly, under the hypothesis of Corollary 2, Thorbergsson's constructions in the paper [7], while false in $\Pi S^n$, are valid in $\bar{\Lambda}^* S^n$ and yield a realization of cycles in the homotopy class of $\bar{\mathrm{A}}^*_{j,k} \bmod \bar{\mathrm{A}}^*_{j,k}$ where all curves have length $<4\pi$. As before, one concludes with the help of the Theorem the existence of $g(n)$ simple closed geodesics.

## References


1. Alber, S.I.: On periodicity problems in the calculus of variations in the large. Uspehi Mat. Nauk **12**. No. 4(76), 57–124 (1975) Russian; Amer. Math. Soc. Transl. (**2**) 14 (1960)
2. Borel, A.: La cohomologie mod 2 des certaines espaces homogènes. Comment. Math. Helv. **27**, 165–197 (1953)
3. Klingenberg, W.: Simple closed geodesics on pinched spheres. J. Diff. Geom. **2**, 225–232 (1968)
4. Klingenberg, W.: Lectures on closed Geodesics. Grundlehren der math. Wissenschaften Bd. **230**. Berlin-Göttingen-Heidelberg: Springer 1978
5. Klingenberg, W.: Riemannian Geometry (1st edition). Bonn, Mathematisches Institut 1980
6. Lyusternik, L.: The topology of function spaces and the calculus of variations in the large. Trudy Mat. Inst. Steklov 19 (1947), (Russian); The Topology of the Calculus of Variations in the Large; Translations of Mathematical Monographs Vol. **16**. Providence R.I.: Amer. Math. Soc. 1966.
7. Thorbergsson, G.: Non-hyperbolic closed geodesics. Math. Scand. **44**, 135–148 (1979)


Received May 9, 1980

J. DIFFERENTIAL GEOMETRY
16 (1981) 179–183

# ON THE MEAN CURVATURE FUNCTION FOR COMPACT SURFACES

H. BLAINE LAWSON, JR.
& RENATO DE AZEVEDO TRIBUZY

*Dedicated to Professor Buchin Su on his 80th birthday*

It is a classical fact that any surface in $\mathbf{R}^3$ is determined up to congruences by its first and second fundamental forms. We shall prove in this article that *compact* surfaces are essentially determined by the first fundamental form and only the trace of the second, that is, by the metric and the mean curvature function. The only possible exception to this phenomenon occurs in the case of constant mean curvature. Of course, it is a long-standing conjecture of Hopf that the only such (compact) surfaces are the round spheres.

An explicit statement of our main result is as follows. Denote by $M^3(c)$ the complete simply-connected 3-manifold of constant sectional curvature $c$.

**Theorem.** *Let $\Sigma$ be a compact oriented surface equipped with a riemannian metric, and let $H\colon \Sigma \to \mathbf{R}$ be a smooth function. If $H$ is not constant, then there exist at most two geometrically distinct isometric immersions of $\Sigma$ into $M^3(c)$ with mean curvature $H$.*

**Remarks.** 1. Two immersions are said to be *geometrically distinct* if they do not differ by an isometry of $M^3(c)$, i.e., by a congruence.

2. The theorem above can be immediately applied to nonorientable surfaces. Here the function $H\colon \Sigma 7 \to \mathbf{R}$ must be replaced by a function $\tilde{H}\colon \tilde{\Sigma} \to \mathbf{R}$ on the 2-sheeted orientable covering surface $\pi\colon \tilde{\Sigma} \to \Sigma$ with the property that $H(\alpha(p)) = -H(p)$ where $\alpha\colon \tilde{\Sigma} \xrightarrow{\approx} \tilde{\Sigma}$ is the deck transformation of the covering $\pi$.

3. The result above represents a generalization to genus greater than one, of a theorem proved in the doctoral dissertation of the second author [5]. The first author insists on stating that the hard part of the proof and the principal ideas originated there.

Received August 3, 1981. The first author was partially supported by NSF Grants MCS77-23579 and INT 77-22241, and the second author by C.N.Pq. (Brasil).

4. In the case that $\Sigma$ is homeomorphic to the sphere $S^2$, the theorem can be strengthened. In that case there exists at most one isometric immersion with a given mean curvature function. This is proved in [5] and follows also from the arguments below.

*Proof of the theorem.* The given metric determines a conformal structure on $\Sigma$, and we shall always work in the corresponding local conformal, or "isothermal", coordinates. With respect to such a local coordinate $z = x_1 + ix_2$, the metric can be written as

$$ds^2 = \lambda^2 |dz|^2.$$

Suppose now that $F: \Sigma \to M^3(c)$ is an isometric immersion with unit normal vector field $\nu$. Let

$$b_{ij} = -\left\langle \nabla_{\frac{\partial}{\partial x_i}} \nu, \frac{\partial}{\partial x_j} \right\rangle$$

for $1 \leqslant i, j \leqslant 2$, denote the components of the second fundamental form of this immersion. Of fundamental importance to this study is the associated quadratic differential

$$Q \equiv \{b_{11} - b_{22} - 2ib_{12}\} dz^2 \equiv f\, dz^2, \tag{1}$$

which is well-defined globally on $\Sigma$. On the metric induced naturally on the bundle $T^{1,0} \otimes T^{1,0}$ we have that

$$\|Q\|^2 = H^2 - 4(K - c), \tag{2}$$

where $H = \lambda^{-2}(b_{11} + b_{22})$ is the mean curvature of the immersion, and $K = \lambda^{-2}(b_{11}b_{22} - b_{12}^2) + c$ is the Gaussian curvature of the surface. In terms of the principal curvatures $k_1$ and $k_2$ of $\Sigma$ we see that

$$\|Q\|^2 = (k_1 - k_2)^2,$$

and so $Q$ vanishes precisely at the umbilic points of the immersion.

We begin by recalling the following well-known consequence of the Mainardi-Codazzi equations (cf. [5]).

**Lemma 5.** *The quadratic form $Q$ is holomorphic if and only if the immersion $F$ has constant mean curvature.*

We now suppose that we are given three isometric immersions $F_k: \Sigma \to \mathbf{R}^3$, $k = 1, 2, 3$, with the same mean curvature function $H$. We let

$$Q_k = f_k(z)\, dz^2, \quad k = 1, 2, 3$$

be the corresponding associated quadratic differentials on $\Sigma$. The following principal results are proved in [5].

**Proposition 6.** *Each of the differences $Q_{ij} \equiv Q_i - Q_j$ for $1 \leqslant i, j \leqslant 3$, is a holomorphic quadratic differential form on $\Sigma$.*

**Theorem 7.** *If the three immersions $F_k$, $k = 1, 2, 3$, are mutually noncongruent, then*

$$\Delta^0 \log(f_k) = \left|\frac{\partial f_k}{\partial \bar{z}}\right|^2 \tag{3}$$

*for each $k$, where $\Delta^0 = 4\,(\partial/\partial z)(\partial/\partial \bar{z})$ is the standard laplacian in the local coordinate $z$.*

From this point on we shall assume that $F_1$, $F_2$ and $F_3$ are mutually noncongruent. However, we shall really only use the fact that (3) holds for the two immersions $F_1$ and $F_2$.

Since $F_1$ and $F_2$ are isometric and have the same mean curvature function, we see from (2) that $\|Q_1\| \equiv \|Q_2\|$. Hence we may write

$$Q_2 = e^{i\theta} Q_1, \tag{4}$$

where $\theta$ is well defined (modulo $2\pi$) outside the zeros of $\|Q_k\|^2 = H^2 - 4(K - c)$. We now consider the holomorphic quadratic form

$$q \equiv Q_1 - Q_2 = (1 - e^{i\theta}) f_1\, dz^2. \tag{5}$$

Clearly, the zeros of $Q_k$ (the umbilic points) are contained in the zeros of $q$. In particular, the zeros of $Q_k$, which we shall denote by $Z = \{p_j\}_{j=1}^n$, are *isolated*.

We now consider the quotient

$$\Psi \equiv \frac{q}{Q_1} = 1 - e^{i\theta},$$

which is well defined on $\Sigma \setminus Z$. Since $q$ is holomorphic, we have from (3) that

$$\Delta \log \Psi = \Delta \log|\Psi| + i\Delta \arg \Psi \leqslant 0, \tag{6}$$

where $\Delta = \lambda^{-2}\Delta^0$ is the Laplace-Beltrami operator on $\Sigma$. Equation (6) can be rewritten by saying that

$$\Delta \log|\Psi| \leqslant 0, \quad \Delta \arg \Psi = 0 \tag{7}$$

on $\Sigma \setminus Z$.

We now observe that since $\Psi$ is not zero in the connected set $\Sigma \setminus Z$, the function $\theta$ cannot be zero (modulo $2\pi$) in this set. Hence we can choose a continuous branch $\theta$: $\Sigma \setminus Z \to (0, 2\pi) \subset \mathbf{R}$. It follows that there exists a continuous branch

$$\arg(\Psi(z)) \in \left(-\frac{\pi}{2}, \frac{\pi}{2}\right), \tag{8}$$

for $z \in \Sigma \setminus Z$. In particular it follows from (7) and (8) that

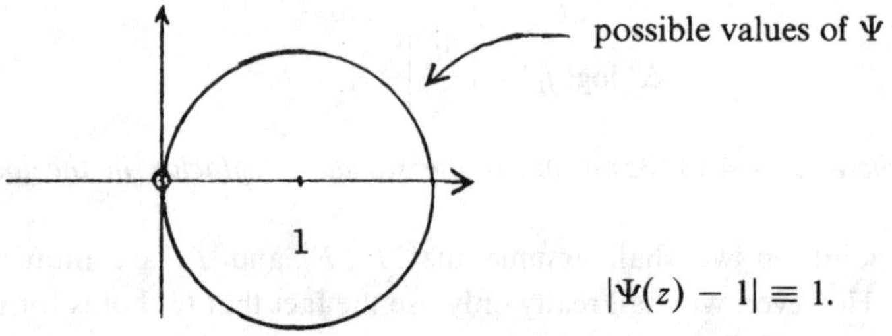


$$|\Psi(z) - 1| \equiv 1.$$

$\arg \Psi$ is a bounded harmonic function on $\Sigma \setminus Z$, where $Z$ consists of a finite set of points. By a classical theorem on removable singularities, $\arg \Psi$ extends to a smooth harmonic function on all of $\Sigma$, and hence $\arg \Psi$ is constant. It follows immediately that $\Psi$ is constant. Consequently, $Q_1$ is holomorphic, and so by Lemma 5, the mean curvature function $H$ is constant. This completes the proof.

## Final comments

1. It should be pointed out that the main result of this paper is definitely global in nature; that is, there exist compact surfaces in $\mathbf{R}^3$ having small neighborhoods which can be continuously deformed through noncongruent isometric embeddings with the same mean curvature function.

2. The local question of isometric immersions with the same mean curvature function into $\mathbf{R}^3$ has been studied in [3], [4] and [2]. In these works it is proved that if a nontrivial family of such immersions does not exist, then there are at most two noncongruent ones. (It follows from [5] that this result is valid also for immersions into $M^3(c)$, any $c$.) A superficial reading of these papers can indicate that in the absence of a nontrivial family, the immersion must be unique. However, in none of these papers do the arguments actually prove this.

3. Complete, simply-connected surfaces of constant mean curvature in $M^3(c)$ always admit 1-parameter families of isometric deformations through noncongruent surfaces with the same constant mean curvature (see [1]).

4. It remains an open question whether there can exist two geometrically distinct isometric immersions $\Sigma \hookrightarrow M^3(c)$ with the same mean curvature function for a compact surface $\Sigma$ of genus $> 0$.

5. For any compact surface $\Sigma$, there do exist families of noncongruent (and *nonisometric*) immersions into $\mathbf{R}^3$ with the same mean curvature function. Such families can be constructed as follows. Let $\gamma$ be a closed curve in the

plane $\mathbf{R}^2 \subset \mathbf{R}^3$, and consider the cylinder $\gamma \times [0, t]$ of height $t$ over $\gamma$. Cap off (smoothly) the bottom of the cylinder with a disk and the top of the cylinder with a surface of desired topological type. These "caps" should be the same, i.e., congruent, for all time $t$. The mean curvature of the annulus at a point $(x, s) \in \gamma \times [0, t]$ is just $\kappa(x) \equiv$ the curvature of the planar curve $\gamma$ at $x$. It is easy to reparameterize these surfaces by a single surface $\Sigma$ in such a manner that the resulting family of immersions $\psi_t$: $\Sigma \to \mathbf{R}^3$ has mean curvature function independent of $t$. (Stretch the parameter along the generators of the cylinder.)

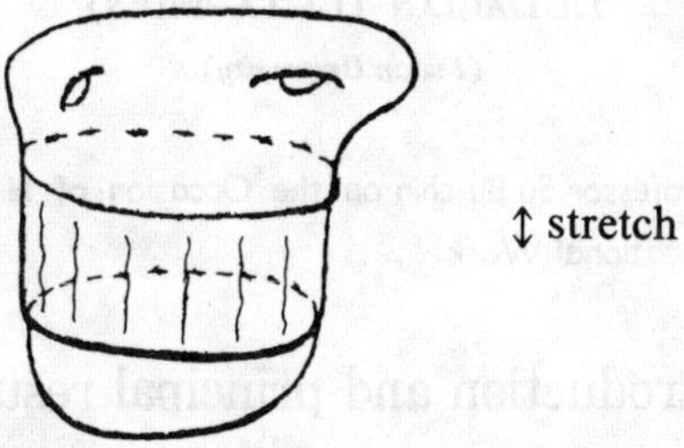


Of course, many such cylinders could be added, giving $k$-fold deformations $\psi_{t_1, \ldots, t_k}$: $\Sigma \to \mathbf{R}^3$ with the same $H$.

## References

[1] H. B. Lawson, Jr., *Complete minimal sufaces in $S^3$*, Ann. of Math. **92** (1970) 335–374.

[2] A. S. Olomouc, *Determination of a surface by its mean curvature*, Casopis Pro Pestovani Matematiky, roc. **103** (1978) Praha, 175–180.

[3] W. Scherrer, *Die Grundgleichungen der Flächentheorie*. II, Comment. Math. Helv. **32** (1957) 73–84.

[4] T. Y. Thomas, *Algebraic determination of the second fundamental form of a surface by its mean curvature*, Bull. Amer. Math. Soc. **51** (1945) 390–399.

[5] R. Tribuzy, *A characterization of tori with constant mean curvature in a space form*, Bol. Soc. Brasil. Mat. **11** (1980) 259–274.

STATE UNIVERSITY OF NEW YORK, STONY BROOK
UNIVERSIDADE DO AMAZONAS, MANAUS, BRAZIL

*Chin. Ann. of Math.*
**3** (4) 1982

# LIMIT BEHAVIORS OF SOLUTIONS FOR SOME PARABOLIC EQUATIONS OF HIGHER ORDER AND THEIR APPLICATIONS TO THE OPTIMAL CONTROL

LI DAQIAN (LI TA–TSIEN)
*(Fudan University)*

Dedicated to Professor Su Bu-chin on the Occasion of his 80th Birthday and his 50th Year of Educational Work

## Introduction and principal results

Let $\Omega$ be a bounded open set in $\mathbf{R}^n$ Containing the origin with a smooth boundary $\Gamma$. In this paper we restrict ourselves to the case $n=2$ or $3$, which is more important in applications.

Let $A$ be a second order self–abjoint elliptic operator with variable coefficients

$$A\varphi=-\sum_{i,j=1}^{n}\frac{\partial}{\partial x_i}\left(a_{ij}(x)\frac{\partial\varphi}{\partial x_j}\right)+c(x)\varphi \tag{1}$$

with $a_{ij}(x)$ and $c(x)$ suitably smooth

$$a_{ij}(x)=a_{ji}(x),\ i,\ j=1,\ \cdots,\ n,\ x\in\Omega \tag{2}$$

and there exists a constant $\alpha>0$ such that

$$\sum_{i,j=1}^{n}a_{ij}(x)\xi_i\xi_j\geqslant\alpha|\xi|^2\quad\forall\xi\in\mathbf{R}^n,\ x\in\Omega. \tag{3}$$

In the first part of this paper we shall prove the following results for limit behaviors of solutions for some parabolic equations of higher order:

**I. 1.** For any $v\in L^2(0,\ T)$, consider the following initial boundary value problem

$$(\mathrm{I})\qquad\begin{cases}\dfrac{\partial y}{\partial t}+Ay=v(t)\delta(x) & \text{in } Q=\Omega\times(0,\ T),\\ y=0 & \text{on } \Sigma=\Gamma\times(0,\ T),\\ y(x,\ 0)=0 & \text{in } \Omega,\end{cases}$$

where $\delta(x)$ is the Dirac mass at the origin. By transposition (cf. [1, 2]), problem (I) admits a unique weak solution $y=y(t;\ v)\in L^2(Q)$.

For any $\varepsilon>0$ fixed, consider the following approximation of problem (I)

Manuscript received June 24, 1981.

$$(\mathrm{I})_\varepsilon \qquad \begin{cases} \dfrac{\partial y_\varepsilon}{\partial t}+\varepsilon A^2 y_\varepsilon+Ay_\varepsilon=v_\varepsilon(t)\delta(x), & \text{in } Q, \\ y_\varepsilon=Ay_\varepsilon=0, & \text{on } \Sigma, \\ y_\varepsilon(x, 0)=0, & \text{in } \Omega. \end{cases}$$

It is well known (cf. [3]) that for any $v_\varepsilon(t)\in L^2(0, T)$, problem $(\mathrm{I})_\varepsilon$ possesses a unique solution $y_\varepsilon=y_\varepsilon(t; v_\varepsilon)$

$$y_\varepsilon\in L^2(0, T;\ H^2(\Omega)\cap H_0^1(\Omega)),\ \frac{\partial y_\varepsilon}{\partial t}\in L^2(0, T;\ H^{-2}(\Omega)) \tag{4}$$

and

$$v_\varepsilon(t)\to\left\{y_\varepsilon,\ \frac{\partial y_\varepsilon}{\partial t}\right\}$$

is a continuous linear mapping from $L^2(0, T)$ to

$$\{L^2(0, T;\ H^2(\Omega))\times L^2(0, T; H^{-2}(\Omega))\}.$$

Hence

$$y_\varepsilon(T;\ v_\varepsilon)\in L^2(\Omega) \tag{5}$$

and

$$L_\varepsilon\colon v_\varepsilon(t)\to y_\varepsilon(T; v_\varepsilon) \tag{6}$$

is a continuous linear mapping from $L^2(0, T)$ to $L^2(\Omega)$.

In I. § 1—2 we shall prove

**Theorem 1** (resp. **Theorem 1 bis**) *As* $\varepsilon\to 0$, *if*

$$v_\varepsilon(t)\to v(t) \tag{7}$$

*in* $L^2[0, T]$ *weakly (resp. strongly), then*

$$y_\varepsilon(t;\ v_\varepsilon)\to y(t;\ v) \tag{8}$$

*in* $L^2(Q)$ *weakly (resp. strongly).*

**Corollary 1. 1.** *In Theorem 1 (resp. Theorem 1 bis) if we suppose further that for* $\varepsilon>0$ *small enough*

$$\{y_\varepsilon(T;\ v_\varepsilon)\} \text{ belongs to a weakly (resp. strongly) compact subset of } L^2(\Omega), \tag{9}$$

*then*

$$y_\varepsilon(T;\ v_\varepsilon)\to y(T;\ v) \text{ in } L^2(\Omega) \text{ weakly (resp. strongly), as } \varepsilon\to 0 \tag{10}$$

*and*

$$v\in \mathscr{U}, \tag{11}$$

*where* $\mathscr{U}$ *is the function space (cf. [2], [7—9])*

$$\mathscr{U}=\{v\,|\,v\in L^2(0, T),\ y(T;\ v)\in L^2(\Omega)\} \tag{12}$$

*provided with the graph norm.*

**I. 2.** For problems (I) and $(\mathrm{I})_\varepsilon$, we consider the corresponding adjoint problems as follows

$$(\mathrm{II}) \qquad \begin{cases} -\dfrac{\partial \varphi}{\partial t}+A\varphi=\psi & \text{in } Q, \\ \varphi=0 & \text{on } \Sigma, \\ \varphi(x, T)=0 & \text{in } \Omega; \end{cases}$$

$$(\mathrm{II})_\varepsilon \quad \begin{cases} -\dfrac{\partial\varphi_\varepsilon}{\partial t}+\varepsilon A^2\varphi_\varepsilon+A\varphi_\varepsilon=\psi_\varepsilon & \text{in } Q, \\ \varphi_\varepsilon=A\varphi_\varepsilon=0 & \text{on } \Sigma, \\ \varphi_\varepsilon(x,\ T)=0 & \text{in } \Omega. \end{cases}$$

Let $\varphi(t;\ \psi)$ and $\varphi_\varepsilon(t;\ \psi_\varepsilon)$ denote the solutions of problem (II) and $(\mathrm{II})_\varepsilon$ respectively, in I. § 1—2 we shall prove

**Theorem 2 (resp. Theorem 2 bis).** *As* $\varepsilon\to 0$, *if*

$$\psi_\varepsilon\to\psi \text{ in } L^2(Q) \text{ strongly (resp. weakly)}, \tag{13}$$

*then*

$$\varphi_\varepsilon(t;\ \psi_\varepsilon)\to\varphi(t;\ \psi) \text{ in } H^{2,1}(Q) \text{ strongly (resp. weakly)}, \tag{14}$$

*where*

$$H^{2,1}(Q)=\left\{\varphi \mid \varphi\in L^2(0,\ T;\ H^2(\Omega)),\ \frac{\partial\varphi}{\partial t}\in L^2(Q)\right\}; \tag{15}$$

*In particular*

$$\varphi_\varepsilon(0,\ t;\ \psi_\varepsilon)\to\varphi(0,\ t;\ \psi) \text{ in } L^2(0,\ T) \text{ strongly (resp. weakly)}. \tag{16}$$

**I. 3.** For any $S(x)\in L^2(\Omega)$, the following problem

$$(\mathrm{III}) \quad \begin{cases} -\dfrac{\partial p}{\partial t}+Ap=0 & \text{in } Q, \\ p=0 & \text{on } \Sigma, \\ p(x,\ T)=S(x) & \text{in } \Omega \end{cases}$$

admits a unique solution $p=p(t;\ S)$ (cf. [3])

$$p\in L^2(0,\ T;\ H_0^1(\Omega)),\ \frac{\partial p}{\partial t}\in L^2(0,\ T;\ H^{-1}(\Omega)). \tag{17}$$

For any $\varepsilon>0$ fixed, construct the following approximation of problem (III)

$$(\mathrm{III})_\varepsilon \quad \begin{cases} -\dfrac{\partial p_\varepsilon}{\partial t}+\varepsilon A^2p_\varepsilon+Ap_\varepsilon=0 & \text{in } Q, \\ p_\varepsilon=Ap_\varepsilon=0 & \text{on } \Sigma, \\ p_\varepsilon(x,\ T)=S_\varepsilon(x) & \text{in } \Omega. \end{cases}$$

It is well known (cf. [3]) that for any $S_\varepsilon(x)\in L^2(\Omega)$, problem $(\mathrm{III})_\varepsilon$ admits a unique solution $p_\varepsilon=p_\varepsilon(t;\ S_\varepsilon)$

$$p_\varepsilon\in L^2(0,\ T;\ H^2(\Omega)\cap H_0^1(\Omega)),\ \frac{\partial p_\varepsilon}{\partial t}\in L^2(0,\ T;\ H^{-2}(\Omega)) \tag{18}$$

and

$$S_\varepsilon(x)\to\left\{p_\varepsilon,\ \frac{\partial p_\varepsilon}{\partial t}\right\}$$

in a continuous linear mapping from $L^2(\Omega)$ to

$$\{L^2(0,\ T;\ H^2(\Omega))\times L^2(0,\ T;\ H^{-2}(\Omega))\}.$$

Hence

$$p_\varepsilon(0,\ t;\ S_\varepsilon)\in L^2(0,\ T) \tag{19}$$

and

$$L_\varepsilon^*:\ S_\varepsilon(x)\to p_\varepsilon(0,\ t;\ S_\varepsilon) \tag{20}$$

is a continuous linear mapping from $L^2(\Omega)$ to $L^2(0,\ T)$.

In I. § 3—4 we shall prove

**Theorem 3 (resp. Theorem 3 bis).** *As $\varepsilon \to 0$, if*

$$S_\varepsilon(x) \to S(x) \ in \ L^2(\Omega) \ weakly \ (resp. \ strongly), \tag{21}$$

*then*

$$p_\varepsilon(t; \ S_\varepsilon) \to p(t; \ S) \ in \ L^2(0, \ T; \ H_0^1(\Omega)) \ weakly \ (resp. \ strongly). \tag{22}$$

**Corollary 3. 1.** *In Theorem 3 (resp. Theorem 3 bis), if we suppose further that for $\varepsilon > 0$ small enough*

$$\{p_\varepsilon(0, \ t; S_\varepsilon)\} \ belongs \ to \ a \ weakly \ (resp. \ strongly) \ compact \ subset \ of \ L^2(0, \ T), \tag{23}$$

*then*

$$p_\varepsilon(0, \ t; \ S_\varepsilon) \to p(0, \ t; \ S) \ in \ L^2(0, \ T) \ weakly \ (resp. \ strongly), \ as \ \varepsilon \to 0 \tag{24}$$

*and*

$$S \in \mathscr{U}^*, \tag{25}$$

*where $\mathscr{U}^*$ is the function space (cf. [2, 7, 9])*

$$\mathscr{U}^* = \{S \mid S \in L^2(\Omega), \ p(0, \ t; \ S) \in L^2(0, \ T)\} \tag{26}$$

*provided with the graph norm.*

In the second part of this paper we shall use the preceding results in the first part to study the following various problems of optimal control:

**II. 1.** By means of Problem (I) we define the cost function

$$J(v) = N \int_0^T v^2 dt + \int_\Omega |y(T; \ v) - Z_d|^2 dx, \ \forall v \in \mathscr{U}, \tag{27}$$

where $Z_d$ is given in $L^2(\Omega)$, $N$ is given $> 0$ and $\mathscr{U}$ is defined by (12). It is easy to see (cf. [5]) that there exists a unique element $u_0 = u_0(t) \in \mathscr{U}$ such that

$$J(u_0) = \inf_{v \in \mathscr{U}} J(v). \tag{28}$$

The corresponding optimality system is the following

$$\text{(IV)} \quad \begin{cases} \dfrac{\partial y}{\partial t} + Ay = -\dfrac{1}{N} p(0, \ t) \delta(x), \ -\dfrac{\partial p}{\partial t} + Ap = 0 & \text{in } Q, \\ y = 0, \ p = 0 & \text{on } \Sigma, \\ y(x, \ 0) = 0, \ p(x, \ T) = y(x, \ T) - Z_d & \text{in } \Omega \end{cases}$$

and the optimal control $u_0$ is given by

$$u_0 = -\frac{1}{N} p(0, \ t). \tag{29}$$

Similarly, by means of problem (I)$_\varepsilon$, we can define the cost function

$$J_\varepsilon(v) = N \int_0^T v^2 dt + \int_\Omega |y_\varepsilon(T; \ v) - Z_d|^2 dx, \ \forall v \in L^2(0, \ T). \tag{30}$$

Noticing (6), it is easily seen that there exists a unique element $u_\varepsilon = u_\varepsilon(t) \in L^2(0, \ T)$ such that

$$J_\varepsilon(u_\varepsilon) = \inf_{v \in L^2(0, T)} J_\varepsilon(v). \tag{31}$$

The corresponding optimality system is the following

$$(\mathrm{IV})_\varepsilon \quad \begin{cases} \dfrac{\partial y_\varepsilon}{\partial t}+\varepsilon A^2 y_\varepsilon+Ay_\varepsilon=-\dfrac{1}{N}p_\varepsilon(0,\ t)\delta(x), \\ -\dfrac{\partial p_\varepsilon}{\partial t}+\varepsilon A^2 p_\varepsilon+Ap_\varepsilon=0 & \text{in } Q, \\ y_\varepsilon=Ay_\varepsilon=0,\ p_\varepsilon=Ap_\varepsilon=0 & \text{on } \Sigma, \\ y_\varepsilon(x,\ 0)=0,\ p_\varepsilon(x,\ T)=y_\varepsilon(x,\ T)-Z_d & \text{in } \Omega \end{cases}$$

and the optimal control $u_\varepsilon$ is given by

$$u_\varepsilon=-\frac{1}{N}p_\varepsilon(0,\ t). \tag{32}$$

In II. § 1 we shall prove

**Theorem 4** *As $\varepsilon\to 0$, we have*

(i) $J_\varepsilon(u_\varepsilon)\to J(u_0)$; (33)

(ii) $u_\varepsilon\to u_0$ *in* $L^2(0,\ T)$ *strongly.* (34)

*Moreover, for the solutions* $\{y_\varepsilon,\ p_\varepsilon\}$ *and* $\{y,\ p\}$ *of the optimality systems* $(\mathrm{IV})_\varepsilon$ *and* (IV) *respectively, we have. as* $\varepsilon\to 0$

$$y_\varepsilon\to y \text{ in } L^2(Q) \text{ strongly}, \tag{35}$$

$$y_\varepsilon(T)\to y(T) \text{ in } L^2(\Omega) \text{ strongly}, \tag{36}$$

$$p_\varepsilon\to p \text{ in } L^2(0,\ T;\ H_0^1(\Omega)) \text{ strongly}, \tag{37}$$

$$p_\varepsilon(0,\ t)\to p(0,\ t) \text{ in } L^2(0,\ T) \text{ strongly}. \tag{38}$$

**II. 2.** By duality, we can define the cost function by means of problem (III) as follows

$$M(S)=N\int_\Omega S^2\,dx+\int_0^T(p(0,\ t;\ S)-Z_d(t))^2\,dt,\ \forall S\in\mathscr{U}^*, \tag{39}$$

where $Z_d(t)$ is given in $L^2(0,\ T)$, $N$ is given$>0$ and $\mathscr{U}^*$ is defined by (26). There exists a unique element $q_0=q_0(x)\in\mathscr{U}^*$ such that

$$M(q_0)=\inf_{S\in\mathscr{U}^*}M(S). \tag{40}$$

The corresponding optimality system is the following

$$(\mathrm{V}) \quad \begin{cases} -\dfrac{\partial p}{\partial t}+Ap=0,\ \dfrac{\partial y}{\partial t}+Ay=(p(0,\ t)-Z_d(t))\delta(x) & \text{in } Q, \\ p=0,\ y=0 & \text{on } \Sigma, \\ p(x,\ T)=-\dfrac{1}{N}y(x,\ T),\ y(x,\ 0)=0 & \text{in } \Omega \end{cases}$$

and the optimal control $q_0$ is given by

$$q_0=-\frac{1}{N}y(x,\ T). \tag{41}$$

Besides, by means of problem $(\mathrm{III})_\varepsilon$ we can also define the cost function

$$M_\varepsilon(S)=N\int_\Omega S^2(x)\,dx+\int_0^T(p_\varepsilon(0,\ t;\ S)-Z_d(t))^2dt,\ \forall S\in L^2(\Omega). \tag{42}$$

Noticing (20), it is easy to see that there exists a unique element $q_\varepsilon=q_\varepsilon(x)\in L^2(\Omega)$ such that

$$M_\varepsilon(q_\varepsilon)=\inf_{S\in L^2(\Omega)} M_\varepsilon(S). \tag{43}$$

The corresponding optimality system is the following

$$(\mathrm{V})_\varepsilon \quad \begin{cases} -\dfrac{\partial p_\varepsilon}{\partial t}+\varepsilon A^2 p_\varepsilon + A p_\varepsilon = 0, & \\ \dfrac{\partial y_\varepsilon}{\partial t}+\varepsilon A^2 y_\varepsilon + A y_\varepsilon = (p_\varepsilon(0,\ t)-Z_d(t))\delta(x) & \text{in } Q, \\ p_\varepsilon = A p_\varepsilon = 0,\ y_\varepsilon = A y_\varepsilon = 0 & \text{on } \Sigma, \\ p_\varepsilon(x,\ T) = -\dfrac{1}{N} y_\varepsilon(x,\ T),\ y_\varepsilon(x,\ 0)=0 & \text{in } \Omega \end{cases}$$

and the optimal control $q_\varepsilon$ is given by

$$q_\varepsilon = -\frac{1}{N} y_\varepsilon(x,\ T). \tag{44}$$

The following theorem can be proved in a similar way as in the proof of Theorem 4.

**Theorem 5** *As $\varepsilon\to 0$, we have*

(i) $M_\varepsilon(q_\varepsilon)\to M(q_0)$, (45)

(ii) $q_\varepsilon\to q_0$ *in $L^2(\Omega)$ strongly.* (46)

*Moreover, for the solutions $\{p_\varepsilon,\ y_\varepsilon\}$ and $\{p,\ y\}$ of the optimality systems* $(\mathrm{V})_\varepsilon$ *and* (V) *respectively, as $\varepsilon\to 0$, we have the same results* (35)—(38).

**II. 3.** Instead of (27) we can also define the cost function by means of (I) as follows

$$J(v)=N\int_0^T v^2\,dt+\int_Q |y(t;\ v)-Z_d|^2\,dx\,dt,\quad \forall v\in L^2(0,\ T), \tag{47}$$

where $Z_d$ is given in $L^2(Q)$. There exists a unique element $u_0=u_0(t)\in L^2(0,\ T)$ such that

$$J(u_0)=\inf_{v\in L^2(0,T)} J(v). \tag{48}$$

The corresponding optimality system is the following

$$(\mathrm{VI}) \quad \begin{cases} \dfrac{\partial y}{\partial t}+Ay=-\dfrac{1}{N}\varphi(0,\ t)\delta(x),\quad -\dfrac{\partial \varphi}{\partial t}+A\varphi = y-Z_d & \text{in } Q, \\ y=0,\ \varphi=0 & \text{on } \Sigma, \\ y(x,\ 0)=0,\ \varphi(x,\ T)=0 & \text{in } \Omega \end{cases}$$

and the optimal control $u_0$ is given by

$$u_0=-\frac{1}{N}\varphi(0,\ t). \tag{49}$$

Similarly, by means of $(\mathrm{I})_\varepsilon$ we can define the cost function

$$J_\varepsilon(v)=N\int_0^T v^2\,dt+\int_Q |y_\varepsilon(t;\ v)-Z_d|^2\,dx\,dt,\quad \forall v\in L^2(0,\ T). \tag{50}$$

There exists a unique element $u_\varepsilon=u_\varepsilon(t)\in L^2(0,\ T)$ such that

$$J_\varepsilon(u_\varepsilon)=\inf_{v\in L^2(0,T)} J_\varepsilon(v). \tag{51}$$

The corresponding optimality system is the following

$$
(\mathrm{VI})_\varepsilon \quad \begin{cases} \dfrac{\partial y_\varepsilon}{\partial t} + \varepsilon A^2 y_\varepsilon + A y_\varepsilon = -\dfrac{1}{N}\varphi_\varepsilon(0, t)\delta(x), \\ -\dfrac{\partial \varphi_\varepsilon}{\partial t} + \varepsilon A^2 \varphi_\varepsilon + A\varphi_\varepsilon = y_\varepsilon - Z_d & \text{in } Q, \\ y_\varepsilon = A y_\varepsilon = 0, \ \varphi_\varepsilon = A\varphi_\varepsilon = 0 & \text{on } \Sigma, \\ y_\varepsilon(x, 0) = 0, \ \varphi_\varepsilon(x, T) = 0 & \text{in } \Omega \end{cases}
$$

and the optimal control $u_\varepsilon$ is given by

$$
u_\varepsilon = -\frac{1}{N}\varphi_\varepsilon(0, t). \tag{52}
$$

In II. § 2, we shall prove

**Theorem 6** *As $\varepsilon \to 0$, We have*

(i) $J_\varepsilon(u_\varepsilon) \to J(u_0)$, (53)

(ii) $u_\varepsilon \to u_0$ *in* $L^2(0, T)$ *strongly.* (54)

*Moreover, for the solutions $\{y_\varepsilon, \varphi_\varepsilon\}$ and $\{y, \varphi\}$ of the optimality systems* $(\mathrm{VI})_\varepsilon$ *and* (VI) *respectively, we have, as $\varepsilon \to 0$*

$$
y_\varepsilon \to y \text{ in } L^2(Q) \text{ strongly}, \tag{55}
$$

$$
\varphi_\varepsilon \to \varphi \text{ in } H^{2,1}(Q) \text{ strongly}, \tag{56}
$$

$$
\varphi_\varepsilon(0, t) \to \varphi(0, t) \text{ in } L^2(0, T) \text{ strongly}. \tag{57}
$$

**II. 4.** By duality, we can define the following cost functions by means of problems (II) and $(\mathrm{II})_\varepsilon$ respectively

$$
M(\psi) = N\int_Q \psi^2\, dx\, dt + \int_0^T (\varphi(0, t; \psi) - Z_d)^2 dt, \ \forall \psi \in L^2(Q) \tag{58}
$$

and

$$
M_\varepsilon(\psi) = N\int_Q \psi^2\, dx\, dt + \int_0^T (\varphi_\varepsilon(0, t; \psi) - Z_d)^2\, dt, \ \forall \psi \in L^2(Q). \tag{59}
$$

We can obtain the similar results as in Theorem 6. The detail is omitted here.

In what follows, the letter $C$ always denotes certain constants independent of $\varepsilon$.

# I. Limit behaviors of solutions for some parabolic equations of higher order

## § 1. Proof of Theorems 1 and 2.

By transposition, the solution $y = y(t; v)$ of (I) is defined by the following Greens formula

$$
\int_Q y\psi\, dx\, dt = \int_0^T \varphi(0, t)v(t)dt, \ \forall \psi \in L^2(Q), \tag{I. 1. 1}
$$

where $\varphi = \varphi(t; \psi)$ is the solution of (II).

Moreover, the solution $y_\varepsilon = y_\varepsilon(t; v_\varepsilon)$ of $(\mathrm{I})_\varepsilon$ satisfies the similar formula

$$
\int_Q y_\varepsilon \psi_\varepsilon\, dx\, dt = \int_0^T \varphi_\varepsilon(0, t)v_\varepsilon(t)dt, \ \forall \psi_\varepsilon \in L^2(Q), \tag{I. 1. 2}
$$

where $\varphi_\varepsilon = \varphi_\varepsilon(t; \psi_\varepsilon)$ is the solution of $(\mathrm{II})_\varepsilon$.

According to (I. 1. 1) and (I. 1. 2), Theorem 1 is a direct consequence of Theorem 2 by duality (cf. [4]).

*Proof of Theorem 2:*

**Lemma 1** (cf. [3]). *For any* $\psi \in L^2(Q)$, *the solution* $\varphi$ *of (II) satisfies*

$$\varphi \in H^{2,1}(Q) \tag{I. 1. 3}$$

*and*

$$\|\varphi(t;\ \psi)\|_{H^{2,1}(Q)} \leqslant C\|\psi\|_{L^2(Q)}. \tag{I. 1. 4}$$

*In particular*

$$\|\varphi(0,\ t;\ \psi)\|_{L^2(0,T)} \leqslant C\|\psi\|_{L^2(Q)}. \tag{I. 1. 5}$$

**Lemma 2.** *For any* $\varepsilon > 0$ *fixed,* $\forall \psi_\varepsilon \in L^2(Q)$, *the solution* $\varphi_\varepsilon$ *of* $(\mathrm{II})_\varepsilon$ *satisfies*

$$\varphi_\varepsilon \in L^2(0,\ T;\ H^4(\Omega)),\ \frac{\partial \varphi_\varepsilon}{\partial t} \in L^2(Q) \tag{I. 1. 6}$$

*and*

$$\|\varphi_\varepsilon(t;\ \psi_\varepsilon)\|_{H^{2,1}(Q)} \leqslant C\|\psi_\varepsilon\|_{L^2(Q)}. \tag{I. 1. 7}$$

*In particular*

$$\|\varphi_\varepsilon(0,\ t;\ \psi_\varepsilon)\|_{L^2(0,T)} \leqslant C\|\psi_\varepsilon\|_{L^2(Q)}. \tag{I. 1. 8}$$

*Proof* Multiplying the equation

$$-\frac{\partial \varphi_\varepsilon}{\partial t} + \varepsilon A^2 \varphi_\varepsilon + A\varphi_\varepsilon = \psi_\varepsilon$$

by $\varphi_\varepsilon$ and $\frac{\partial \varphi_\varepsilon}{\partial t}$ respectively and integrating by parts, Lemma 2 follows from the classical theorem of regularity of solutions for linear elliptic equations.

From Lemmas 1 and 2 we obtain

**Lemma 3.** *It is sufficient to prove Theorem 2 for* $\psi_\varepsilon \equiv \psi$ *independent of* $\varepsilon$ *and* $\psi \in \mathscr{D}(Q)$, *where* $\mathscr{D}(Q)$ *is the space of infinitely differentiable functions with compact support.*

Hence, it remains only to prove

**Lemma 4.** *Suppose that* $\varphi_\varepsilon$ *is the solution of*

$$\begin{cases} -\dfrac{\partial \varphi_\varepsilon}{\partial t} + \varepsilon A^2 \varphi_\varepsilon + A\varphi_\varepsilon = \psi & \text{in } Q, \\ \varphi_\varepsilon = A\varphi_\varepsilon = 0 & \text{on } \Sigma, \\ \varphi_\varepsilon(x,\ T) = 0 & \text{in } \Omega, \end{cases} \tag{I. 1. 9}$$

*where* $\psi \in \mathscr{D}(Q)$, *then as* $\varepsilon \to 0$, *we have*

$$\varphi_\varepsilon \to \varphi \text{ in } H^{2,1}(Q) \text{ strongly}. \tag{I. 1. 10}$$

*In particular*

$$\varphi_\varepsilon(0,\ t) \to \varphi(0,\ t) \text{ in } L^2(0,\ T) \text{ strongly}. \tag{I. 1. 11}$$

*Proof* Since $\psi \in \mathscr{D}(Q)$, $\varphi$ and $\varphi_\varepsilon$ are all regular. According to Iemma 2, we have

$$\|\varphi_\varepsilon\|_{H^{2,1}(Q)} \leqslant C, \tag{I. 1. 12}$$

$$\|\varphi_\varepsilon(0,\ t)\|_{L^2(0,T)} \leqslant C. \tag{I. 1. 13}$$

By regularity, we have also

$$\left\|\frac{\partial \varphi_\varepsilon}{\partial t}\right\|_{H^{2,1}(Q)} \leqslant C,\ \left\|\frac{\partial^2 \varphi_\varepsilon}{\partial t^2}\right\|_{H^{2,1}(Q)} \leqslant C, \tag{I. 1. 14}$$

$$\left\| \frac{\partial \varphi_s}{\partial t}(0, t) \right\|_{L^2(0,T)} \leqslant C. \tag{I. 1. 15}$$

Hence

$$\|\varphi_s\|_{H^1(Q)} \leqslant C, \tag{I. 1. 16}$$

$$\|\varphi_s(0, t)\|_{H^1(0,T)} \leqslant C. \tag{I. 1. 17}$$

Moreover, since $A\varphi_s$ is also the solution of (I. 1. 9) in which $\psi$ is replaced by $A\psi$, we have

$$\|A\varphi_s\|_{H^{2,1}(Q)} \leqslant C, \quad \left\| \frac{\partial A\varphi_s}{\partial t} \right\|_{H^{2,1}(Q)} \leqslant C, \tag{I. 1. 18}$$

hence

$$\|A\varphi_s\|_{H^1(Q)} \leqslant C. \tag{I. 1. 19}$$

According to (I. 1. 16)、(I. 1. 17) and (I. 1. 19), by compactness from $\{\varphi_s\}$ we can extract a subsequence $\{\varphi_\eta\}$ such that

$$\varphi_\eta \to \Phi \text{ in } H^1(Q) \text{ strongly}, \tag{I. 1. 20}$$

$$\varphi_\eta(0, t) \to \Phi(0, t) \text{ in } L^2(0, T) \text{ strongly}, \tag{I. 1. 21}$$

$$A\varphi_\eta \to A\Phi \text{ in } L^2(Q) \text{ strongly}. \tag{I. 1. 22}$$

Then, according to the classical theorem of reqularity of solutions for linear elliptic equations, it follows from (I. 1. 20) and (I. 1. 22) that

$$\varphi_\eta \to \Phi \text{ in } H^{2,1}(Q) \text{ strongly}. \tag{I. 1. 23}$$

Hence, it remains only to prove

$$\Phi = \varphi. \tag{I. 1. 24}$$

Passing to the limit in (I. 1. 9), from (I. 1. 23) we get

$$\begin{cases} -\dfrac{\partial \Phi}{\partial t} + A\Phi = \psi & \text{in } Q, \\ \Phi = 0 & \text{on } \Sigma \\ \Phi(x, T) = 0 & \text{in } \Omega, \end{cases}$$

so (I. 1. 24) holds.

Thus, Theorem 2 (hence Theorem 1) is proved.

## § 2. Proof of Theorems 1 bis and 2 bis.

**Lemma 5.** *Theorem 1 bis is a direct consequence of Theorem 2 bis.*

*Proof* This Lemma can be obtained by duality. Here we give a direct proof. Taking $\psi = y$ and $\psi_s = y$ in (I. 1. 1) and (I. 1. 2) respectively, we get

$$\int_Q y^2\,dx\,dt = \int_0^T \varphi(0, t; y)\,v(t)\,dt, \tag{I. 2. 1}$$

$$\int_Q y_s y\,dx\,dt = \int_0^T \varphi_s(0, t; y)\,v_s(t)\,dt. \tag{I. 2. 2}$$

Using Theorem 2 we have

$$\int_Q y_s y\,dx\,dt \to \int_Q y^2\,dx\,dt, \text{ as } s \to 0. \tag{I. 2. 3}$$

Now taking $\psi = y$ and $\psi_s = y_s$ in (I. 1. 1) and (I. 1. 2) respectively, we get (I. 2. 1) and

$$\int_Q y_\varepsilon^2 \, dx \, dt = \int_0^T \varphi_\varepsilon(0, \ t; \ y_\varepsilon) v_\varepsilon(t) dt. \tag{I. 2. 4}$$

Hence under the hypotheses of Theorem 1 bis, if Theorem 2 bis holds, then using Theorem 1 we have

$$\int_Q y_\varepsilon^2 \, dx \, dt \to \int_Q y^2 \, dx \, dt, \quad \text{as } \varepsilon \to 0. \tag{I. 2. 5}$$

Thus, Theorem 1 bis follows from (I. 2. 3) and (I. 2. 5).

*Proof of Theorem 2 bis:*

According to Theorem 2, it is easy to see that as $\varepsilon \to 0$, if

$$\psi_\varepsilon \to 0 \text{ in } L^2(Q) \text{ weakly}, \tag{I. 2. 6}$$

then

$$\varphi_\varepsilon(t; \ \psi_\varepsilon) \to 0 \text{ in } H^{2,1}(Q) \text{ weakly}. \tag{I. 2. 7}$$

By Lemma 2, from $\{\psi_\varepsilon\}$ we can extract a subsequence $\{\psi_\eta\}$ such that

$$\varphi_\eta \to \Phi \text{ in } L^2(0, \ T; \ H^2(\Omega)) \text{ weakly},$$

$$\frac{\partial \varphi_\eta}{\partial t} \to \frac{\partial \Phi}{\partial t} \text{ in } L^2(Q) \text{ weakly},$$

$$A\varphi_\eta \to A\Phi \text{ in } L^2(Q) \text{ weakly},$$

$$A^2\varphi_\eta \to A^2\Phi \text{ in } L^2(0, \ T; \ H^{-2}(\Omega)) \text{ weakly}.$$

Thus, passing to the limit in $(II)_\eta$, we get

$$\begin{cases} -\dfrac{\partial \Phi}{\partial t} + A\Phi = 0 & \text{in } Q, \\ \Phi = 0 & \text{on } \Sigma, \\ \Phi(x, \ T) = 0 & \text{in } \Omega, \end{cases}$$

hence $\Phi = 0$, namely (I. 2. 7) holds.

Theorem 2 bis (hence Theorem 1 bis) is proved.

*Proof of corollary 1. 1:* Since $y_\varepsilon(t; \ v_\varepsilon)$ converges to $y \ (t; \ v)$ in $L^2(Q)$ as $\varepsilon \to 0$, it follows from $(I)_\varepsilon$ that

$$\frac{\partial y_\varepsilon}{\partial t}(t; \ v_\varepsilon) \to \frac{\partial y}{\partial t}(t; \ v), \text{ in } L^2(0, \ T; \ H^{-k}(\Omega)) \ (k > 0, \text{ suitable integer}),$$

hence          $y_\varepsilon(T; \ v_\varepsilon) \to y(T; \ v)$, in $H^{-s}(\Omega) \ (s > 0, \text{ suitable integer})$,

from this the corollary follows easily.

### § 3. Proof of Theorem 3

It is easy to verify the following lemmas.

**Lemma 6.** *For any $S(x) \in L^2(\Omega)$, the solution $p$ of (III) satisfies*

$$\|p(t; \ S)\|_{L^2(0, T; \ H^{\frac{1}{2}}(\Omega))} \leqslant C \|S\|_{L^2(\Omega)}. \tag{I. 3. 1}$$

**Lemma 7.** *For any $S_\varepsilon(x) \in L^2(\Omega)$, the solution $p_\varepsilon$ of (III)$_\varepsilon$ satisfies*

$$\|p_\varepsilon(t; \ S_\varepsilon)\|_{L^2(0, \ T; \ H^{\frac{1}{2}}(\Omega))} \leqslant C \|S_\varepsilon\|_{L^2(\Omega)}. \tag{I. 3. 2}$$

*Proof of Theorem 3* By lemma 7 we can extract from $\{S_\varepsilon\}$ a subsequence $\{S_\eta\}$ such that

$$p_\eta(t;\ S_\eta) \to P \text{ in } L^2(0,\ T;\ H_0^1(\Omega)) \text{ weakly}, \tag{I. 3. 3}$$

then

$$\begin{cases} Ap_\eta \to AP, \text{ in } L^2(0,\ T;\ H^{-1}(\Omega)) \text{ weakly}, \\ A^2p_\eta \to A^2P, \text{ in } L^2(0,\ T;\ H^{-3}(\Omega)) \text{ weakly}, \\ \dfrac{\partial p_\eta}{\partial t} \to \dfrac{\partial P}{\partial t}, \text{ in } L^2(0,\ T;\ H^{-3}(\Omega)) \text{ weakly}. \end{cases} \tag{I. 3. 4}$$

Passing to the limit in $(\mathrm{III})_\eta$ we get

$$\begin{cases} -\dfrac{\partial P}{\partial t} + AP = 0 & \text{in } Q, \\ P = 0 & \text{on } \Sigma, \\ P(x,\ T) = S(x) & \text{in } \Omega, \end{cases}$$

hence

$$P = p(t;\ S). \tag{I. 3. 5}$$

From this, Theorem 3 follows easily.

*Proof of Corollary* 3. 1 It is well known [cf. [2, 7, 9]) that for any $S(x) \in L^2(\Omega)$ we have

$$p(0,\ t;\ S) \in \mathscr{U}', \tag{I. 3. 6}$$

and

$$\int_0^T p(0,\ t;\ S)v(t)\,dt = \int_\Omega y(T;\ v)S(x)dx,\ \forall v \in \mathscr{U}, \tag{I. 3. 7}$$

where $\mathscr{U}'$ denotes the dual of $\mathscr{U}$ (when $L^2(0,\ T)$ is identified with its dual), $y(T;\ v)$ is given by (I) and in the left hand side of (I. 3. 7) the integral denotes the duality between $\mathscr{U}'$ and $\mathscr{U}$.

Besides, from $(\mathrm{I})_\varepsilon$ and $(\mathrm{III})_\varepsilon$ we can also get

$$\int_0^T p_\varepsilon(0,\ t;\ S_\varepsilon)v(t)dt = \int_\Omega y_\varepsilon(T;\ v)S_\varepsilon(x)dx,\ \forall v \in L^2(0,\ T). \tag{I. 3. 8}$$

In order to prove Corollary 3.1 we shall use the following Lemma 8 whose proof will be given at the end of this section.

**Lemma 8.** *For any $v(t) \in \mathscr{D}(0,\ T)$ fixed, we have*

$$y_\varepsilon(T;\ v) \to y(T;\ v) \text{ in } L^2(\Omega) \text{ strongly, as } \varepsilon \to 0, \tag{I. 3. 9}$$

*where $y_\varepsilon$ and $y$ are defined by $(\mathrm{I})_\varepsilon$ and (I) respectively.*

Suppose (if necessary, extract a subsequence)

$$p_\varepsilon(0,\ t;\ S_\varepsilon) \to d_0(t) \text{ in } L^2(0,\ T) \text{ weakly (resp. strongly)}, \tag{I. 3. 10}$$

for any $v(t) \in \mathscr{D}(0,\ T)$ fixed, passing to the limit in (I. 3. 8), by Lemma 8 it follows from (21) and (I. 3. 10) that

$$\int_0^T d_0(t)v(t)dt = \int_\Omega y(T;\ v)S(x)dx,\ \forall v \in \mathscr{D}(0,\ T), \tag{I. 3. 11}$$

hence, according to (I. 3. 7) we get

$$d_0(t) = p(0,\ t;\ S) \text{ in } \mathscr{D}'(0,\ T). \tag{I. 3. 12}$$

Noticing (cf. [2, 7, 8])

$$\mathscr{D}(0,\ T) \subset \mathscr{U} \subset L^2(0,\ T) \subset \mathscr{U}' \subset \mathscr{D}'(0,\ T); \tag{I. 3. 13}$$

(24) and (25) follow from (I. 3. 12). Corollary 3.1 is proved.

*Proof of Lemma* 8 It follows from Theorem 1 bis that

$$y_\varepsilon(t;\ v)\to y(t;\ v) \text{ in } L^2(Q) \text{ strongly, as } \varepsilon\to 0. \tag{I. 3. 14}$$

Since $v(t)\in\mathscr{D}(0,\ T)$, by regularity we have also

$$\frac{\partial y_\varepsilon}{\partial t}(t;\ v)\to\frac{\partial y}{\partial t}(t;\ v) \text{ in } L^2(Q) \text{ strongly, as } \varepsilon\to 0. \tag{I. 3. 15}$$

Hence (I. 3. 9) follows from (I. 3. 14) and (I. 3. 15).

### § 4. Proof of Theorem 3 bis

From Lemmas 6 and 7, we get

**Lemma 9.** *It is sufficient to prove Theorem 3 bis for $S_\varepsilon\equiv S$ independent of $\varepsilon$ and $S\in\mathscr{D}(\Omega)$.*

Hence, it remains only to prove

**Lemma 10.** *Let $p_\varepsilon$ be the solution of*

$$\begin{cases} -\dfrac{\partial p_\varepsilon}{\partial t}+\varepsilon A^2 p_\varepsilon+Ap_\varepsilon=0, & \text{in } Q,\\ p_\varepsilon=Ap_\varepsilon=0, & \text{on } \Sigma,\\ p_\varepsilon(x,\ T)=S(x), & \text{in } \Omega \end{cases} \tag{I. 4. 1}$$

*in which $S(x)\in\mathscr{D}(\Omega)$, then as $\varepsilon\to 0$, we have*

$$p_\varepsilon\to p(t;\ S) \text{ in } L^2(0,\ T;\ H_0^1(\Omega)) \text{ strongly.} \tag{I. 4. 2}$$

*Proof* From Lemma 7 we have

$$\|p_\varepsilon\|_{L^2(0,T;H_0^1(\Omega))}\leqslant C. \tag{I. 4. 3}$$

Since $Ap_\varepsilon$ is also the solution of (I. 4. 1) in which $S$ is seplaced by $AS$, we have

$$\|Ap_\varepsilon\|_{L^2(0,T;H_0^1(\Omega))}\leqslant C. \tag{I. 4. 4}$$

Then, according to the classical theorem of regularity of solutions for linear elliptic equations, we get

$$\|p_\varepsilon\|_{L^2(0,T;H^3(\Omega))}\leqslant C. \tag{I. 4. 5}$$

By regularity, we have also

$$\left\|\frac{\partial p_\varepsilon}{\partial t}\right\|_{L^2(0,T;H^1(\Omega))}\leqslant C. \tag{I. 4. 6}$$

Hence, it follows from (I. 4. 5) and (I. 4. 6) that $p_\varepsilon$ belongs to a strongly compact subset of $L^2(0,\ T;\ H_0^1(\Omega))$, then Lemma 10 holds.

Theorem 3 bis is proved.

## II. Applications to the optimal control

### § 1. Proof of Theorem 4

1. Since

$$\begin{aligned} 0\leqslant J_\varepsilon(u_\varepsilon)&=N\int_0^T u_\varepsilon^2\,dt+\int_\Omega |y_\varepsilon(T;\ u_\varepsilon)-Z_d|^2\,dx\\ &=\inf_{v\in L^2(0,T)} J_\varepsilon(v)\leqslant J_\varepsilon(0)=C_0, \end{aligned} \tag{II. 1. 1}$$

where

$$C_0=\int_\Omega Z_d^2 dx. \tag{II. 1. 2}$$

we have

$$\|u_s(t)\|_{L^2(0,T)}\leqslant C, \tag{II. 1. 3}$$

$$\|y_s(T;\ u_s)\|_{L^2(\Omega)}\leqslant C. \tag{II. 1. 4}$$

Hence, we can extract from $\{u_s\}$ a subsequence $\{u_\eta\}$ such that, as $\eta\to 0$,

$$u_\eta(t)\to w_0 \text{ in } L^2(0,\ T) \text{ weakly}, \tag{II. 1. 5}$$

$$y_\eta(T;\ u_\eta)\to Y \text{ in } L^2(\Omega) \text{ weakly}. \tag{II. 1. 6}$$

By Corollary 1.1 we get

$$y_\eta(T;\ u_\eta)\to y(T;\ w_0) \text{ in } L^2(\Omega) \text{ weakly}, \tag{II. 1. 7}$$

and

$$w_0\in \mathscr{U}. \tag{II. 1. 8}$$

According to the inferior semi-continuity for the weak topology, we can pass to the limit in (II. 1. 1) and obtain that

$$\varliminf_{\eta\to 0} J_\eta(u_\eta)\geqslant N\int_0^T w_0^2 dt+\int_\Omega |y(T;\ w_0)-Z_d|^2 dx=J(w_0)\geqslant J(u_0), \tag{II. 1. 9}$$

hence

$$\varliminf_{s\to 0} J_s(u_s)\geqslant J(u_0). \tag{II. 1. 10}$$

2. By means of the duality, we are going to prove

$$\varlimsup_{s\to 0} J_s(u_s)\leqslant J(u_0). \tag{II. 1. 11}$$

Set

$$F(v)=N\int_0^T v^2 dt,\ \forall v\in \mathscr{U}, \tag{II. 1. 12}$$

$$G(q)=\int_\Omega |q-Z_d|^2 dx,\ \forall q\in L^2(\Omega), \tag{II. 1. 13}$$

we introduce the corresponding conjugate functionals

$$F^*(w)=\begin{cases}\dfrac{1}{4N}\displaystyle\int_0^T w^2 dt, & \text{if } w\in L^2(0,\ T),\\ +\infty, & \text{if } w\in \mathscr{U}'\backslash L^2(0,\ T),\end{cases} \tag{II. 1. 14}$$

$$G^*(S)=\int_\Omega\left(\frac{S^2}{4}+SZ_d\right)dx,\ \forall S\in L^2(\Omega), \tag{II. 1. 15}$$

where $\mathscr{U}'$ is the dual of $\mathscr{U}$.

Let $L$ be the continuous linear mapping from $\mathscr{U}$ to $L^2(\Omega)$, defined by

$$L:\ v\to y(T;\ v), \tag{II. 1. 16}$$

where $y(T;\ v)$ is given by problem (I). The corresponding adjoint mapping $L^*$ can be defined by

$$L^*:\ S\to p(0,\ t;\ S), \tag{II. 1. 17}$$

where $p$ is the solution of problem (III). It is well known that $L^*$ is a continuous linear mapping from $L^2(\Omega)$ to $\mathscr{U}'$ (cf. [2, 7, 9]).

According to Rockafellar's duality Theorem (cf. [6]), we have

$$J(u_0)=\inf_{v\in\mathscr{U}} J(v)=\inf_{v\in\mathscr{U}}(F(v)+G(Lv))$$
$$=-\inf_{S\in L^2(\Omega)}(F^*(L^*S)+G^*(-S))=-\inf_{S\in\mathscr{U}^*} M(S)+C_0, \tag{II. 1. 18}$$

where

$$M(S)=\frac{1}{4N}\int_0^T p(0,\ t;\ S)^2\,dt+\int_\Omega\left(\frac{S}{2}-Z_d\right)^2 dx, \tag{II. 1. 19}$$

$\mathscr{U}^*$ is defined by (26) and $C_0$ is given by (I. 1. 2).

Obviously, there exists a unique element $q_0\in\mathscr{U}^*$ such that

$$M(q_0)=\inf_{S\in\mathscr{U}^*} M(S). \tag{II. 1. 20}$$

Hence

$$J(u_0)=-M(q_0)+C_0. \tag{II. 1. 21}$$

In a similar way, for any $s>0$ fixed, set

$$\tilde{F}(v)=N\int_0^T v^2\,dt,\ \forall v\in L^2(0,\ T) \tag{II. 1. 22}$$

and

$$\tilde{F}^*(w)=\frac{1}{4N}\int_0^T w^2\,dt,\ \ \forall w\in L^2(0,\ T), \tag{II. 1. 23}$$

according to Rockafellar's duality Theorem we have

$$J_s(u_s)=\inf_{v\in L^2(0,T)} J_s(v)=\inf_{v\in L^2(0,T)}(\tilde{F}(v)+G(L_s v))$$
$$=-\inf_{S\in L^2(\Omega)}(\tilde{F}^*(L_s^*S)+G^*(-S))=-\inf_{S\in L^2(\Omega)} M_s(S)+C_0, \tag{II. 1. 24}$$

where $L_s$ and $L_s^*$ are defined by (6) and (20) respectively

$$M_s(S)=\frac{1}{4N}\int_0^T p_s(0,\ t;\ S)^2\,dt+\int_\Omega\left(\frac{S}{2}-Z_d\right)^2 dx \tag{II. 1. 25}$$

and $C_0$ is given by (II. 1. 2). Besides, in (II. 1. 25), $p_s(0,\ t;\ S)$ is given by problem $(\mathrm{III})_s$ with $S_s\equiv S$.

There exists a unique element $q_s\in L^2(\Omega)$ such that

$$M_s(q_s)=\inf_{S\in L^2(\Omega)} M_s(S), \tag{II. 1. 26}$$

hence

$$J_s(u_s)=-M_s(q_s)+C_0. \tag{II. 1. 27}$$

Thus, in order to get (II. 1. 11) it is sufficient to prove

$$\varliminf_{s\to 0} M_s(q_s)\geqslant M(q_0). \tag{II. 1. 28}$$

3. Proof of (II.1.28)

Since

$$0\leqslant M_s(q_s)=\frac{1}{4N}\int_0^T p_s(0,\ t;\ q_s)^2\,dt+\int_\Omega\left(\frac{q_s}{2}-Z_d\right)^2 dx$$
$$=\inf_{S\in L^2(\Omega)} M_s(S)\leqslant M_s(0)=C_0, \tag{II. 1. 29}$$

we have

$$\|q_s\|_{L^2(\Omega)}\leqslant C, \tag{II. 1. 30}$$
$$\|p_s(0,\ t;\ q_s)\|_{L^2(0,T)}\leqslant C. \tag{II. 1. 31}$$

Hence, we can extract from $\{q_s\}$ a subsequence $\{q_\eta\}$ such that as $\eta\to 0$,

$$q_\eta\to S_0 \text{ in } L^2(\Omega) \text{ weakly}, \tag{II. 1. 32}$$

$$p_\eta(0,\ t;\ q_\eta)\rightarrow d_0 \text{ in } L^2(0,\ T) \text{ weakly.} \tag{II. 1. 33}$$

By Corollary 3.1, we have

$$d_0=p(0,\ t;\ S_0) \tag{II. 1. 34}$$

and

$$S_0\in \mathscr{U}^*, \tag{II. 1. 35}$$

hence

$$p_\eta(0,\ t;\ q_\eta)\rightarrow p(0,\ t;\ S_0) \text{ in } L^2(0,\ T) \text{ weakly.} \tag{II. 1. 36}$$

According to the inferior semi-continuity for the weak topology, we can pass to the limit in (II. 1. 29) and obtain that

$$\varliminf_{\eta\to 0} M_\eta(q_\eta)\geqslant\frac{1}{4N}\int_0^T p(0,\ t;\ S_0)^2dt+\int_\Omega\left(\frac{S_0}{2}-Z_d\right)^2dx$$
$$=M(S_0)\geqslant M(q_0), \tag{II. 1. 37}$$

hence (II.1.28) follows immediately.

4. From (II. 1. 5)、(II. 1. 9)—(II. 1. 11) we obtain (33) and

$$u_\varepsilon\rightarrow u_0 \text{ in } L^2(0,\ T) \text{ weakly,} \tag{II. 1. 38}$$

hence it follows from Corollary 1.1 and (II. 1. 4) that

$$y_\varepsilon(T)\rightarrow y(T) \text{ in } L^2(\Omega) \text{ weakly.} \tag{II. 1. 39}$$

Thus, according to the inferior semi-continuity for the weak topology, it follows from (33) that

$$\|u_\varepsilon\|^2_{L^2(0,T)}\rightarrow\|u_0\|^2_{L^2(0,T)}, \tag{II. 1. 40}$$

$$\|y_\varepsilon(T)\|^2_{L^2(\Omega)}\rightarrow\|y(T)\|^2_{L^2(\Omega)}, \tag{II. 1. 41}$$

hence (34) and (36) hold. Moreover, from Theorem 1 bis and (29)、(32) we can get (35) and (38). Finally, (37) follows from Theorem 3 bis.

## § 2. Proof of Theorem 6

The proof is similar to the proof of Theorem 4, but we must set

$$F(v)=N\int_0^T v^2\,dt,\ \forall v\in L^2(0,\ T), \tag{II. 2. 1}$$

$$G(q)=\int_Q |q-Z_d|^2\,dx\,dt,\ \forall q\in L^2(Q), \tag{II. 2. 2}$$

$$F^*(w)=\frac{1}{4N}\int_0^T w^2\,dt,\ \forall w\in L^2(0,\ T), \tag{II. 2. 3}$$

$$G^*(\psi)=\int_Q\left(\frac{\psi^2}{4}+\psi Z_d\right)dx\,dt,\ \forall\psi\in L^2(Q) \tag{II. 2. 4}$$

and

$$L:\ \ v\rightarrow y(t;\ v), \tag{II. 2. 5}$$

$$L^*:\ \ \psi\rightarrow\varphi(0,\ t;\ \psi), \tag{II. 2. 6}$$

$$L_\varepsilon:\ \ v\rightarrow y_\varepsilon(t;\ v), \tag{II. 2. 7}$$

$$L^*_\varepsilon:\ \ \psi\rightarrow\varphi_\varepsilon(0,\ t;\ \psi). \tag{II. 2. 8}$$

Moreover, instead of using Theorems 3 and 3 bis, we must use Theorems 2 and 2 bis. The detail is omitted here.

# III. Remarks

The previous results remain still valid if the boundary condition on $\Sigma$ is changed by any one of the following boundary conditions:

For the original problem:

1° $$\frac{\partial y}{\partial \nu_A}+\lambda(x)y=0; \tag{III. 1}$$

or

2° $y|_\Gamma=c(t)$ (unknown function of t) and

$$\int_\Gamma\left(\frac{\partial y}{\partial \nu_A}+\lambda(x)y\right)dS=0 \text{ for a. e. } t\in(0,\ T); \tag{III. 2}$$

or

3° $\left.\frac{\partial y}{\partial \nu_A}\right|_\Gamma=c(t)$ (unknown function of $t$) and

$$\int_\Gamma\left(y+\lambda(x)\frac{\partial y}{\partial \nu_A}\right)dS=0 \text{ for a. e. } t\in(0,\ T). \tag{III. 3}$$

For the approximate problem:

1° $$\frac{\partial y_s}{\partial \nu_A}+\lambda(x)y_s=\frac{\partial Ay_s}{\partial \nu_A}+\lambda(x)Ay_s=0; \tag{III. 1$'$}$$

or

2° $y_s|_\Gamma=c_s(t)$, $Ay_s|_\Gamma=d_s(t)$ (unknown functions of $t$) and

$$\int_\Gamma\left(\frac{\partial y_s}{\partial \nu_A}+\lambda(x)y_s\right)dS=\int_\Gamma\left(\frac{\partial Ay_s}{\partial \nu_A}+\lambda(x)Ay_s\right)dS=0 \tag{III. 2$'$}$$

for a. e. $t\in(0,\ T)$;

or

3° $\left.\frac{\partial y_s}{\partial \nu_A}\right|_\Gamma=c_s(t)$, $\left.\frac{\partial Ay_s}{\partial \nu_A}\right|_\Gamma=d_s(t)$ (unknown functions of $t$) and

$$\int_\Gamma\left(y_s+\lambda(x)\frac{\partial y_s}{\partial \nu_A}\right)dS=\int_\Gamma\left(Ay_s+\lambda(x)\frac{\partial Ay_s}{\partial \nu_A}\right)dS=0 \tag{III. 3$'$}$$

for a. e. $t\in(0,\ T)$.

Where $\lambda(x)\geqslant 0$ is a smooth function on $\Gamma$

$$\frac{\partial y}{\partial \nu_A}=\sum_{i,j=1}^{n}a_{ij}\nu_i\frac{\partial y}{\partial x_j}, \tag{III. 4}$$

in which $\nu=(\nu_1,\ \cdots,\ \nu_n)$ is the unit normal oriented towards the exterior of $\Omega$.

# Acknowledgement

The author would like to thank Professor Lions, J. L. for his kind suggestion and very helpful discussions.

## References

[1] Lions, J. L. & Magenes, E., Problémes aux limites non homogènes et applications, volume 1, Dunod, Paris, 1968.

[2] Lions, J. L., Nouveaux espaces et ensembles fonctionnels intervenant en contrôle optimal, Cours au Collège de France, 1979; *C. R. A. S.*, **289**, *série* **A**, (1979), 315.

[3] Lions, J. L. & Magenes, E., Problémes aux limites non homogènes et applications, volume 2, Dunod, Paris, 1968.

[4] Damlamian, A. & Li Ta-tsien, Comportement limite des solutions de certains problèmes mixtes pour des équations paraboliques, *C. R. A. S.*, **290**, *série* **A**, (1980), 957.

[5] Lions, J. L., Contrôle optimal de systèmes gouvernés par des èquations aux dérivées partielles, Dunod, Gauthier-Villars, Paris, 1968.

[6] Ekeland, I. & Temam, R., Convex analysis and variational problems, North-Holland Publishing Company, 1976.

[7] Lions, J. L., Function spaces and optimal control of distributed systems, *Lectures Notes, U. F. R. J.*, Rio de Janeiro, 1980.

[8] Li Ta-tsien, Propriétés d'espaces fonctionnels intervenant en contrôle optimal, *C. R. A. S.*, **289**, *série* **A**, (1979), 687.

[9] Lions, J. L., Some methods in the mathematical analysis of systems and their control, *Science Press*, Peking, 1981.

[10] Li Ta-tsien, Comportements limites des solutions de certaines éqnations paraboliques d'ordre supérieur et leurs applications au contrôle optimal, *C. R. A. S.*, **293**, *série* **A**, (1981), 205.

*Chin. Ann. of Math.*
3 (4) 1982

# A NEW ATTEMPT ON GOLDBACH CONJECTURE

PAN CHENGDONG
(*Shandong University*)

Dedicated to Professor Su Bu-chinY on the Occasion of his 80th Birthday and his 50th Year of Educational Work

Let $N$ be a large even integer and $D(N)$ denote the number of the ways of representing $N$ as a sum of two primes, that is

$$D(N)=\sum_{N=p_1+p_2} 1. \tag{1}$$

By cycle method, we can derive that

$$D(N)=\mathfrak{S}(N)\frac{N}{\log^2 N}+R, \tag{2}$$

where

$$\mathfrak{S}(N)=2\prod_{p>2}\left(1-\frac{1}{(p-1)^2}\right)\prod_{p|N,p>2}\left(1+\frac{1}{p-2}\right)$$

$$R=\left(\sum_{q>Q}\frac{\mu^2(q)}{\phi^2(q)}C_q(-N)\right)\frac{N}{\log^2 N}+\int_E S^2(\alpha, N)e^{-2\pi i\alpha N}d\alpha \tag{3}$$

$$S(\alpha, N)=\sum_{p<N}e^{2\pi i\alpha p},\quad C_q(-N)=\sum_{h=1}^{q}e^{\frac{-2\pi iNh}{q}},$$

$Q=\log^{16}N$ and the $E$ denotes the supplement interval as usual. This suggests us to conjecture that the main term of $D(N)$ is $\mathfrak{S}(N)\dfrac{N}{\log^2 N}$, that is

$$D(N)\sim\mathfrak{S}(N)\frac{N}{\log^2 N}. \tag{4}$$

It is well known that the difficulty in proving this conjecture is to deal with the integral in the remainder term $R$. So far as we know, up to now, the cycle method might be the unique approach* which suggests us to conjecture (4) is true. In this paper, we shall give another method which also suggests us to conjecture (4) is true . It seems to be more direct and elementary than the cycle method.

For convenience, we consider

$$\hat{D}(N)=\sum_{N=d+d'}\Lambda(d)\Lambda(d')=\sum_{d<N}\Lambda(d)\Lambda(N-d)$$

in place of $D(N)$. It is easy to see that

$$D(N)=\frac{\hat{D}(N)}{\log^2 N}\left[1+O\left(\frac{\log\log N}{\log N}\right)\right]+O\left(\frac{N}{\log^3 N}\right).$$

Now We shall prove the following theorems:

Manuscript received February. 2, 1982.

* Recently Prof. Hua, L. K. has proposed a different new method on this line, however not yet published.

**Theorem I.** *Let N be a large even integer. Then for*

$$Q=\sqrt{N}\log^{-20}N,$$

*we have*

$$\hat{D}(N)=\mathfrak{S}(N)N+\hat{R}, \tag{5}$$

*where* $\mathfrak{S}(N)$ *is defined by* (3)

$$\hat{R}=R_1+R_2+R_3+O(N\log^{-1}N), \tag{6}$$

*and*

$$R_1=\sum_{n<N}\Big(\sum_{\substack{d_1\mid n\\ d_1<Q}}a(d_1)\Big)\Big(\sum_{\substack{d_2\mid N-n\\ (d_2,N)=1\\ d_2>Q}}a(d_2)\Big),$$

$$R_2=\sum_{n<N}\Big(\sum_{\substack{d_1\mid n\\ d_1>Q}}a(d_1)\Big)\Big(\sum_{\substack{d_2\mid N-n\\ (d_2,N)=1\\ d_2<Q}}a(d_2)\Big)$$

$$R_3=\sum_{n<N}\Big(\sum_{\substack{d_1\mid n\\ d_1>Q}}a(d_1)\Big)\Big(\sum_{\substack{d_2\mid N-n\\ (d_2,N)=1\\ d_2>Q}}a(d_2)\Big),$$

$$a(m)=\mu(m)\log m.$$

**Theorem II.** *By means of Bombieri Theorem, we have*

$$R_1=R_2=O(N\log^{-1}N). \tag{7}$$

First of all, we prove some lemmas as follows:

**Lemma 1.** *Let m be a positive integer, and* $m\leqslant N^{c_1}$. *Then for*

$$\sigma\geqslant 1-\frac{c_2}{\sqrt{\log N}}\geqslant 1/2,$$

*we have*

$$\prod_{p\mid m}\Big(1-\frac{1}{p^s}\Big)^{-1}\ll\log^{c_3}N. \tag{8}$$

*Proof* Put $T=e^{\sqrt{\log N}}$, we have

$$\Big|\prod_{p\mid m}\Big(1-\frac{1}{p^s}\Big)^{-1}\Big|\leqslant\prod_{p\mid m}\Big(1-\frac{1}{p^\sigma}\Big)^{-1}=\prod_{p\mid m}\Big(1+\frac{1}{p^\sigma-1}\Big)$$

and

$$\log\prod_{p\mid m}\Big(1+\frac{1}{p^\sigma-1}\Big)\leqslant\sum_{p\mid m}\frac{1}{p^\sigma-1}\ll\sum_{p\mid m}\frac{1}{p^\sigma}$$

$$=\sum_{\substack{p\mid m\\ p<T}}\frac{1}{p^\sigma}+\sum_{\substack{p\mid m\\ p>T}}\frac{1}{p^\sigma}=\Sigma_1+\Sigma_2.$$

Furthermore, we have

$$\Sigma_1\ll\log\log N$$

and

$$\Sigma_2\ll T^{-1/2}\log N\ll 1.$$

Since $\sigma\geqslant 1/2$, summing the above up, the Lemma is proved.

**Lemma 2.** *Let m be a positive integer* $m\leqslant N^{c_1}$. *Then we have*

$$\sum_{\substack{d<N\\ (d,m)=1}}\frac{\mu(d)}{d}\ll e^{-c_3\sqrt{\log N}} \tag{9}$$

*and*

$$\sum_{\substack{d<N\\ (d,m)=1}}\frac{\mu(d)}{d}\log d=-\frac{m}{\phi(m)}+O(e^{-c_4\sqrt{\log N}}). \tag{10}$$

*Proof* Put $X=N+1/2$ and

$$F(s)=\prod_{p|m}\left(1-\frac{1}{p^s}\right)\zeta(s).$$

Then
$$\sum_{\substack{d<N\\(d,m)=1}}\frac{\mu(d)}{d}=\frac{1}{2\pi i}\int_{b-iT}^{b+iT}\frac{1}{F(1+w)}\frac{X^w}{w}dw+O\left(\frac{\log N}{T}\right),$$

where
$$b=\frac{1}{\log X},\quad T=e^{\sqrt{\log X}}.$$

Moving the line of the integral to $[c-iT,\ c+iT]$, $c=-\frac{c_5}{\sqrt{\log X}}$, and using Lemma 1, we have

$$\sum_{\substack{d<N\\(d,m)=1}}\frac{\mu(d)}{d}\ll e^{-c_6\sqrt{\log N}}.$$

It is easy to prove by the method of Abel Summation that

$$\sum_{\substack{d=1\\(d,m)=1}}^{\infty}\frac{\mu(d)}{d}\log d=\sum_{\substack{d<X\\(d,m)=1}}\frac{\mu(d)}{d}\log d+O(e^{-c_6\sqrt{\log X}}) \tag{11}$$

and
$$\sum_{\substack{d=1\\(d,m)=1}}^{\infty}\frac{\mu(d)}{d}\log d=\lim_{\sigma\to1+}\sum_{\substack{d=1\\(d,m)=1}}^{\infty}\frac{\mu(d)}{d^\sigma}\log d=-\left(\frac{1}{F(s)}\right)'_{s=1}. \tag{12}$$

From (11), (12) and

$$\left(\frac{1}{F(s)}\right)'_{s=1}=\prod_{p|m}\left(1-\frac{1}{p}\right)^{-1}=\frac{m}{\phi(m)}, \tag{13}$$

(10) is derived at once.

**Lemma 3.** *We have*

$$\sum_{\substack{n<N\\(n,m)=1}}\frac{\mu(n)\log n}{\phi(n)}=-\mathfrak{S}(m)+O(e^{-c_6\sqrt{\log N}}).$$

*Proof* We have

$$\sum_{\substack{d<N\\(d,m)=1}}\frac{\mu(d)\log d}{\phi(d)}=\sum_{\substack{d<N\\(d,m)=1}}\frac{\mu(d)\log d}{d}\sum_{t|d}\frac{\mu^2(t)}{\phi(t)}=\sum_{\substack{t<N\\(t,m)=1}}\frac{\mu^2(t)}{\phi(t)}\sum_{\substack{d<N\\(d,m)=1\\t|d}}\frac{\mu(d)\log d}{d}$$

$$=\sum_{\substack{t<N\\(t,m)=1}}\frac{\mu^2(t)}{\phi(t)}\sum_{\substack{v<N/t\\(v,m)=1}}\frac{\mu(vt)\log vt}{vt}$$

$$=\sum_{\substack{t<N\\(t,m)=1}}\frac{\mu^2(t)}{\phi(t)}\frac{\mu(t)}{t}\sum_{\substack{v<N/t\\(v,mt)=1}}\frac{\mu(v)}{v}(\log v+\log t)$$

$$=\sum_{\substack{t<N\\(t,m)=1}}\frac{\mu(t)}{t\phi(t)}\sum_{\substack{v<N/t\\(v,mt)=1}}\frac{\mu(v)}{v}\log v+\sum_{\substack{t<N\\(t,m)=1}}\frac{\mu(t)\log t}{t\phi(t)}\sum_{\substack{v<N/t\\(v,mt)=1}}\frac{\mu(v)}{v}$$

$$=\Sigma_1+\Sigma_2.$$

By (10)

$$\Sigma_1=\sum_{\substack{t<\sqrt{N}\\(t,m)=1}}\frac{\mu(t)}{t\phi(t)}\sum_{\substack{v<N/t\\(v,mt)=1}}\frac{\mu(v)}{v}\log v+\sum_{\sqrt{N}<t<N}=-\sum_{\substack{t<\sqrt{N}\\(t,m)=1}}\frac{\mu(t)}{t\phi(t)}\frac{rt}{\phi(rt)}+O(e^{-c_7\sqrt{\log N}})$$

$$=-\frac{m}{\phi(m)}\sum_{\substack{t=1\\(t,m)=1}}^{\infty}\frac{\mu(t)}{\phi^2(t)}+O(e^{-c_8\sqrt{\log N}})=-\mathfrak{S}(m)+O(e^{-c_8\sqrt{\log N}}).$$

Similarly, by (9)

$$\Sigma_2=\sum_{\substack{t<\sqrt{N}\\(t,m)=1}}\frac{\mu(t)\log t}{t\phi(t)}\sum_{\substack{v<N/t\\(v,mt)=1}}\frac{\mu(v)}{v}+\sum_{\substack{\sqrt{N}<t<N\\(t,m)=1}}\sum_{\substack{v<N/t\\(v,mt)=1}}\frac{\mu(v)}{v}=O(e^{-c_3\sqrt{\log N}}).$$

Summing the above up, the Lemma is proved.

*The proof of Theorem I.* we have

$$\begin{aligned}\hat{D}(N)&=-\sum_{n<N}\Lambda(n)\sum_{d|N-n}a(d)\\&=-\sum_{n<N}\Lambda(n)\sum_{\substack{d|N-n\\(d,N)=1}}a(d)-\sum_{n<N}\Lambda(n)\sum_{\substack{d|N-n\\(d,N)>1}}a(d)=I_1+I_2.\end{aligned}\tag{14}$$

It is evident that

$$I_2=O(N^{\frac{2}{3}})\tag{15}$$

and

$$\begin{aligned}I_1&=\sum_{n<N}\sum_{d_1|n}a(d_1)\sum_{\substack{d_2|N-n\\(d_2,N)=1}}a(d_2)\\&=\sum_{n<N}\Bigg(\sum_{\substack{d_1|n\\d_1<Q}}a(d_1)+\sum_{\substack{d_1|n\\d_1>Q}}a(d_1)\Bigg)\Bigg(\sum_{\substack{d_2|N-n\\(d_2,N)=1\\d_2<Q}}a(d_2)+\sum_{\substack{d_2|N-n\\(d_2,N)=1\\d_2>Q}}a(d_2)\Bigg)\\&=\Sigma_1+R_1+R_2+R_3,\end{aligned}\tag{16}$$

where

$$\Sigma_1=\sum_{n<N}\sum_{\substack{d_1|n\\d_1<Q}}a(d_1)\sum_{\substack{d_2|N-n\\d_2<Q\\(d_2,N)=1}}a(d_2).$$

It is easily seen that

$$\begin{aligned}\Sigma_1&=\sum_{n<N}\sum_{\substack{d_1|n\\d_1<Q}}a(d_1)\sum_{\substack{d_2|N-n\\d_2<Q\\(d_2,N)=1}}a(d_2)=\sum_{\substack{d_2<Q\\(d_2,N)=1}}a(d_2)\sum_{\substack{d_1<Q\\(d_1,d_2)=1}}a(d_1)\sum_{\substack{d_1n<N\\d_1n\equiv N(d_2)}}1\\&=N\sum_{\substack{d_2<Q\\(d_2,N)=1}}\frac{a(d_2)}{d_2}\sum_{\substack{d_1<Q\\(d_1,d_2)=1}}\frac{a(d_1)}{d_1}+O(Q^2\log^2 N).\end{aligned}\tag{17}$$

From (17) and Lemma 2 and Lemma 3, we have

$$\begin{aligned}\Sigma_1&=N\sum_{\substack{d_2<Q\\(d_2,N)=1}}\frac{\mu(d_2)\log d_2}{d_2}\sum_{\substack{d_1<Q\\(d_1,d_2)=1}}\frac{\mu(d_1)\log d_1}{d_1}+O(N\log^{-1}N)\\&=-N\sum_{\substack{d_2<Q\\(d_2,N)=1}}\frac{\mu(d_2)\log d_2}{\phi(d_2)}+O(N\log^{-1}N)\\&=\mathfrak{S}(N)N+O(N\log^{-1}N).\end{aligned}$$

The Theorem is completed.

Now we are going to prove Theorem II.

*The proof of Theorem II.*

$$\begin{aligned}R_1&=\sum_{n<N}\sum_{\substack{d_1|n\\d_1<Q}}\mu(d_1)\log d_1\sum_{\substack{d_2|N-n\\(d_2,N)=1\\d_2>Q}}\mu(d_2)\log d_2\\&=\sum_{d_1<Q}\mu(d_1)\log d_1\Bigg(\sum_{\substack{n<N\\n\equiv o(d_1)}}\sum_{\substack{d_2|N-n\\(d_2,N)=1\\d_2>Q}}\mu(d_2)\log d_2\Bigg)\\&=\sum_{d_1<Q}\mu(d_1)\log d_1\Bigg(\sum_{\substack{n<N\\n_1\equiv o(d)}}\Lambda(N-n)-\sum_{\substack{n<N\\n\equiv o(d_1)}}\sum_{\substack{d_2|N-n\\(d_2,N)=1\\d_2<Q}}\mu(d_2)\log d_2\Bigg)+O(N\log^{-1}N)\end{aligned}$$

$$= \sum_{d_1<Q} \mu(d_1)\log d_1 \left( \sum_{\substack{n<N \\ n\equiv N(d_1)}} \Lambda(n) - \frac{N}{d_1} \sum_{\substack{d_2<Q \\ (d_2,d_1)=1}} \frac{\mu(d_2)\log d_2}{d_2} \right) + O(N\log^{-1}N)$$

$$= \sum_{d_1<Q} \mu(d_1)\log d_1 \left( \sum_{\substack{u<N \\ n\equiv N(d_1)}} \Lambda(n) - \frac{N}{\phi(d_1)} \right) + O(N\log^{-1}N)$$

$$= \sum_{\substack{d_1<Q \\ (d_1,N)=1}} \mu(d_1)\log d_1 \left( \sum_{\substack{n<N \\ n\equiv N(d_1)}} \Lambda(n) - \frac{N}{\phi(d_1)} \right) + O(N\log^{-1}N)$$

$$= O(N\log^{-1}N).$$

Similarly, We have

$$R_2 = O(N\log^{-1}N).$$

The Theorem is completed.

*Chin. Ann. of Math.*
**3** (4) 1982

# SOME NONEXISTENCE THEOREMS ON STABLE HARMONIC MAPPINGS

PAN YANGLIAN
*(Institute of Mathematics, Fudan University)*

Dedicated to Professor Su Bu-chin on the Occasion of his 80th Birthday and his 50th Year of Educational Work

As well known, a harmonic mapping is a critical point of the energy integral Therefore, from the viewpoint of variational calculus, it is natural to study the stable harmonic mapping, i.e. the harmonic mapping with non-negative second variation.

Xin, Y. L.[1] proved the nonexistence of nonconstant stable harmonic mapping from Euclidean sphere $S^n$ $(n>2)$ to any Riemannian manifold, and generalized the result to the case of $S^n\times S^m$ $(n>2,\ m>2)$. All these theorems generalized the results of Eells, J., Sampson J. H. and Smith, R. T.

Noting the fact that $S^n$ is a 1-codimensional submanifold in Euclidean space $E^{n+1}$, in this paper we study the stable harmonic mapping from a compact submanifold with codimension $p$ in Euclidean space $E^{n+p}$ to any Riemannian manifold. We obtain several nonexistence theorems which generalize the results in[1].

**1. Basic formulas** Let $M$ and $N$ be Riemannian manifolds with dimensions $n$ and $m$ respectively. $M$ is compact without boundary. And $\nabla$, $\nabla'$ represent the Riemannian connections of $M$, $N$ respectively.

Suppose that $\phi$: $M\to N$ is a differentiable mapping, with $\phi_*:TM\to TN$ as its induced mapping, where $TM$, $TN$ are the tangent bundles of $M$, $N$ respectively. As known to all, the induced bundle $E=\phi^{-1}\,TN\to M$ possesses the induced Riemannian connection as follows

$$\tilde{\nabla}_X S=\nabla'_{\phi_* X}S, \tag{1.1}$$

where $X\in TM$, $S\in\Gamma(E)$.

$E(\phi)$, the energy of the mapping $\phi$, is defined by the formula

$$E(\phi)=\int_M e(\phi)*1=\frac{1}{2}\int_M\langle\phi_*e_i,\ \phi_*e_i\rangle_N*1, \tag{1.2}$$

where $e(\phi)$ is the energy density, $\{e_i\}$ $(i,\ j,\ k,\ \cdots,\ =1,\ \cdots,\ n)$ an orthonormal basis in $M$, $\langle,\rangle_N$ the Riemannian metric in $N$ and $*1$ the volume form of $M$. We follow

Manuscript received May 6, 1981.

the summation convention throughout this paper.

Set

$$\tau(\phi)=\tilde{\nabla}_{e_i}\phi_* e_i-\phi_*\nabla_{e_i}e_i, \tag{1.3}$$

$\phi$ is called a harmonic mapping if $\tau(\phi)=0$.

For any vector field $V$ along $\phi$, we have a 1-parametric family of mappings $\phi_t$: $M\to N$, $\phi_t(p)=\exp_{\phi(p)}(tV(p))$. The first variational formula for the correponding energy functional $E(\phi)$ is

$$\frac{d}{dt}E(\phi_t)\,|_{t=0}=-\int_M\langle V,\ \tau(\phi)\rangle_N*1. \tag{1.4}$$

Hence, the harmonic mapping is the critical point of the energy functional.

As known to all, the second variation of the energy functional is given as follows[3]

$$\frac{d^2}{dt^2}E(\phi_t)\,|_{t=0}=-\int_M\langle V,\ \tilde{\nabla}*\tilde{\nabla}V+R^N(\phi_*e_i,\ V)\phi_*e_i\rangle_N*1, \tag{1.5}$$

where $\tilde{\nabla}*\tilde{\nabla}$ is the trace Laplacian with respect to $\tilde{\nabla}$, and $R^N$ is the curvature operator of $N$.

Therefore, the index form for harmonic mapping is

$$I(V,\ W)=\int_M\langle-\tilde{\nabla}*\tilde{\nabla}V-R^N(\phi_*e_i,\ V)\phi_*e_i,\ W\rangle_N*1. \tag{1.6}$$

The harmonic mapping $\phi$ is called stable, if $I(V,\ V)\geqslant 0$ for any vector field $V$ along $\phi$.

Consider $\phi_*$ as $\phi^{-1}TN$ valued 1-form $d\phi$ i.e.

$$d\phi(X)=\phi_*X. \tag{1.7}$$

When mapping $\phi$ is harmonic, it follows from Weitzenböck formula that

$$-\tilde{\nabla}*\tilde{\nabla}\,d\phi+S=0, \tag{1.8}$$

where

$$S(v)=-R^N(\phi_*e_i,\ \phi_*v)\phi_*e_i+\phi_*(\mathrm{Ric}^M(v)),\ v\in TM. \tag{1.9}$$

Thus

$$R^N(\phi_*e_i,\ \phi_*v)\phi_*e_i=-(\tilde{\nabla}*\tilde{\nabla}\,d\phi)(v)+\phi_*(\mathrm{Ric}^M(v)). \tag{1.10}$$

Substitutng (1.10) into (1.6), we have

$$I(\phi_*v,\ \phi_*w)=\int_M\langle-\tilde{\nabla}*\tilde{\nabla}\phi_*v+(\tilde{\nabla}*\tilde{\nabla}\,d\phi)(v)-\phi_*(\mathrm{Ric}^M(v)),\ \phi_*w\rangle_N*1. \tag{1.11}$$

On the other hand, for any fixed point $P$, choose $\{e_i\}$ such that $\nabla_{e_i}e_j\,|_P=0$. Then

$$\tilde{\nabla}*\tilde{\nabla}\,d\phi=\tilde{\nabla}_{e_i}\tilde{\nabla}_{e_i}d\phi.$$

Therefore

$$\begin{aligned}-\tilde{\nabla}*\tilde{\nabla}\phi_*v+(\tilde{\nabla}*\tilde{\nabla}\,d\phi)(v)&=-\tilde{\nabla}*\tilde{\nabla}\phi_*v+(\tilde{\nabla}_{e_i}\tilde{\nabla}_{e_i}d\phi)(v)\\&=-\tilde{\nabla}*\tilde{\nabla}\phi_*v+\tilde{\nabla}_{e_i}((\tilde{\nabla}_{e_i}d\phi)v)-(\tilde{\nabla}_{e_i}d\phi)(\nabla_{e_i}v)\\&=d\phi(\nabla_{e_i}\nabla_{e_i}v)-2\tilde{\nabla}_{e_i}(d\phi(\nabla_{e_i}v)).\end{aligned} \tag{1.12}$$

Hence

$$I(\phi_*v,\ \phi_*w)=\int_M\langle d\phi(\nabla_{e_i}\nabla_{e_i}v)-2\tilde{\nabla}_{e_i}(d\phi(\nabla_{e_i}v))-\phi_*(\mathrm{Ric}^M(v)),\ \phi_*w\rangle_N*1. \tag{1.13}$$

**2. The main theorems** Suppose that $M$ is a $p$-codimensional compact submanifold without boundary in Euclidean space $E^{n+p}$ and $\{e_\alpha\}$ $(\alpha, \beta, \gamma, \cdots = n+1, \cdots, n+p)$ is an orthonormal basis in the normal bundle of $M$. Choose an orthonormal basis of $M$ such that $\nabla_{e_i} e_j|_P = 0$ is satisfied.

Denote the second fundamental form of $M$ by $h$

$$h(X, Y) = h^\alpha(X, Y)e_\alpha, \; h^\alpha(e_i, e_j) = h^\alpha_{ij}. \tag{2.1}$$

Gauss-Codazzi equations are

$$\overline{\nabla}_X Y = \nabla_X Y + h^\alpha(X, Y)e_\alpha, \tag{2.2}$$

$$\overline{\nabla}_X e_\alpha = -h^\alpha(X, e_i)e_i + \nabla^\perp_X e_\alpha, \; \nabla^\perp_{e_i} e_\alpha = \mu_{i\alpha\beta} e_\beta, \; \mu_{i\alpha\beta} + \mu_{i\beta\alpha} = 0, \tag{2.3}$$

where $\overline{\nabla}$ denotes the connection in $E^{n+p}$ and $\nabla^\perp$ the connection in the normal bundle of $M$.

Set $L = \{a|_M$: $a$ is a constant vector in $E^{n+p}\}$. If $v \in L$, then $v = a - \langle a, e_\alpha\rangle e_\alpha$, where $\langle\ ,\ \rangle$ is the inner product in $E^{n+p}$. Using (2.2) and (2.3), we have

$$\nabla_{e_i} v = \overline{\nabla}_{e_i} v - h(e_i, v) = \langle a, e_\alpha\rangle h^\alpha_{ij} e_j, \tag{2.4}$$

$$\nabla_{e_i}(\nabla_{e_i} v) = -\langle a, e_k\rangle h^\alpha_{ik} h^\alpha_{ij} e_j + \langle a, e_\beta\rangle \mu_{i\alpha\beta} h^\alpha_{ij} e_j + \langle a, e_\alpha\rangle (\nabla_{e_i} h^\alpha_{ij}) e_j, \tag{2.5}$$

$$\begin{aligned}\widetilde{\nabla}_{e_i}(d\phi(\nabla_{e_i} v)) = &-\langle a, e_k\rangle h^\alpha_{ik} h^\alpha_{ij} \phi_* e_j + \langle a, e_\beta\rangle \mu_{i\alpha\beta} h^\alpha_{ij} \phi_* e_j + \langle a, e_\alpha\rangle (\widetilde{\nabla}_{e_i} h^\alpha_{ij}) \phi_* e_j \\ &+ \langle a, e_\alpha\rangle h^\alpha_{ij} \widetilde{\nabla}_{\phi_* e_i} \phi_* e_j.\end{aligned} \tag{2.6}$$

Thus

$$\begin{aligned}I(\phi_* v, \phi_* v) = \int_M \langle &\langle a, e_k\rangle h^\alpha_{ik} h^\alpha_{ij} \phi_* e_j - \langle a, e_\beta\rangle \mu_{i\alpha\beta} h^\alpha_{ij} \phi_* e_j - \langle a, e_\alpha\rangle (\nabla_{e_i} h^\alpha_{ij}) \phi_* e_j \\ &- 2\langle a, e_\alpha\rangle h^\alpha_{ij} \widetilde{\nabla}_{\phi_* e_i} \phi_* e_j - \phi_*(\mathrm{Ric}^M(v)), \; \langle a, e_i\rangle \phi_* e_i\rangle_N *1.\end{aligned} \tag{2.7}$$

From Gauss formula it follows that

$$\mathrm{Ric}^M(v) = \langle a, e_i\rangle (h^\alpha_{ll} h^\alpha_{ij} - h^\alpha_{ik} h^\alpha_{kj}) e_j. \tag{2.8}$$

Substituting (2.8) into (2.7), we have

$$\begin{aligned}I(\phi_* v, \phi_* v) = \int_M \langle &2\langle a, e_k\rangle h^\alpha_{ik} h^\alpha_{ij} \phi_* e_j - \langle a, e_i\rangle h^\alpha_{ll} h^\alpha_{ij} \phi_* e_j - \langle a, e_\beta\rangle \mu_{i\alpha\beta} h^\alpha_{ij} \phi_* e_j \\ &- \langle a, e_\alpha\rangle (\nabla_{e_i} h^\alpha_{ij}) \phi_* e_j - 2\langle a, e_\alpha\rangle h^\alpha_{ij} \widetilde{\nabla}_{\phi_* e_i} \phi_* e_j, \; \langle a, e_i\rangle \phi_* e_i\rangle_N *1.\end{aligned} \tag{2.9}$$

Denote $I(\phi_* v, \phi_* w) = \int_M F(\phi_* v, \phi_* w) *1$, $v, w \in L$. Then $\mathrm{tr} I = \int_M \mathrm{tr} F *1$. Because trace is independent of the choice of orthonormal basis, we can pointwise take $\{e_i, e_\alpha\}$ as the basis of $L$ to compute $\mathrm{tr} F$. Thus

$$\mathrm{tr} I = \int_M (2h^\alpha_{ik} h^\alpha_{ij} - h^\alpha_{ll} h^\alpha_{ij}) \langle \phi_* e_j, \phi_* e_k\rangle_N *1. \tag{2.10}$$

**Theorem 1.** *Let $M^n \to E^{n+p}$ be a compact submanifold without boundary in Euclidean space $E^{n+p}$. If there exists a negative constant $B$ such that*

$$2\langle h(e_i, e_k), h(e_i, e_j)\rangle - \langle h(e_i, e_i), h(e_k, e_j)\rangle \leqslant B\delta_{jk}, \tag{2.11}$$

*then there is no nonconstant stable harmonic mapping from $M^n$ to any Riemannian manifold.*

*Proof* By (2.10), we see that

$$\operatorname{tr} I \leqslant B\int_M \langle \phi_* e_j,\ \phi_* e_j\rangle_N *1 = 2BE(\phi) \leqslant 0,$$

and the equality holds if and only if $E(\phi)=0$, i. e. $\phi$ is a constant mapping.

**Theorem 2.** *Let $M^n$ be a closed convex hypersurface in Euclidean space $E^{n+1}$. If each principal curvature of $M^n$ is less than half the sum of all principal curvatures, then there is no nonconstant stable harmonic mapping from $M^n$ to any Riemannian manifold.*

*Proof* Let $\lambda_1, \cdots, \lambda_n$ be $n$ principal curvatures, and $e_1, \cdots, e_n$ eigenvectors of $h$. Set $H=\sum_{i=1}^{n}\lambda_i$. From (2.10), we obtain

$$\operatorname{tr} I = \int_M (2\lambda_k^2 - H\lambda_k)\langle \phi_* e_k,\ \phi^* e_k\rangle_N *1 = \int_M (2\lambda_k - H)\lambda_k \langle \phi_* e_k,\ \phi_* e_k\rangle_N *1 \leqslant BE(\phi),$$

where $B$ is a negative constant, and the equality holds if and only if $E(\phi)=0$, i. e. $\phi$ is a constant mapping.

**Remark.** Theorem 2 obviously includes Xin's result as a special case.

Applying Theorem 1, we obtain

**Theorem 3.** *If* $\min(n_1, \cdots, n_q) > 2$, *then there is no nonconstant stable harmonic mapping from the product of Euclidean spheres $S^{n_1}\times\cdots\times S^{n_q}$ to any Riemannian manifold.*

*Proof* Noting that $S^{n_1}\times\cdots\times S^{n_q}$ can be imbedded canonically into $E^{n_1+1}\times\cdots\times E^{n_q+1}$, in this case, we may take $B=2-\min(n_1, \cdots, n_q)$ in (2.11).

The author would like to thank Professor Hu Hesheng for her encouragement and help.

## References

[1] Xin, Y. L., Some results on stable harmonic maps, *Duke Math. J.*, **47**:3 (1980).
[2] Smith, R. T., The second variation formula for harmonic mappings, *Proc. Amer. Math. Soc.* **47** (1975).
[3] Eells, J. & Sampson J. H., Harmonic mappings of Riemannian manifolds, *Amer. J. Math.*, **86** (1964).

*Chin. Ann. of Math.*
3 (4) 1982.

# A NOTE ON THE APPROXIMATE SOLUTION OF THE CAUCHY PROBLEM BY NUMBER-THEORETIC NETS

WANG YUAN
(*Institute of Mathematics, Academia Sinica*)

Dedicated to Professor Su Bu-chin On the Occaion of his 80th Birthday and his 50th Year of Educational Work

## § 1. Introduction

We use $\boldsymbol{x}=(x_1, \cdots, x_s)$ to denote a vector with real coeffients and $\boldsymbol{m}=(m_1, \cdots, m_s)$, $\boldsymbol{l}=(l_1, \cdots, l_s)$ and $\boldsymbol{a}=(a_1, \cdots, a_s)$ the vectors with integral components. We use the notations $\bar{x}=\max(1, |x|)$, $\|\boldsymbol{m}\|=\bar{m}_1\cdots\bar{m}_s$, $(\boldsymbol{m}, \boldsymbol{x})=\sum_{i=1}^{s} m_i x_i$ the scalar product of $\boldsymbol{m}$ and $\boldsymbol{x}$ and $Q(\boldsymbol{x})$ a polynomial of $\boldsymbol{x}$. We also use $C(\xi, \cdots, \eta)$ to denote a positive constant depending on $\xi, \cdots, \eta$ only, but not always with the same value.

Consider the problem of approximate solution of the equation

$$\frac{\partial u}{\partial t}=Q\left(\frac{\partial}{\partial x_1}, \cdots, \frac{\partial}{\partial x_s}\right)u, \quad 0\leqslant t\leqslant T, \quad -\infty<x_\nu<\infty\,(1\leqslant\nu\leqslant s) \tag{1}$$

with the initial condition

$$u(0, \boldsymbol{x})=\varphi(\boldsymbol{x})=\sum C(\boldsymbol{m})e^{2\pi i(\boldsymbol{m},\boldsymbol{x})},$$

where the Fourier coeffients $C(\boldsymbol{m})$ satisfy

$$|C(\boldsymbol{m})|\leqslant C/\|\boldsymbol{m}\|^{\alpha}$$

in which $C(>0)$ and $\alpha(>1)$ are two constants.

We use $p$ to denote prime number and $N=[p^{\frac{2\alpha}{4\alpha-1}}(\ln p)^{\frac{-(2\alpha-1)(s-1)}{4\alpha-1}}]$, where $[x]$ denotes the integral part of $x$. We also use the following notations:

1° $f(t, \boldsymbol{x})^T$ denotes the set of numbers $f\left(t, \frac{\boldsymbol{a}k}{p}\right)$, $1\leqslant k\leqslant p$,

2° $\Gamma f^T=\sum_{\|\boldsymbol{m}\|\leqslant N} \tilde{C}(t, \boldsymbol{m})e^{2\pi i(\boldsymbol{m},\boldsymbol{x})}$,

where

$$\tilde{C}(t, \boldsymbol{m})=\frac{1}{p}\sum_{k=1}^{p} f\left(t, \frac{\boldsymbol{a}k}{p}\right)e^{-2\pi i(\boldsymbol{a},\boldsymbol{m})k/p},$$

3° $D^T_{r_1,\cdots,r_s}f^T=\left(\frac{\partial^r}{\partial x_1^{r_1}\cdots\partial x_s^{r_s}}\Gamma f^T\right)^T$,

Manuscript received March 30, 1981.

where $r_1+\cdots+r_s=r$, $r_i\geqslant 0(1\leqslant i\leqslant s)$,

4° $\|f\|^2=\int_{G_s}|f|^2 d\boldsymbol{x}$,

where $G_s$ denotes the $s$-dimensional unit cube $0\leqslant x_i\leqslant 1$ $(1\leqslant i\leqslant s)$.

**Theorem 1.** *Suppose that $Q(\boldsymbol{x})$ is a polynomial such that the solution of (1) satisfies $\|u(t, \boldsymbol{x})\|\leqslant c(s)\|\varphi(\boldsymbol{x})\|$.*) Then for any given $p$, there exists an $\boldsymbol{a}=\boldsymbol{a}(p)$ such that*

$$R=\|u(t, \boldsymbol{x})-\Gamma v(t, \boldsymbol{x})^T\|\leqslant Cc(\alpha, s)p^{\frac{-\alpha(2\alpha-1)}{4\alpha-1}}(\ln p)^{\frac{2\alpha^2(s-1)}{4\alpha-1}}, \tag{2}$$

*Where $v(t, \boldsymbol{x})^T$ denotes the solution of the system of the ordinary differential equation*

$$\frac{dv(t, \boldsymbol{x})^T}{dt}=Q(D^T_{1,0,\cdots,0}, \cdots, D^T_{0,\cdots,0,1})v(t, \boldsymbol{x})^T \tag{3}$$

*with initial condition*

$$v(0, \boldsymbol{x})=D^T_{0,\cdots,0}\varphi(\boldsymbol{x})^T.$$

This gives a modification of a result due to Рабенький В. С.[1] which will be obtained, if the right hand side of (2) is replaced by $Cc(\alpha, s)p^{\frac{1-\alpha}{2}}(\ln p)^{\frac{(\alpha+1)(s-1)}{2}}$.

If $p$ and $\boldsymbol{a}(p)$ in Theorem 1 are changed by $F_{n+1}$ and $(1, F_n)$ respectively for the case $s=2$, where $F_n=\frac{1}{\sqrt{s}}\left(\left(\frac{1+\sqrt{s}}{2}\right)^n-\left(\frac{1-\sqrt{s}}{2}\right)^n\right)$, $(n=1, 2, \cdots)$ denote the Fibonacci sequence, then the right hand side of (2) may be improved slightly by $Cc(\alpha)F_n^{\frac{-\alpha(2\alpha-1)}{4\alpha-1}}(\ln 3F_n)^{\frac{3\alpha-1}{4\alpha-1}}$.

The vectoro $\boldsymbol{a}$ is called a good lattice point modulo $p$ by Hlawka, E. or an optimal coefficient modulo $p$ by Коробов, Н. М. and a table of good lattice points is contained in many books for the purpose of practical use, for example the book of Hua Loo Keng and Wang Yuan[2]

## § 2. Several lemmas.

**Lemma 1.** *For any given $p$, there exists $\boldsymbol{a}$ such that any non-zero solution $\boldsymbol{l}$ of the congruence*

$$(\boldsymbol{a}, \boldsymbol{l})\equiv 0 \pmod p$$

*satisfies*

$$\|\boldsymbol{l}\|>c(s)p/(\ln p)^{s-1} \tag{4}$$

*and*

$$\sum_{(\boldsymbol{a},\boldsymbol{l})\equiv 0 (\mathrm{mod}\ p)}{}' \frac{1}{\|\boldsymbol{l}\|^\alpha}\leqslant c(\alpha, s)p^{-\alpha}(\ln p)^{\alpha(s-1)}, \tag{5}$$

*where $\sum'$ denotes a sum with an exception $\boldsymbol{l}=\boldsymbol{0}$,* (Cf, Бахвалов, Н. С. [3])

**Lemma 2.** *Suppose that $\|\boldsymbol{l}\|\geqslant 3^s$ and that $1\leqslant M\leqslant\|\boldsymbol{l}\|/3^s$. Then*

$$\sum_{\|\boldsymbol{m}\|\leqslant M}\frac{1}{\|\boldsymbol{l}+\boldsymbol{m}\|^\alpha}<c(\alpha, s)M^\alpha\|\boldsymbol{l}\|^{-\alpha}$$

(Cf. Wang Yuan [4]).

*) For example, $Q(\boldsymbol{x})$ is a positive definite quadratic form.

In the following, the vector $\boldsymbol{a}$ is taken such that (4) and (5) are satisfied.

**Lemma 3.** *We have*

$$\Gamma(e^{2\pi i(\boldsymbol{l},\boldsymbol{x})})^T=\sum_{\substack{\|\boldsymbol{m}\|<N\\(\boldsymbol{a},\boldsymbol{l}-\boldsymbol{m})\equiv 0(\mathrm{mod}\ p)}} e^{2\pi i(\boldsymbol{m},\boldsymbol{x})}.$$

*In particular,*

$$\Gamma(e^{2\pi i(\boldsymbol{l},\boldsymbol{x})})^T=e^{2\pi i(\boldsymbol{l},\boldsymbol{x})}$$

*for* $\|\boldsymbol{l}\|\leqslant N$ *and* $p>c(s)$.

*Proof*

$$\Gamma(e^{2\pi i(\boldsymbol{l},\boldsymbol{x})})^T=\sum_{\|\boldsymbol{m}\|\leqslant N}\frac{1}{p}\sum_{k=1}^{p}e^{2\pi i(\boldsymbol{a},\boldsymbol{l})k/p}e^{-2\pi i(\boldsymbol{a},\boldsymbol{m})k/p}e^{2\pi i(\boldsymbol{m},\boldsymbol{x})}=\sum_{\substack{\|\boldsymbol{m}\|<N\\(\boldsymbol{a},\boldsymbol{l}-\boldsymbol{m})\equiv 0(\mathrm{mod}\ p)}} e^{2\pi i(\boldsymbol{m},\boldsymbol{x})}$$

The Lemma is proved.

**Lemma 4.** *Suppose that* $\varphi(\boldsymbol{x})=e^{2\pi i(\boldsymbol{m},\boldsymbol{x})}$, *where* $\|\boldsymbol{m}\|\leqslant N$. *Then* $R=0$.

*Proof* Suppose that

$$u(t,\ \boldsymbol{x})=u(t)e^{2\pi i(\boldsymbol{m},\boldsymbol{x})}$$

and

$$v(t,\ \boldsymbol{x})^T=v(t)(e^{2\pi i(\boldsymbol{m},\boldsymbol{x})})^T,$$

where $u(0)=v(0)=1$. Substituting into (1) and (3), we have

$$\frac{\partial u}{\partial t}=u'(t)e^{2\pi i(\boldsymbol{m},\boldsymbol{x})}=Qu=u(t)Q(2\pi i\boldsymbol{m})e^{2\pi i(\boldsymbol{m},\boldsymbol{x})}$$

and

$$v'(t)(e^{2\pi i(\boldsymbol{m},\boldsymbol{x})})^T=v(t)Q(2\pi i\boldsymbol{m})(e^{2\pi i(\boldsymbol{m},\boldsymbol{x})})^T$$

by Lemma 3. Hence

$$u'(t)=u(t)Q(2\pi i\boldsymbol{m})$$

and

$$v'(t)=v(t)Q(2\pi i\boldsymbol{m}).$$

Since $u(t)$ and $v(t)$ satisfy the same ordinary differential equation with the same initial value, therefore $u(t)\equiv v(t)$ and the Lemma follows.

**Lemma 5.** *Let*

$$\varphi_2(\boldsymbol{x})=\sum_{\|\boldsymbol{m}\|>N}C(\boldsymbol{m})e^{2\pi i(\boldsymbol{m},\boldsymbol{x})}.$$

*Then*

$$\|\Gamma\varphi_2(\boldsymbol{x})^T\|\leqslant Cc(\alpha,\ s)p^{\frac{-\alpha(2\alpha-1)}{4\alpha-1}}(\ln p)^{\frac{4\alpha^2(s-1)}{4\alpha-1}}$$

*Proof* It follws from Lemma 3 that

$$\|\Gamma\varphi_2(\boldsymbol{x})^T\|^2=\int_{G_s}\Big|\sum_{\|\boldsymbol{l}\|>N}C(\boldsymbol{l})\Gamma(e^{2\pi i(\boldsymbol{l},\boldsymbol{x})})^T\Big|^2d\boldsymbol{x}$$

$$=\int_{G_s}\Big|\sum_{\|\boldsymbol{m}\|\leqslant N}\sum_{\substack{\|\boldsymbol{l}\|>N\\(\boldsymbol{a},\boldsymbol{l}-\boldsymbol{m})\equiv 0(\mathrm{mod}\ p)}}C(\boldsymbol{l})e^{2\pi i(\boldsymbol{m},\boldsymbol{x})}\Big|^2d\boldsymbol{x}=\sum_{\|\boldsymbol{m}\|\leqslant N}\Big(\sum_{\substack{\|\boldsymbol{l}\|>N\\(\boldsymbol{a},\boldsymbol{l}-\boldsymbol{m})\equiv 0(\mathrm{mod}\ p)}}C(\boldsymbol{l})\Big)^2.$$

Let $\boldsymbol{l}-\boldsymbol{m}=\boldsymbol{n}$. Then

$$\|\Gamma\varphi_2(\boldsymbol{x})^T\|^2\leqslant C^2\sum_{\|\boldsymbol{m}\|\leqslant N}\Big(\sum_{(\boldsymbol{a},\boldsymbol{n})\equiv 0(\mathrm{mod}\ p)}{}'\frac{1}{\|\boldsymbol{n}+\boldsymbol{m}\|^{\alpha}}\Big)^2$$

$$=C^2\sum_{\|\boldsymbol{m}\|\leqslant N}\sum_{(\boldsymbol{a},\boldsymbol{n})\equiv 0(\mathrm{mod}\ p)}{}'\Big(\frac{\|\boldsymbol{n}\|}{\|\boldsymbol{m}\|\,\|\boldsymbol{n}+\boldsymbol{m}\|}\Big)^{\alpha}\frac{\|\boldsymbol{m}\|^{\alpha}}{\|\boldsymbol{n}\|^{\alpha}}\sum_{(\boldsymbol{a},\boldsymbol{l})\equiv 0(\mathrm{mod}\ p)}{}'\frac{1}{\|\boldsymbol{l}+\boldsymbol{m}\|^{\alpha}}.$$

Since

$$\frac{\|\boldsymbol{n}\|}{\|\boldsymbol{m}\|\|\boldsymbol{n}+\boldsymbol{m}\|}\leqslant 2^s$$

and we may suppose that $p>c(s)$, by Lemmas 1 and 2 we have

$$\|\Gamma\varphi_2(\boldsymbol{x})^T\|^2\leqslant C^2c(\alpha, s)N^{2\alpha}p^{-2\alpha}(\ln p)^{2\alpha(s-1)}$$
$$\leqslant C^2c(\alpha, s)p^{\frac{-2\alpha(2\alpha-1)}{4\alpha-1}}(\ln p)^{\frac{4\alpha(s-1)}{4\alpha-1}}.$$

The Lemma is proved.

**Lemma 6.** *If* $v(0, \boldsymbol{x})^T=(\Gamma\varphi_2(\boldsymbol{x})^T)^T$, *then*

$$\|\Gamma v(t, \boldsymbol{x})^T\|\leqslant Cc(\alpha, s)p^{\frac{-\alpha(2\alpha-1)}{4\alpha-1}}(\ln p)^{\frac{2\alpha^2(s-1)}{4\alpha-1}}.$$

*Proof* For any given $\boldsymbol{l}$, we shall prove that the congruence

$$(\boldsymbol{a}, \boldsymbol{l}-\boldsymbol{m})\equiv 0 \pmod p \tag{6}$$

has at most 1 solution $\boldsymbol{m}$ satisfying

$$\|\boldsymbol{m}\|\leqslant N.$$

In fact, if there are two different vectors $\boldsymbol{m}$ and $\boldsymbol{m}'$ satisfying (6), then $\boldsymbol{m}-\boldsymbol{m}'\neq\boldsymbol{0}$,

$$(\boldsymbol{a}, \boldsymbol{m}-\boldsymbol{m}')\equiv 0 \pmod p$$

and

$$\|\boldsymbol{m}-\boldsymbol{m}'\|\leqslant 2^sN$$

which leads to a contradiction with Lemma 1. Hence by Lemma 3, we have

$$\Gamma(e^{2\pi i(\boldsymbol{l},\boldsymbol{x})})^T=0 \text{ or } e^{2\pi i(\boldsymbol{m},\boldsymbol{x})},$$

where $\|\boldsymbol{m}\|\leqslant N$. Consequently, it follows from Lemma 4 that the solution $u(t, \boldsymbol{x})$ of the partial differential equation

$$\begin{cases}\dfrac{\partial u}{\partial t}=Q\Big(\dfrac{\partial}{\partial x_1}, \cdots, \dfrac{\partial}{\partial x^s}\Big)u,\\ u(0, \boldsymbol{x})=\Gamma\varphi_2(\boldsymbol{x})^T\end{cases}$$

and $\Gamma v(t, \boldsymbol{x})^T$ are identical, where $v(t, \boldsymbol{x})^T$ is the solution of the ordinary differential equation

$$\begin{cases}\dfrac{dv(t, \boldsymbol{x})^T}{dt}=Q(D^T_{1,0,\cdots,0}, \cdots, D^T_{0,\cdots,0,1})v(t, \boldsymbol{x})^T,\\ v(0, \boldsymbol{x})^T=(\Gamma\varphi_2(\boldsymbol{x})^T)^T.\end{cases}$$

Hence by Lemma 5, we have

$$\|\Gamma v(t, \boldsymbol{x})^T\|=\|v(t, \boldsymbol{x})\|\leqslant c(s)\|u(0, \boldsymbol{x})\|\leqslant c(s)\|\Gamma\varphi_2(\boldsymbol{x})^T\|$$
$$\leqslant Cc(\alpha, s)p^{\frac{-\alpha(2\alpha-1)}{4\alpha-1}}(\ln p)^{\frac{2\alpha^2(s-1)}{4\alpha-1}}.$$

The Lemma is proved.

## § 3. The proof of Theorem 1.

Let

$$\varphi(\boldsymbol{x})=\varphi_1(\boldsymbol{x})+\varphi_2(\boldsymbol{x}),$$

where

$$\varphi_1(\boldsymbol{x})=\sum_{\|\boldsymbol{m}\|\leqslant N}C(\boldsymbol{m})e^{2\pi i(\boldsymbol{m},\boldsymbol{x})}$$

and

$$\varphi_2(\boldsymbol{x})=\sum_{\|\boldsymbol{m}\|>N} C(\boldsymbol{m})e^{2\pi i(\boldsymbol{m},\boldsymbol{x})}.$$

Let $u_1(t, \boldsymbol{x})$ and $u_2(t, \boldsymbol{x})$ denote the solutions of the equation (1) with the initial conditions $u_1(0, \boldsymbol{x})=\varphi_1(\boldsymbol{x})$ and $u_2(0, \boldsymbol{x})=\varphi_2(\boldsymbol{x})$ respectively. Further let $v_1(t, \boldsymbol{x})^T$ and $v_2(t, \boldsymbol{x})^T$ be the solutions of the equation (3) with the initial conditions $v_1(0, \boldsymbol{x})^T = D^T_{0,\cdots,0}\varphi_1(\boldsymbol{x})^T$ and $v_2(0, \boldsymbol{x})^T=D^T_{0,\cdots,0}\varphi_2(\boldsymbol{x})^T$ respectively. Then

$$u(t, \boldsymbol{x})=u_1(t, \boldsymbol{x})+u_2(t, \boldsymbol{x})$$

and

$$v_1(t, \boldsymbol{x})^T=v_1(t, \boldsymbol{x})^T+v_2(t, \boldsymbol{x})^T.$$

It follows that

$$\|u_1(t, \boldsymbol{x})-\Gamma v_1(t, \boldsymbol{x})^T\|=0$$

by Lemma 4 and that

$$\|\Gamma v_2(t, \boldsymbol{x})^T\|\leqslant Cc(\alpha, s)p^{\frac{-\alpha(2\alpha-1)}{4\alpha-1}}(\ln p)^{\frac{2\alpha^2(s-1)}{4\alpha-1}}$$

by Lemma 5. Since

$$\|u_2(t, \boldsymbol{x})\|\leqslant\|u_2(0, \boldsymbol{x})\|=\left(\int_{G_s}|\varphi_2(\boldsymbol{x})|^2d\boldsymbol{x}\right)^{1/2}\leqslant C\left(\sum_{\|\boldsymbol{m}\|>N}\frac{1}{\|\boldsymbol{m}\|^{2\alpha}}\right)^{1/2}$$
$$\leqslant Cc(\alpha, s)N^{-\frac{2\alpha-1}{2}}(\ln p)^{\frac{s-1}{2}}\leqslant Cc(\alpha, s)p^{-\frac{\alpha(2\alpha-1)}{4\alpha-1}}(\ln p)^{\frac{2\alpha^2(s-1)}{4\alpha-1}},$$

we have

$$\|u(t, \boldsymbol{x})-\Gamma v(t, \boldsymbol{x}\|^T\|^2=\|u_1(t, \boldsymbol{x})+u_2(t, \boldsymbol{x})-\Gamma v_1(t, \boldsymbol{x})^T-\Gamma v_2(t, \boldsymbol{x})^T\|^2$$
$$\leqslant 3(\|u_1(t, \boldsymbol{x})-\Gamma v_1(t, \boldsymbol{x})^T\|+\|u_2(t, \boldsymbol{x})\|^2+\|\Gamma v_2(t, \boldsymbol{x})^T\|^2)$$
$$\leqslant C^2c(\alpha, s)p^{\frac{-2\alpha(2\alpha-1)}{4\alpha-1}}(\ln p)^{\frac{4\alpha^2(s-1)}{4\alpha-1}}.$$

The Theorem Ts proved.

## References

[1] Рябенький, В. С., Об одном способе получения разностных схем и об использовании теоретико-числовых сеток для Решенмя Задачи Коши Мегодом Конечных Разностей, *Тру. Мат. ИНСТ. АН СССР*, Том LX, (1961), 232—237.

[2] Hua Loo Keng and Wang Yuan, Applications of number theory to numerical analysis, *Sci. Press* (Beijing), 1978.

[3] Бахвалов, Н. С., Приближенном Вычислении Кратных Ингегралов, *Вес. МГУ*, **4** (1959), 3—18.

[4] Wang Yuan, A note on interpolation of a certain class of functions, *Sci. Sin.*, **10** (1960), 632—636

J. DIFFERENTIAL GEOMETRY
16 (1981) 551–557

# HYPER-$q$-CONVEX DOMAINS IN KÄHLER MANIFOLDS

H. WU

*Dedicated to Professor Buchin Su on his 80th birthday*

**1.** In a recent paper [2], C. Badji used harmonic theory to prove that if $D$ is a $C^\infty$ strongly pseudoconvex domain in a Kähler manifold of nonnegative bisectional curvature, then the Dolbeault group $H^{1,1}(D)$ vanishes. Now it is implicit in the arguments of [4] that under the same hypothesis, $-\log \rho$ ($\rho =$ distance to the boundary $\partial D$) is strictly plurisubharmonic so that $D$ is in fact a Stein manifold and hence $H^{p,q}(D) = 0$ for all $q \geqslant 1$. With the availability of [7], even more general statements can be made under weaker hypotheses. It therefore seems worth while to explicitly write down some of these theorems for future references. Deferring the technical definitions to the next section, we may state the main theorem as follows.

**Theorem.** *Let $M$ be an $n$-dimensional Kähler manifold (not necessarily complete), and $D$ be a relatively compact domain in $M$. Then $D$ is strongly $q$-pseudoconvex ($1 \leqslant q \leqslant n$), if for some neighborhood $W$ of $\partial D$ in $M$ the bisectional curvature of $M$ is $q$-positive in $W \cap D$, and $D$ is weakly hyper-$q$-convex. Furthermore, $D$ is $q$-complete if any of the following holds*:

(i) *The bisectional curvature of $M$ is $q$-nonnegative in all of $D$ and is $q$-positive in $W \cap D$ for some neighborhood $W$ of $\partial D$ in $M$, and $D$ is weakly hyper-$q$-convex.*

(ii) *The bisectional curvature of $M$ is $q$-nonnegative in all of $D$, and $D$ is $C^\infty$ hyper-$q$-convex.*

(iii) *The bisectional curvature of $M$ is $q$-nonnegative in all of $D$, $D$ is weakly hyper-$q$-convex, and there exists a continuous function $f$: $D \to \mathbf{R}$ which is in $\Psi(q)$ on $D$.*

**Corollary.** *Let $M$ be a Kähler manifold whose bisectional curvature is $q$-nonnegative. Then $M$ has no exceptional analytic sets of dimension not less than $q$.*

Received November 14, 1981. Work under partial support by the National Science Foundation Grant MC 577-23579.

In this theorem there is no need to make any assumption on the bisectional curvature of $M$ outside $D$. However, in part (iii) above the existence of such an $f$ would follow, if $M$ is complete and its bisectional curvature is everywhere $q$-nonnegative and is $q$-positive outside a compact set [7, Theorem 3]. Note that many variations on this theorem are possible. Indeed, the crucial argument of such a theorem in its various guises always involves showing that a certain sum of second variations of arc length is positive (cf. [4, pp. 177–178] or [7, Lemma 6]). Since the second variation formula is itself the sum of a boundary term involving the Levi form of $\partial D$ and an integral involving the bisectional curvature of $D$, suitable assumptions on $\partial D$ and on the bisectional curvature of $D$ balancing one against the other will always insure the positivity of the resulting sum. Thus while [4, Theorem 1] assumes the pseudoconvexity of $D$ and the *positivity* of the bisectional curvature in $D$, it is obvious that it could have assumed instead the *strict* pseudoconvexity of $\partial D$ and the nonnegativity of the bisectional curvature. This observation explains the remark in the opening paragraph as well as the proofs to be given below. We leave the precise enumeration of the other possibilities to the reader.

The above theorem has been known to the author for some time (when $q = 1$ it was of course also known to R. E. Greene), but the thought of actually writing down the details came only after the receipt of the Badji preprint [2]. The author wishes to thank Dr. Badji for the courtesy of sending this preprint. In the meantime, the author received the preprint [5] which contains among other things the case $q = 1$ in (ii) and (iii) of the above theorem as well as the case $q = 1$ in the Corollary. It should be made clear however that the Corollary was added only after reading [5], and that its proof uses the argument in [5].

**2.** We first recall some definitions, A $C^\infty$ function $\tau$ on an $n$-dimension complex manifold $M$ is *strongly q-pseudoconvex* if its Levi form $L\tau$ has at least $n - q + 1$ positive eigenvalues at each point of $M$; $M$ is *strongly q-pseudoconvex* (resp. *q-complete*) if it possesses a $C^\infty$ exhaustion function which is strongly $q$-pseudoconvex outside a compact set (resp. strongly $q$-pseudoconvex everywhere), [1]. A domain $D$ in $M$ is said to have $C^\infty$ *boundary* $\partial D$ if $\partial D$ is an imbedded $C^\infty$ real hypersurface of $M$; in this case, $D$ is said to be a $C^\infty$ *domain*. Let $M$ be Kähler. Then $D$ or $\partial D$ is said to be $C^\infty$ *hyper-q-convex* if $D$ is $C^\infty$, and each $x \in \partial D$ admits a local defining function $\phi$ of $\partial D$ at $x$ (i.e., locally $\partial D = \phi^{-1}(0)$, $\phi|_D < 0$ and $|d\phi(x)| = 1$) such that the eigenvalues $\lambda_1, \cdots, \lambda_{n-1}$ of the restriction of the Levi form $L\phi$ to the maximal complex subspace of the tangent space $T_x(\partial D)$ of $\partial D$ at $x$ satisfy $\Sigma_{i=1}^{q} \lambda_{j_i} > 0$ for all $1 \leqslant j_i \leqslant n - 1$, [3]. It is natural to consider the case of a $C^\infty$ domain $D$ which merely satisfies $\Sigma_{i=1}^{q} \lambda_{j_i} \geqslant 0$ for all $1 \leqslant j_i \leqslant n - 1$; for a reason which will become obvious, we adopt the ad hoc terminology that such a $D$ or $\partial D$ is $C^\infty$ *$q$-convex*. Next, recall the class of continuous functions $\Psi(q)$ from [7]. A set of vectors $\{Z_1,\cdots,Z_q\}$ in $T_xM$ is *$\varepsilon$-orthonormal* if $|G(Z_i, Z_j) - \delta_{ij}| < \varepsilon$ for $i, j = 1,\cdots,q$, where $G$ is the Hermitian inner product on $T_xM$ given by the Kähler metric. Given $K \subset M$ and positive constants $\varepsilon$ and $\eta$, define $\mathfrak{L}(K, \varepsilon, \eta)$ to be the set of all $C^\infty$ functions $f$ defined on $K$ such that if $\{Z_1,\cdots,Z_q\}$ is an $\varepsilon$-orthonormal set in $T_xM$ $(x \in K)$, then $\Sigma_{i=1}^q Lf(Z_i, Z_i) \geqslant \eta$. Now let $U$ be an open set in $M$. Then a function $F \in \Psi(q; U)$ iff for each compact subset $K$ of $U$, there exist positive constants $\varepsilon$ and $\eta$ and a sequence $\{f_i\} \subset \mathfrak{L}(K, \varepsilon, \eta)$ such that $f_i$ converges uniformly to $F$ on $K$. If $U = M$ or there is no danger of confusion, we will simply write $\Psi(q)$ in place of $\Psi(q; U)$. It follows easily from the considerations in [7, §2] that this definition of $\Psi(q)$ coincides with that given in [7]. Note that by definition, $\Psi(q)$ consists of continuous functions, and that if $\mathcal{C}^\infty$ (or in case there is any confusion, $\mathcal{C}^\infty(U)$) denotes the $C^\infty$ functions on $U$, then $\mathcal{C}^\infty \cap \Psi(q; U)$ consists of exactly those $C^\infty$ functions $f$ on $U$ such that the sum of any $q$ eigenvalues of $Lf$ at any point of $U$ is positive. In particular, any function in $\mathcal{C}^\infty \cap \Psi(q)$ is strongly $q$-pseudoconvex. Moreover, [7, Proposition 1] shows that $\mathcal{C}^\infty \cap \Psi(q)$ is dense in $\Psi(q)$ in the $C^0$-topology.

A domain $D$ in a Kähler manifold $M$ is said to be *weakly hyper-$q$-convex* if for every $x \in \partial D$ there exists a neighborhood $V_x$ of $x$ in $M$ such that $V_x \cap D$ admits an exhaustion function which belongs to $\Psi(q; V_x \cap D)$. The following lemma gives the relationship among the various domains; its proof will be given at the end of §3.

**Lemma.** *Let $D$ be a $C^\infty$ domain in a Kähler manifold $M$. Then the following hold*:

(i) *If $D$ is $C^\infty$ hyper-$q$-convex, then it is weakly hyper-$q$-convex.*

(ii) *Suppose in addition the bisectional curvature is $q$-nonnegative in a neighborhood of $\partial D$, then $D$ is $C^\infty$ $q$-convex iff it is weakly hyper-$q$-convex.*

We conclude this section by defining the $q$-positivity of the bisectional curvature. Let $M$ be Kähler and let $x \in M$. If $X$ and $Y$ are nonzero vectors in $T_xM$, the bisectional curvature determined by $X$ and $Y$ is by definition $H(X, Y) \equiv R(X, JX, Y, JY)/(|X|^2 \cdot |Y|^2)$, where $R$ is the curvature tensor of $M$. The bisectional curvature of $M$ is *$q$-nonnegative* (resp., *$q$-positive*) in an open set $U$ if for every $x \in U$ and for every orthonormal basis $\{e_1, Je_k,\cdots,e_n, Je_n\}$ of $T_xM$ and $0 \neq X \in T_xM$, $\Sigma_{i=1}^q H(X, e_i) \geqslant 0$ (resp., $\Sigma_{i=1}^q H(X, e_i) > 0$). See [7] for further details.

**3.** We now supply the proofs of the preceding theorem, corollary, and lemma. The reader is assumed to be acquainted with [4] and [7].

We first prove the first assertion of the theorem concerning the strong $q$-pseudoconvexity of $D$. Since $D$ is weakly hyper-$q$-convex, each $x \in \partial D$ has a neighborhood $V_x$ such that $V_x \cap D$ admits an exhaustion function $\phi$ which belongs to $\Psi(q; V_x \cap D)$. By the density $\mathcal{C}^\infty \cap \Psi(q; V_x \cap D)$ in $\Psi(q; V_x \cap D)$, we may further assume that $\phi$ is $\mathcal{C}^\infty$. If $\{r_i\}$ is a sequence of regular values of $\phi$ such that $r_i \uparrow \infty$, then $\{\phi^{-1}(r_i)\}$ is a sequence of $\mathcal{C}^\infty$ real hypersurfaces in $V_x \cap D$ which are $\mathcal{C}^\infty$ hyper-$q$-convex and approximate $V_x \cap \partial D$ from within. The proof of Theorem 1(A) in [4] coupled with the technique in §3 of [7] now yields the following assertion: Let $\rho: D \to [0, \infty)$ denote the distance from the boundary $\partial D$. Then there exist a neighborhood $W$ of $\partial D$ in $M$ and a $\mathcal{C}^\infty$ increasing convex function $\chi: (-\infty, 0) \to \mathbf{R}$ such that $\chi(-\rho) \uparrow \infty$ near $\partial D$ and $\chi(-\rho) \in \Psi(q; W \cap D)$. Again using the density of $\mathcal{C}^\infty \cap \Psi(q; W \cap D)$ in $\Psi(q; W \cap D)$ in the $C^0$-topology, we obtain a $\tau \in \mathcal{C}^\infty \cap \Psi(q; W \cap D)$ such that $\tau \uparrow \infty$ uniformly near $\partial D$. By shrinking $W$ if necessary, we may assume $\tau$ is a $\mathcal{C}^\infty$ function defined on $D$. This proves that $D$ is strongly $q$-pseudoconvex.

Continuing with the same notation, we shall go on to prove part (i) of the theorem. Indeed, let $\{t_i\}$ be a sequence of regular values of $\tau$ such that $t_i \uparrow \infty$. Then $\{\tau^{-1}(t_i)\}$ is a sequence of $\mathcal{C}^\infty$ real hypersurfaces in $D$, which are $\mathcal{C}^\infty$ hyper-$q$-convex and uniformly approximate $\partial D$ from within. The arguments used for the proofs of Theorems 2 and 4 in [7] are now applicable; they prove that for some $\mathcal{C}^\infty$ increasing convex function $\chi_0: (-\infty, 0) \to \mathbf{R}$ such that $\chi_0 \uparrow \infty$ near 0, the function $\chi_0(-\rho)$, is an exhaustion function of $D$ and is in $\Psi(q; D)$ because the bisectional curvature is now everywhere $q$-nonnegative in $D$. The density of $\mathcal{C}^\infty \cap \Psi(q; D)$ in $\Psi(q; D)$ allows us to replace $\chi_0(-\rho)$ by a $\mathcal{C}^\infty$ exhaustion function of $D$, which belongs to $\Psi(q; D)$. This proves (i).

The proof of (ii) is essentially identical with the proof of Theorem 2 of [7], the only necessary change being in the proof of Lemma 6 of [7]; the latter has to do with showing a certain sum of second variations of arc length is positive. That this is so is guaranteed by the $\mathcal{C}^\infty$ hyper-$q$-convexity of $\partial D$ (in the presense of everywhere $q$-nonnegative bisectional curvature; see [7, (15)]). See the discussion after the corollary in §1.

Finally to prove (iii), we have to invoke the generalized Levi form $Pf$ of a continuous function $F$, [6], [7]. Under the assumption of (iii), the by-now familiar arguments of [4] and [7] show that for some $\mathcal{C}^\infty$ increasing convex function $\chi_1: (-\infty, 0) \to \mathbf{R}$, $\chi_1(-\rho)$ is an exhaustion function of $D$, and the following holds for $\chi_1(-\rho)$. Let $\delta$ be a given continuous positive function on $D$. Then there exists a continuous positive function $\varepsilon_1$ on $D$ such that whenever $x \in D$ and $\{Z_1, \cdots, Z_q\}$ is any $\varepsilon_1(x)$-orthonormal set in $T_xM$, $\Sigma_{i=1}^q P(\chi_1(-\rho))(x, Z_i) > -\delta(x)$. By hypothesis, there exists an $f \in \Psi(q; D)$. By the density of $\mathcal{C}^\infty \cap \Psi(q; D)$ in $\Psi(q; D)$, $f$ may be assumed to be actually $\mathcal{C}^\infty$. Replacing $f$ by $e^f$ if necessary, we may also assume $f > 0$. Now let $\delta$ be any continuous positive function in $D$ such that for some positive continuous function $\varepsilon_2$ on $D$, $\Sigma_{i=1}^q Lf(Z_i, Z_i) > 2\delta(x)$ for all $x \in D$ and all $\varepsilon_2(x)$-orthonormal sets $\{Z_1, \cdots, Z_q\}$ in $T_xM$. Consequently, $\tau \equiv \chi_1(-\rho) + f$ is an exhaustion function of $D$, and if we denote $\min\{\varepsilon_1, \varepsilon_2\}$ by $\varepsilon$, then $\Sigma_{i=1}^q P\tau(x, Z_i) > \delta(x)$ for all $x \in D$ and all $\varepsilon(x)$-orthonormal sets $\{Z_1, \cdots, Z_q\}$ in $T_xM$. In particular, $\tau \in \Psi(q; D)$. By the usual reasoning, the $q$-completeness of $D$ follows. q.e.d.

Next, we prove the corollary using the idea of [5]. If the corollary is false, there would exist a compact subvariety $S$ in $M$ of dimension $s > q$ and a holomorphic map $\pi: M \to M'$ into a complex space $M'$ such that $\pi(S)$ is a point $x' \in M'$ and $\pi: M - S \to M' - \{x'\}$ is biholomorphic. Let $\Sigma'$ be the boundary of some $\varepsilon$-ball $B'$ relative to some coordinate system centered at $x' \in M'$. $B'$ is then a $\mathcal{C}^\infty$ strictly pseudoconvex domain. Let $B = \pi^{-1}(B')$ and $\Sigma = \pi^{-1}(\Sigma')$. Then $B$ is a $\mathcal{C}^\infty$ strictly pseudoconvex domain because $\pi$ is biholomorphic in a neighborhood of $\Sigma$. By part (ii) of the theorem, there is an exhaustion function $\tau$ of $B$, which belongs to $\Psi(q; B)$. By the density of $\mathcal{C}^\infty \cap \Psi(q; B)$ in $\Psi(q; B)$, we may assume $\tau$ is $\mathcal{C}^\infty$. Since $s \geqslant q$, it is clear that the restriction $\tau|_S$ is a $\mathcal{C}^\infty$ strictly subharmonic function on the regular points $\mathcal{R}S$ of $S$ and is a continuous function on $S$. Moreover, since $S$ is compact, $\Delta\tau$ is bounded below by a positive constant on $\mathcal{R}S$, where $\Delta$ denotes the Laplacian of the Kähler manifold $\mathcal{R}S$. Since $\tau|_S$ must attain an absolute maximum on $S$, a standard argument shows that such an $S$ does not exist. q.e.d.

We finally give the proof of the lemma. To prove part (1), let $x \in \partial D$ and let $\phi$ be a local defining function at $X$ satisfying the hyper-$q$-convex condition at $X$. By a standard argument, there exists a $\mathcal{C}^\infty$ strictly increasing, strictly convex function $\chi(t)$ such that $\chi(\phi)$ has the property that $\Sigma_{i=1}^q L\chi(\phi)(Z_i, Z_i) > 0$ for any orthonormal basis $\{Z_1, JZ_1, \cdots, Z_n, JZ_n\}$ in $T_xM$. Thus for a sufficiently small neighborhood $V_x$ of $x$, $\chi(\phi) \in \mathcal{C}^\infty \cap \Psi(q; V_x)$. We may assume that relative to some coordinate system $\{z_1, \cdots, z_n\}$ centered at $x$, $V_x$ is given by $\{\Sigma_{i=1}^n |z_i|^2 < b\}$ for some $b > 0$. Let $\tau_1 = -1/\chi(\phi)$ with $\chi(0) = 0$, and let $\tau_2 = 1/(b - \Sigma_{i=1}^n |z_i|^2)$. Both $\tau_1$ and $\tau_2$ are in $\mathcal{C}^\infty \cap \Psi(q; V_x \cap D)$ so that also $\max\{\tau_1, \tau_2\} \equiv \tau$ belongs to $\Psi(q; V_x \cap D)$ (cf. [7, Lemma 2(d)]). $\tau$ is clearly an exhaustion function on $V_x \cap D$. This proves (i).

To prove (ii), first assume $D$ is $\mathcal{C}^\infty$ $q$-convex. We know from the proof of the theorem that, in the presence of $q$-nonnegative bisectional curvature near $\partial D$, the following holds. There exist a neighborhood $W$ of $\partial D$ in $M$ and a $\mathcal{C}^\infty$ increasing convex function $\chi: (-\infty, 0) \to \mathbf{R}$ such that $\chi(-\rho) \uparrow \infty$ near $\partial D$ ($\rho$

denotes the distance from $\partial D$ as usual) and such that if $\delta$ is a given continuous positive function on $W \cap D$, then there exists a continuous positive function $\varepsilon_1$ on $W \cap D$ with the property that whenever $x \in W \cap D$ and $\{Z_k, \cdots, Z_q\}$ is any $\varepsilon_1(x)$-orthonormal set in $T_x M$, $\Sigma_{i=1}^q P\chi(-\rho)(x, Z_i) > -\delta(x)$. Now fix an $x_0 \in \partial D$ and let $\{z_1, \cdots, z_n\}$ be a coordinate system centered at $x_0$. Choose a positive number $b$ so small that if $V_0 \equiv \{\Sigma_{i=1}^n |z_i|^2 < b\}$, then $V_0 \cap D \subset W \cap D$. Let $\delta$ be a positive continuous function on $V_0$, so that for some positive continuous function $\varepsilon_2$ on $V_0$, every $\varepsilon_2(x)$-orthonormal set $\{Z_1, \cdots, Z_q\}$ in $T_x M$ $(x \in V_0)$ satisfies $\Sigma_{i=1}^q Lf(Z_i, Z_i) > 2\delta(x)$, where $f \equiv 1/(b - \Sigma_{i=1}^n |z_i|^2)$. Let $\tau_1 = \chi(-\rho) + f$. Then $\tau_1 \in \Psi(q; V_0 \cap D)$ and $\tau_1 \uparrow \infty$ near $\partial D$. Therefore the function $\tau \equiv \max\{f, \tau_1\}$ is an exhaustion function of $V_0 \cap D$, which belongs to $\Psi(q; V_0 \cap D)$. This shows $D$ is weakly hyper-$q$-convex. To prove the converse, let $D$ be $C^\infty$ and weakly hyper-$q$-convex. Fix an $x_0 \in \partial D$, and let $V_0$ be a neighborhood of $x_0$ such that on $V_0 \cap D$ there exists an exhaustion function $\phi$ of $V_0 \cap D$, which is in $\Psi(q; \phi)$. Let $\{r_i\}$ be a sequence of regular values of $\phi$ such that $r_i \uparrow \infty$, and let $D_i \equiv \{\phi < r_i\}$. Thus each $D_i \subset\subset V_0 \cap D$, and each $D_i$ is $\mathcal{C}^\infty$ hyper-$q$-convex; furthermore, each $\partial D_i$ approximates $\partial D \cap V_0$ (we ignore the portion of $\partial D_i$ which approximates $\partial V_0 \cap D$). In $D_i$, let $\rho_i$ denote the distance from $\partial D_i$, and let $\rho$ be the usual function on $D$ denoting the distance from $\partial D$. Let $V_1$ be a sufficiently small neighborhood of $x_0$ such that $V_1 \subset V_0$ and such that for all $y \in V_1 \cap D$, $\rho(y)$ is realized as the length of a unique geodesic from $y$ to $\partial D \cap V_1$. Fix such a $y \in V_1 \cap D$, and consider only $i$ so large that $y \in V_1 \cap D_i$. For each such $i$, there exist a $p_i \in \partial D$ and a geodesic of unit speed $\zeta_i$ joining $y$ to $p_i$ with length $\rho_i(y)$. Elementary considerations show that, after passing to a subsequence if necessary, $p_i \in V_1 \cap \partial D_i$ for all large $i$ and that $p_i$ converges to some $p \in V_1 \cap \partial D$, and $\zeta_i$ converges to a minimizing geodesic $\zeta$ joining $y$ to $p$. We may as well assume at this point that $V_0$ is so small that the bisectional curvature is $q$-nonnegative in $V_0$, and $\rho$ is $\mathcal{C}^\infty$ in $V_0$. Now at $y$, let $C_i$ (resp., $C$) be the complex subspace of dimension $n-1$ in $T_y M$ orthogonal to $\dot{\zeta}_i$ (resp., $\dot{\zeta}$). Then the standard second variation argument (see especially the proof of Lemma 6 in [7]) shows that for any orthonormal basis $\{e_1, Je_1, \cdots, e_{n-1}, Je_{n-1}\}$ in $C_i$, $-\Sigma_{j=1}^q P\rho_i(e_j e_j) > 0$. Since the $C_i$'s converge to $C$ and $\rho_i$ converges uniformly in a neighborhood of $y$ to $\rho$, we obtain in the limit: $-\Sigma_{j=1}^q P\rho(e_j, e_j) \geqslant 0$ for all orthonormal bases $\{e_1, Je_1, \cdots, e_{n-1}, Je_{n-1}\}$ in $C$. This is equivalent to $\Sigma_{j=1}^q L(-\rho)(e_j, e_j) \geqslant 0$ at $y$, since $\rho$ is $\mathcal{C}^\infty$ near $y$. By letting $y$ approach $x_0$, we obtain the following: Let $\tilde{\rho}: V_1 \to \mathbf{R}$ be the function $\tilde{\rho} = -\rho$ on $V_1 \cap D$ and $\tilde{\rho} =$ the distance from $\partial D$ on $V_1 - D$. Then $\tilde{\rho}$ is a $\mathcal{C}^\infty$ local defining function of $\partial D$ at $x_0$ such that $\Sigma_{j=1}^q L\tilde{\rho}(e_j, e_j) \geqslant 0$ at $x_0$ for any orthonormal basis $\{e_1, Je_1, \cdots, e_{n-1}, Je_{n-1}\}$ of $T_{x_0}M$, which lies in $\partial D$. This is clearly equivalent to the $C^\infty$ $q$-convexity of $\partial D$ at $x_0$.

## Bibliography

[1] A. Andreotti & H. Grauert, *Théorèmes de finitude pour la cohomologie des espaces complexes*, Bull. Soc. Math. France **90** (1962) 193–259.

[2] C. Badji, *Sur la $\bar{\partial}$-cohomologies de bidegre* (1, 1) *de certaines variétés Kähleriennes à bord, à courbure holomorphe bisectionnelle non negative: Un théorème de nullité*, Proc. Sem. Complex Analysis, Intern. Centre Theor. Phys., Trieste, 1980.

[3] H. Grauert & O. Riemenschneider, *Kählersche Mannifaltigkeiten mit hyper-q-konvexen Rand*, Problems in Analysis, A Sympos. in Honor of Salomon Bochner, Princeton University Press, Princeton, 1970, 61–79.

[4] R. E. Greene & H. Wu, *On Kähler manifolds of positive bisectional curvature and a theorem of Hartogs*, Abh. Math. Sem. Univ. Hamburg **47** (1978) 171–185.

[5] A. Kasue, *A note on Kähler manifolds of nonnegative holomorphic bisectional curvature*, to appear in Osaka J. Math.

[6] H. Wu, *An elementary method in the study of nonnegative curvature*, Acta Math. **142** (1979) 57–78.

[7] ———, *On certain Kähler manifolds which are q-complete*, to appear.

UNIVERSITY OF CALIFORNIA, BERKELEY

*Chin. Ann. of Math.*
3 (4) 1982

# ON THE REPRESENTATIONS OF THE LOCAL CURRENT ALGEBRA AND THE GROUP OF DIFFEOMORPHISMS (II)

XIA DAOXING
(*Institute of Mathmetics, Fudan University*)

Dedicated to Professor Su Bu-chin on the Occasion of his 80th Birthday and his 50th Year of Educational Work

**1.** Let $X$ be a $k$-dimensional connected $C^\infty$-manifold. Following [5, 6], we consider the group Diff$(X)$ of all $C^\infty$-bijections $\varphi$ which are identical mappings outside some compact sets $K_\varphi$. The group Diff$(X)$ is a topological group if it is endowed by the usual Schwartz's topology. The unitary representations of this group is closely connected with the theory of quasi-invariant measures, statistical mechanics, the representation of the local current algebra in the quantum theory of fields, etc. (cf. [1—5]). In [6] there is some series of elementary unitary representations of the group Diff$(X)$. The aim of the present paper is to find another series of elementary unitary representations by means of the tangent bundle.

In § 1, the preliminary of the tangent bundle connected with the representations is discussed. In § 2, the representations connected with finite configuations are given. In § 3, the representations connected with the infinite configuations are given. In § 4, the representations connected with the Poisson measures are discussed.

Let $C^\infty(p)$ be the family of all $C^\infty$-functions defined on a neighbourhood of the point $p\in X$, where the different functions in $C^\infty(p)$ may have different domains of definition, and $s$ be a fixed natural number. Now we consider the following linear functional $t(\cdot)$ on $C^\infty(p)$, which depands on the partial derivatives of $f\in C^\infty(p)$ up to the order $s$ only, namely, if $x=(x_1, x_2, \cdots, x_k)$ is the local coordinate in the neighbourhood of the point $p$, with coordinate $x^0$ at $p$, then there are real numbers $t^{i_1 i_2\cdots i_k}$ such that

$$t(f)=\sum_{1\leqslant i_1+\cdots+i_k\leqslant s} t^{i_1\cdots i_k}\frac{\partial^{i_1+\cdots+i_k}}{\partial x_1^{i_1}\cdots\partial x_k^{i_k}} f(p(x))\Big|_{x=x^0} \tag{1}$$

for $f\in C^\infty(p)$. The vecter spaces of all linear functionals of the type (1) is denoted by $\mathscr{T}_s(p)$. The dimension of $\mathscr{T}_s(p)$ is

$$N_s=\sum_{1\leqslant k\leqslant s} n(n+1)\cdots(n+k)/k!.$$

Let $\mathscr{T}(p)$ be the tangent space of $X$ at the point $p$, $\otimes^k\mathscr{T}(p)$ be the space of all

Manuscript received February 2, 1981.

contravariant vectors of order $k$. Obviously,

$$\mathscr{T}_s(p) = \bigoplus_{k=1}^{s} \otimes^k \mathscr{T}(p).$$

Let $B^{(s)}$ be $N_s$-dimensional Euclidean space, $X^{(s)} = \{(p, t) \mid p \in X, t \in \mathscr{T}_s(p)\}$ be projection

$$\pi:\ (p, t) \mapsto p, \quad (p, t) \in X^{(s)},$$

$G^{(s)}$ be the group of all non-singular linear transforms in $B^{(s)}$, and $(X^{(s)}, B^{(s)}, \pi, G)$ be the fiber bundle like the tangent bundle. We choose a basis $t_m$, $m=1, 2, \cdots, N$ of $\mathscr{T}_s(p)$ arbitrarily, then we choose the cooresponding dual basis $f_m$, $m=1, 2, \cdots, N$, such that

$$t_l(f_m) = \delta_{lm}.$$

We can choose the neighbourhood $O(p)$ of the point $p$ sufficiently small such that $O(p)$ is diffeomorphic to a sphere in the $k$-dimensional Euclidean space with the local coordinate $x = (x_1, \cdots, x_k)$ and there exists the dual basis $t_l^{(q)}$, $l=1, 2, \cdots, N_s$ is $\mathscr{T}_s(q)$ for any $q \in O(p)$ such that

$$t_l^{(q)}(f_m) = \delta_{lm}.$$

If $t \in \mathscr{T}_s(q)$, there is a set of numbers $\eta_m$ such that $t = \sum \eta_m t_m^{(q)}$, hence we have an open set $O_p^{(s)} = \{(q, t) \mid q \in O(p), t \in \mathscr{T}_s(q)\}$ of $X$, in which the local coordinate is

$$(x_1, \cdots, x_k, \eta_1, \cdots, \eta_{N_s}\},$$

where $(x_1, \cdots, x_k)$ is the local coordinate of $q$. Of course, $X^{(s)}$ is also a manifold.

If $\psi \in \mathrm{Diff}(X)$, $p \in X$, $t \in \mathscr{T}_s(p)$, then $(d\psi)t$ is an element in $\mathscr{T}_s(\psi(p))$ satisfying

$$((d\psi)t)(f) = t(f \circ \psi), \quad f \in C^\infty(\psi(p)),$$

and $d\psi$: $t \mapsto (d\psi)t$ is a linear transformation on $\mathscr{T}_s(p)$. Hence we can define a $C^\infty$-diffeomorphism $\tilde{\psi}$ as follows

$$\tilde{\psi}(p, t) = (\psi(p), (d\psi)t), \quad (p, t) \in X^{(s)}.$$

Let $(\mathrm{Diff}(X))^\sim = \{\tilde{\psi} \mid \psi \in \mathrm{Diff}(X)\}$.

If $p \in X$, $D(p) = \{\psi \mid \psi(p) = p, \psi \in \mathrm{Diff}(X)\}$ and $dD(p) = \{d\psi \mid \psi \in D(p)\}$, then $dD(p)$ is obviously a group of linear transformations in $\mathscr{T}_s(p)$. The space $\mathscr{T}_s(p)$ can be decomposed into multually disjoint, invariant and transitive with respect to the group $dD(p)$ sets $W_j(p)$, $j=0, 1, \cdots$ with $W_0(p) = \{0\}$. For any $q \in X$, if $\psi \in \mathrm{Diff}(X)$ and $\psi(p) = q$, then $(d\psi)W_j(p)$ is invariant and transtive with respect to $dD(q)$. The set $(d\psi)W_j(p)$ is denote by $W_j(q)$. Hence the submanifold

$$X_j^{(s)} = \{(p, t) \mid p \in X, t \in W_j(p)\}, \quad j=1, 2, \cdots$$

of $X^{(s)}$ is invariant and transitive with respective to $(\mathrm{Diff}(X))^\sim$ (actually the restriction of $(\mathrm{Diff}(X))^\sim$ in $X_j^{(s)}$). The trivial case is $W_0(p) = \{0\}$ and $X_0^{(s)} = X$.

For any $t_0 \in W_j(p)$, the isotropic group at $t_0$ is $dD(p, t_0) = \{d\psi \mid \psi \in D(p), (d\psi)t_0 = t_0\}$ The submanifold is diffeomorphic to the manifold $dD(p)/dD(p, t_0)$, which is a left coset. Thus every $W_j(p)$ is a $C^\infty$-manifold, for $j \neq 0$.

Suppose that $m$ is a smooth measure in $X$, Diff $(X, m)$ is the subgroup of all diffeomorphisms $\psi$ in Diff$(X)$ satisfying

$$\psi m = m.$$

According to the measure $m$, we can choose the neighbourhood of $p$ and the local coordinate there such that

$$dm(p(x)) = \prod_{j=1}^{k} dx_j.$$

Let $D(p, m)$ be the group of all diffeomorphisms $\psi$ in a certain Diff$(O(p), m)$ and $dD(p, m) = \{d\psi \mid \psi \in D(p, m)\}$. In the following we only consider the case that $W_j(p)$ is also transitive with respect to $dD(p, m)$ and there is a smooth measure in it, which is aslo invariant with respective to $dD(p, m)$. In this non-trival case, the manifold is called suitable. We shall then fixed a suitable $W_j(p)$ and denote it by $T(p)$. This does exist, for example, if $T(p) = \mathscr{T}(p) - \{0\}$ and the smooth measure $\nu$ in it is

$$d\nu(t) = \prod_{j=1}^{k} dt^j,$$

where $t = \sum_{j=1}^{k} t^j \frac{\partial}{\partial x_j}$. If $T(p)$ is suitable, and $\psi \in$ Diff$(X)$, then $T(\psi(p)) = (d\psi)T(p)$ is also suitable. The manifold

$$(X, T) = \{(p, t) \mid p \in X, t \in T(p)\}$$

is also denoted by $\widetilde{X}$. If $\tilde{\nu}$ is any smooth measure in $T$, then there is a measure $\xi$ in $X$ which is equivalent to the product measure $m \times \tilde{\nu}$. The measure $\xi$ is denote by $\tilde{m}$ when $\tilde{\nu}$ is $\nu$. In the following sections we shall construct the unitary representations by means of $\widetilde{X}$ and $(\text{Diff}(X))^{\sim}$.

**2.** By the method similar to that in § 1 in [6], we construct the unitary representations of Diff$\widetilde{X}$. For the convienence of the reader we shall give the details.

Let $\widetilde{X}^n$ be the topological product of $n$-copies of $\widetilde{X}$, $\xi_n = \xi \times \cdots \times \xi$, where $\xi$ is a smooth measure in $\widetilde{X}$. Let $L^2_{\xi_n}(\widetilde{X}^n, W)$ be the Hilbert space of all $W$-valued, measurable and square integrable functions $F$ on $\widetilde{X}^n$ with norm

$$\|F\|^2 = \int \|F(q_1, q_2, \cdots, q_n)\|^2_W d\xi(q_1) \cdots d\xi(q_n) < +\infty,$$

where $W$ is a Hilbert space.

Now we construct a unitary represtation $U$ of Diff$(X)$ in $L^2_{\xi_n}(\widetilde{X}^n, W)$ as follows

$$(U(\psi)F)(q_1, \cdots, q_n) = \prod_{j=1}^{n} J_\psi^{1/2}(q_j) F(\tilde{\psi}^{-1}q_1, \cdots, \tilde{\psi}^{-1}q_n),$$

where $J_\psi(q) = d\xi(\tilde{\psi}^{-1}q)/d\xi(q)$. In particular, when $\xi = m \times \tilde{\nu}$, we have

$$J_\psi(p, t) = \frac{dm(\psi^{-1}p)}{dm(p)} \frac{d\tilde{\nu}((d\psi^{-1})t)}{d\tilde{\nu}(t)}, \quad t \in T(p).$$

If $\xi = m \times \nu$ and $\psi \in$ Diff$(X, m)$, then

$$(U(\psi)F)(q_1, \cdots, q_n) = F(\tilde{\psi}^{-1}q_1, \cdots, \tilde{\psi}^{-1}q_n).$$

Let $\rho$ be an irreducible unitary representation of the symmetric group of order $n$

in $W$. Let $H_{n,\rho}$ be the subspace of all functions $F$ in $L^2_{\xi_n}(\widetilde{X}^n, W)$ satisfying

$$F(q_{\sigma(1)}, \cdots, q_{\sigma(n)}) = \rho(\sigma)^{-1} F(q_1, \cdots, q_n), \quad \sigma \in S_n.$$

The restriction of $U$ in $H_{n,\rho}$ is denoted by $V^{\rho,T,\xi}$ or simplely by $V^\rho$. This is also a unitatry representation of Diff$(X)$. In particular, $V^\rho$ coincides with that in [6], when $T = \{0\}$. But in general these two representations are different.

**Theorem 1.** *If* dim $X > 1$, *then the restriction of* $V^{\rho,T,\widetilde{m}}$ *in* Diff $(X)$ *is irreducible.*

*Proof* By the similar method in the proof of the Theorem 2 of § 1 in [6], we replace the Lemma 1 there by the following lemma

**Lemma 1** *If* $p_1, \cdots, p_n$ *are n different points,* $t_j \in T(p_j)$, $j = 1, 2, \cdots, m$ *then there exist the neighbourhoods* $O_j$ *of the points* $(p_j, t_j)$ *in* $\widetilde{X}$, $j = 1, 2, \cdots, n$, *such that*

(1) *the closure* $\overline{O}_j$ of $O_j$ *is* $C^\infty$*-diffeomorphic to a closed sphere,* $\overline{O}_i \cap \overline{O}_j = \phi$ *for* $i \neq j$, *and* $\widetilde{m}(O_1) = \cdots = \widetilde{m}(O_n)$.

(2) *for any permutation* $k_1, k_2, \cdots, k_n$ *of* 1, 2, $\cdots$, *n, there exists a* $\psi \in$ Diff$(X, m)$ *such that* $\overline{\psi}(\overline{O}_i) = \overline{O}_{k_i}$, $i = 1, 2, \cdots, n$.

*Proof* Without lose of generality, we may suppose that $X$ is an open sphere in the Euclidean space, and $m$ is the Lebesgue measure. From [6], there exists $\psi_{ij} \in$ Diff $(X, m)$ for any two points $x_i$, $x_j (x_i \neq x_j)$, such that (1) $\psi_{ij} D^\varepsilon_{x_i} = D^\varepsilon_{x_j}$, $\psi_{ij} D^\varepsilon_{x_j} = D^\varepsilon_{x_i}$ for a certain sufficient small positive number $\varepsilon$, where $D^\varepsilon_x$ is an open sphere with center $x$ and radius $\varepsilon$, (2) $\psi_{ij}$ is an identical mapping in a certain neighbourhood of every $x_k$ for $k \neq i$, $k \neq j$.

Now we construct a mapping $\varphi_{ij} \in$ Diff$(X, m)$ satisfying the following condition: there exists a small positive number $\varepsilon$ such that $\varphi_{ij} D^\varepsilon_{x_i} = D^\varepsilon_{x_j}$, $\varphi_{ij} D^\varepsilon_{x_j} = D^\varepsilon_{x_j}$ and

$$(d\varphi_{ij})(d\psi_{ij}) t_i = t_j, \ (d\varphi_{ij})(d\psi_{ij}) t_j = t_i.$$

In fact, by the transitivity of $T$, there is a mapping $\varphi \in D(x_i, m)$ such that $(d\varphi) \cdot (d\psi_{ij}) t_j = t_i$. Let

$$\hat{\varphi} = \psi_{ij}^{-1} \circ \varphi^{-1} \circ \psi_{ij}^{-1}.$$

Obviously, $\hat{\varphi} \in D(x_j, m)$ and $(d\hat{\varphi})(d\psi_{ij}) t_i = t_j$. We can modify the mappings $\varphi$ and $\hat{\varphi}$ in the neighbourhood of $x_j$ and extend them suitablely so that $\varphi_{ij}$ satisties the above conditions. From § 1, we know that the mapping $\varphi_{ij} \circ \psi_{ij}$ preserves the measure $\widetilde{m}$.

We take a suitable neighbourhood $O_i$ of $(x_i, t_i)$ such that $\pi O_i = D^\varepsilon_{x_i}$. Let $\widetilde{\varphi}_{ij} \circ \widetilde{\psi}_{ij} O_i = O_j$. Hence

$$\widetilde{\varphi}_{ij} \circ \widetilde{\psi}_{ij} O_j = \widetilde{\hat{\varphi}} \circ \widetilde{\psi}_{ij} O_j = (\widetilde{\varphi} \circ \widetilde{\psi}_{ij})^{-1} O_j = O_i.$$

**Theorem 2.** *If* dim $X > 1$, *and* $\xi$ *is an arbitrary smooth measure in* $\widetilde{X}$, *then* $V^{\rho,T,\xi}$ *is an irreducible unitary represention of* Diff$(X)$.

*Proof* We construct a unitary operator $\Xi$ from $L^2_{\xi_n}(\widetilde{X}^n, W)$ onto $L^2_{\widetilde{m}_n}(\widetilde{X}^n, W)$ as follows

$$\Xi: F(q_1, \cdots, q_n) \mapsto F(q_1, \cdots, q_n) \prod_{j=1}^{n} \left( \frac{d\xi(q_j)}{d\widetilde{m}(q_j)} \right)^{1/2}.$$

Then $V^{\rho,T,\xi}=\Xi^{-1}V^{\rho,T,\tilde{m}}\Xi$. However, $V^{\rho,T,\tilde{m}}$ is an irreducible unitary representation of Diff$(X)$ by Theorem 1. Thus $V^{\rho,T,\xi}$ is irreducible unitary representation also.

**3**. Let $B_{\tilde{X}}$ be the set of all finite configurations in $X$, $\Gamma_{\tilde{X}}$ be the set of all infinite and locally finite configurations in $\tilde{X}$, $\mu$ be a measure in $\Gamma_{\tilde{X}}$ which is quasi-invariant with respect to $(\text{diff}(X))^{\sim}$, and$L^2_\mu(\Gamma_{\tilde{X}})$ be the Hilbert space of all measurable and $\mu$-square integrable functions on $\Gamma_{\tilde{X}}$. We construct the unitary representation $U_\mu$ defined by

$$(U_\mu(\psi)F)(\gamma)=J_\psi^{1/2}(\gamma)F(\gamma),\quad \gamma\in\Gamma_{\tilde{X}},\quad F\in L^2_\mu(\Gamma_{\tilde{X}}),$$

where $J_\psi(\gamma)=d\mu(\tilde{\psi}^{-1}\gamma)/d\mu(\gamma)$.

**Theorem 3.** *If $\mu$ is a quasi-invariant and ergodic measure in $\Gamma_{\tilde{X}}$ with respect to* $(\text{Diff}(X))^{\sim}$, *$\rho$ is an irreducible unitary representation of the symmectric group $S_n$, then $U_\mu\otimes V^{\rho,T,\xi}$ is also an irreducible representation of* Diff$(X)$.

Let $\rho$ be an irreducible representation of $S_n$ in $W$, $n=0, 1, 2, \cdots$. We use the similar notation $\Gamma_{\tilde{X},n}$ as in [6]. Suppose that $\tilde{\mu}$ is the Campbell measure of $\mu$ in $\Gamma_{\tilde{X},n}$. Let $L^2_{\tilde{\mu}}(\Gamma_{\tilde{X},n}, W)$ be the Hilbert space of all the $W$-valued measurable and square integrable functions $F$ on $\Gamma_{\tilde{X},n}$, with

$$\|F\|^2=\int_{\Gamma_{\tilde{X}n}}\|F(c)\|^2_W\,d\tilde{\mu}(c)<+\infty.$$

We construct the unitary representation

$$(U(\psi)F)(\gamma, q_1, \cdots, q_n)=J_\psi^{1/2}(\gamma)F(\tilde{\psi}^{-1}\gamma, \tilde{\psi}^{-1}q_1, \cdots, \tilde{\psi}^{-1}q_n).$$

Let $H_{\mu,n,\rho}$ be the subspace of all functions $F$ in $L^2_{\tilde{\mu}}(\Gamma_{\tilde{X},n}, W)$ which satisfies the condition

$$F(\gamma;\ q_{\sigma(1)}, \cdots, q_{\sigma(n)})=\rho(\sigma)^{-1}F(\gamma; q_1, \cdots, q_n),\quad \sigma\in S_n,$$

and $\tilde{U}^\rho_\mu$ be the restriction of the representation of $U(\psi)$ in the space $H_{\mu,n,\rho}$.

**Theorem 4.** *If $\mu$ is an ergodic and quasi-invariant measure in $\Gamma_{\tilde{X}}$ with respact to* $(\text{Diff}(X))^{\sim}$ *and $\rho$ is an irreducible unitary representbtion of $S_n$, then the unitatry representation of $\tilde{U}^\rho_\mu$ is irreducible.*

The proof of these two theorems is similar to that in § 3 of [6].

**4**. In this section we consider the unitatry represntation constructed by the Poisson measure.

Let $\Delta_{\tilde{X}}=B_{\tilde{X}}\cup\Gamma_{\tilde{X}}$. We add the point $O$ to the set $T(p)$ and define $\nu(\{0\})=0$. Let $\tilde{\Delta}_{\tilde{X}}$ be the set of all those configurations in $\Delta_{\tilde{X}}$ of which there are only finit points $(p, t)$ with have non-vanishing $t$. Let $\xi$ be a smooth measure on $X$,, $\lambda>0$ be a parameter, and $\mu^\xi_\lambda$ be the Poisson measure with the parameter $\lambda$ cooresponding to the smooth measure $\xi$.

**Theorem 5.** *If $\xi=m\times\tilde{\nu}$ is equivalentk to a smooth measure and $\tilde{\nu}\sim\nu, \tilde{\nu}(T)<+\infty$, then the Poisson measure $\mu^\xi_\lambda$ is concentrated in $\tilde{\Delta}_{\tilde{X}}$ and quasi-invariant with respect to* $(\text{Diff}(X))^{\sim}$ *with the Radon-Nikodym's derivative*

$$\frac{d\mu_\lambda^\xi(\tilde{\psi}^{-1}\nu)}{d\mu_\lambda^\xi(\nu)} = \prod_{(p,t)\in\nu} \frac{dm(\tilde{\psi}^{-1}p)}{dm(p)} \frac{d\tilde{\nu}((d\psi^{-1})t)}{d\tilde{\nu}(t)}, \quad \nu\in\tilde{\Delta}_{\tilde{X}} \tag{2}$$

*for* $\psi\in \mathrm{Diff}(X)$.

*Proof* We have to consider the case of $m(X)=+\infty$ only, since in the opposite case the measure $\mu_\lambda^\xi$ is concentrated in the set $B_{\tilde{X}}$. We constuct a sub-manifold $Y\subset X$ with $m(Y)<+\infty$. Let $\tilde{Y}=\{(p, t)\,|\,p\in X,\ t\in T(p),\ t\neq 0\}$. By the definition of the Poisson measure, we have

$$\mu_\lambda^\xi\{\nu\in \Delta_{\tilde{X}}\,|\,|\nu\cap\tilde{Y}|=n\} = \frac{(\lambda m(Y)\tilde{\nu}(T))^n}{n!}\, e^{-\lambda m(Y)\tilde{\nu}(T)}.$$

Hence

$$\mu_\lambda^\xi(\tilde{\Delta}_{\tilde{X}}) \geqslant \sum_{n=1}^{\infty} \mu_\lambda^\xi(\{\nu\in \Delta_{\tilde{X}}\,|\,|\gamma\cap\tilde{Y}|=n\}) = (e^{\lambda m(Y)\tilde{\nu}(T)}-1)e^{-\lambda m(Y)\tilde{\nu}(T)}.$$

When the submanifold $Y$ varies and $m(Y)\to\infty$, we have $\mu_\lambda^\xi(\tilde{\Delta}_{\tilde{X}})=1$. The other part of Theorem 5 can be proved by this equality.

**Theorem 6.** *If* $m(X)=+\infty$, $\tilde{\nu}(T)<+\infty$ *and* $\xi=m\times\tilde{\nu}$, *then the cooresponding Poisson measure* $\mu_\lambda^\xi$ *is concentrated in* $\tilde{\Gamma}_{\tilde{X}}=\tilde{\Delta}_{\tilde{X}}\cap\Gamma_{\tilde{X}}$, *quasi-invariant and ergodic with respect to the group* $(\mathrm{Diff}(X))^{\sim}$.

## References


[1] Goldin, G. A., Grodnik, J., Powers, R. & Sharp, D. H., *Jour. Math. Phys.*, **15** (1974), 88.

[2] Goldin, G. A., Menikoff, R, Sharp, D, H., Particle Statistics from induced representations of a local current group. (preprints).

[3] Matthes, K., Kerstan, J., Mecke, J., Infinite Divisibie Point Processes, John Wiley Sons, (1978)

[4] Menikoff, R., *Jour. Math. Phys.* **15** (1974), 1138.

[5] Xia, Daoxing (夏道行), On the represtatiots of the local current algebra and the group of diffeomorphisms (I), *Sci. Sinica* (1979) *Special Issue (II)*, 249—260.

[6] Верщик, А. М., Гельфанд, И. М., Граев, Й. М., *Представления Группы Диффеоморфизмов, УМН*, **30** (1975), Вып. 6, 2—50.

J. DIFFERENTIAL GEOMETRY
16 (1981) 153–159

# THE CHARACTERISTIC NUMBERS OF 4-DIMENSIONAL KÄHLER MANIFOLDS

Y. L. XIN

*Dedicated to the author's teacher Professor Buchin Su*

## 1. Introduction

There have been many results about the relation between the curvature of a Riemannian manifold $M$ and its characteristic numbers. S. S. Chern and J. Milnor [3] proved that a 4-dimensional manifold with sectional curvature everywhere of the same sign has nonnegative Euler number. M. Berger [1] and N. Hitchin [6] considered the case of an Einstein manifold. H. Donnelly [4] obtained inequalities involving the Euler number and the Pontrjagin number of Einstein Kähler manifolds. S. T. Yau [11] and A. Polombo [8] generalized Gray-Hitchin-Thorpe [5], [6], [9] inequality to $k$-Ricci pinched manifolds and considered the $k$-sectionally pinched case.

In the present paper the similar problem for $k$-Ricci pinched Kähler manifold is considered, and a generalization of Donnelly's inequalities is obtained (Theorem 1).

On the other hand R. Bishop and S. I. Goldberg [2] proved that a 4-dimensional Kähler manifold with holomorphic sectional curvature everywhere of the same sign has nonnegative Euler number. This result is improved in Theorem 2 of this paper.

Thus the main results are the following two theorems.

**Theorem 1.** *Let $M$ be a compact 4-dimensional Kähler manifold with Euler number $\chi$ and Pontrjagin number $p$. If $M$ is $k$-th Ricci pinched with $k \geqslant \sqrt{2}/2$, then the inequalities*

$$\chi + \frac{3 - 5k^2}{2k^2} p \geqslant 0, \tag{1}$$

Received, November 1, 1979, and, in revised form, April 14, 1980. The author would like to thank Professors C. N. Yang, C. H. Gu and A. Phillips for their help, and is grateful to the State University of New York at Stony Brook and University of California at Berkeley for their hospitality.

(2) $$\chi + \frac{1}{2}p \geqslant 0$$

*are valid. Furthermore, if the equality in* (1) *occurs, then M must be in one of the following three cases:*

(i) *M has constant holomorphic curvature,*

(ii) *the universal covering manifold of M is a $K_3$ surface,*

(iii) *M is flat.*

*If the equality in* (2) *occurs, then M must be in one of cases (ii) and (iii) above.*

**Theorem 2.** *Let M be a compact* 4*-dimensional Kähler manifold with Euler number $\chi$ and Pontrjagin number p. If M is $\lambda$-holomorphically pinched with $\lambda \geqslant 0$, then*

$$\chi + \frac{1}{2}p \geqslant 0, \quad \chi + \min\left(\frac{1-2\lambda-5\lambda^2}{6\lambda^2}, \frac{\lambda^2}{\lambda^2-4}\right)p \geqslant 0 \quad \text{for } \frac{1}{4} \leqslant \lambda \leqslant 1,$$

*and, otherwise*

(3) $$\chi + \frac{\lambda^2}{1-4\lambda^2}p \geqslant 0, \quad \chi + \frac{\lambda^2}{\lambda^2-4}p \geqslant 0.$$

We should point out that A. Polombo [7] has obtained similar results, which however do not cover the above theorems.

## 2. Preliminary notation

First of all, we construct a special Hermitian basis at any point $p$ in a 4-dimensional Kähler manifold $M$. Let $e_1$ and $e_2$ be unit eigenvectors of the Ricci curvature such that it reaches its maximum and minimum respectively. It is clear that $e_1$ and $e_2$ are mutually perpendicular. Therefore using the canonical almost complex structure $J$ we obtain a Hermitian basis $\{e_1, Je_1, e_2, Je_2\}$ which diagonalizes the Ricci curvature tensor. In this case $R_{11} = R_{22}$ and $R_{33} = R_{44}$.

From the author's previous paper [10], we have the Euler number $\chi$ and the Pontrjagin number $p$ for any 4-dimensional Kähler manifold:

(4) $$\chi = \frac{1}{8\pi^2}\int_M \left(|W^-|^2 + \frac{S^2}{12} - 2|R^{+-}|^2\right)\Omega,$$

(5) $$p = \frac{1}{4\pi^2}\int_M \left(\frac{S^2}{24} - |W^-|^2\right)\Omega,$$

where $W^-$ is the antiself dual part of the conformal curvature tensor, $R^{+-}$ is the part of the Riemannian curvature tensor which is self dual on the first two indices as well as antiself dual on the last two [10], $S$ is the scalar curvature, and $\Omega$ is the volume form of the manifold.

Equivalently, (4) and (5) can be expressed in another form:

$$\chi = \frac{1}{8\pi^2}\int_M\left(|R^{--}|^2 + \frac{1}{16}S^2 - 2|R^{+-}|^2\right)\Omega, \tag{4'}$$

$$p = \frac{1}{4\pi^2}\int_M\left(\frac{S^2}{16} - |R^{--}|^2\right)\Omega, \tag{5'}$$

where $R^{--}$ is the part of the Riemannian curvature tensor and is antiself dual on both pairs of indices.

By directly computing, we have

$$|R^{+-}|^2 = \frac{1}{4}(R_{1212} - R_{3434})^2, \tag{6}$$

$$|R^{+-}|^2 = \frac{1}{16}(R_{11} + R_{22} - R_{33} - R_{44})^2 = \frac{1}{4}(R_{11} - R_{33})^2 \tag{6'}$$

under the special Hermitian basis $\{e_1, Je_1, e_2, Je_2\}$.

Let $X$ and $Y$ be perpendicular unit tangent vectors of $M$ at any point $p$, such that $\langle X, JY\rangle = 0$. Then we have the formula [2]

$$\begin{aligned} K(X, Y) + K(X, JY) = \frac{1}{4}[&H(X + JY) + H(X - JY) + H(X + Y) \\ &+ H(X - Y) - H(X) - H(Y)], \end{aligned} \tag{7}$$

where $K(X, Y)$ is the sectional curvature of the plane spanned by $X$, $Y$, and $H(X) = K(X, JX)$. By (7) we obtain the components of the Ricci curvature tensor:

$$\begin{aligned} R_{11} &= K(e_1, Je_1) + K(e_1, e_2) + K(e_1, Je_2) \\ &= H(e_1) + \frac{1}{4}[H(e_1 + Je_2) + H(e_1 - Je_2) + H(e_1 + e_2) \\ &\qquad + H(e_1 - e_2) - H(e_1) - H(e_2)], \\ R_{33} &= H(e_2) + \frac{1}{4}[H(e_1 + Je_2) + H(e_1 - Je_2) + H(e_1 + e_2) \\ &\qquad + H(e_1 - e_2) - H(e_1) + H(e_2)], \end{aligned}$$

from which it follows that

$$\begin{aligned} S = {}&H(e_1 + Je_2) + H(e_1 - Je_2) + H(e_1 + e_2) \\ &+ H(e_1 - e_2) + H(e_1) + H(e_2), \end{aligned} \tag{8}$$

and (6) can be written in the following form:

$$|R^{+-}|^2 = \frac{1}{4}[H(e_1) - H(e_2)]^2. \tag{9}$$

By definition a $k$-Ricci pinched manifold is one in which there is a number $k > 0$ such that

$$\frac{1}{4}|S| \geqslant k|R_{ii}| \tag{10}$$

for all $i$. It is easy to see $k \leqslant 1$. If the equality in (10) occurs, then either $k = 1$ or $S = 0$. Both conditions imply that the manifold is an Einstein manifold; furthermore in the second case it must be Ricci flat.

## 3. Proof of Theorem 1

From the pinching condition (10), we have

$$R_{11}^2 + R_{33}^2 \leqslant \frac{S^2}{8k^2}.$$

Substituting the above inequality into (6′) yields the following:

$$|R^{+-}|^2 = \frac{1}{4}(R_{11} - R_{33})^2 = \tfrac{1}{2}(R_{11}^2 + R_{33}^2) - \frac{1}{4}(R_{11} + R_{33})^2$$

$$\leqslant \frac{S^2}{16K^2} - \frac{S^2}{16} = \frac{1 - K^2}{16K^2} S^2. \tag{11}$$

If the equality holds above, then the equality also holds in (10) for $k$-Ricci pinched manifolds. Thus the equality in (11) occurs iff $k = 1$ or $S = 0$.

From (4), (5) and (11), we have

$$\chi + bp \geqslant \frac{1}{8\pi^2} \int_M \left[ (1 - 2b)|W^-|^2 + \frac{5k^2 - 3 + 2bk^2}{24k^2} S^2 \right] \Omega \tag{12}$$

for any real $b$. Taking $b = \frac{1}{2}(3 - 5k^2)/k^2$, we reduce (12) to

$$\chi + \frac{3 - 5k^2}{2k^2} p \geqslant \frac{3}{8\pi^2} \int_M \frac{2k^2 - 1}{k^2} |W^-|^2 \Omega, \tag{12′}$$

which gives (1) when $K \geqslant \sqrt{2}/2$. The equality in (1) occurs only if one of the following conditions holds:

(i) $K = 1$, $|W^-| = 0$ and $S \neq 0$;
(ii) $S = 0$, $k^2 = \frac{1}{2}$ and $|W^-| \neq 0$;
(iii) $S = 0$, $|W^-| = 0$.

Under the first condition $M$ has constant holomorphic curvature [10]. The second condition means that the universal covering of $M$ is a $K_3$ surface [6]. When $S = 0$ and $|W^-|^2 = 0$, then $\chi = 0$, which forces $M$ to be flat [1]. Taking $b = \frac{1}{2}$, from (12) we have (2) provided $k \geqslant \sqrt{2}/2$. If the equality holds in (2), then $M$ satisfies either (ii) or (iii) above. The same discussion as above would not be repeated.

## 4. Proof of Theorem 2.

If $M$ is a $\lambda$-holomorphically pinched Kähler manifold with $\lambda > 0$, then there is a constant $A > 0$ such that

$$\lambda A \leqslant H(X) \leqslant A, \tag{13}$$

for any $X \in T_p(M)$.

The pinching condition (13) and (8) give the inequality

$$6\lambda A \leqslant S \leqslant 6A. \tag{14}$$

From (9) and (13) we have

$$|R^{+-}|^2 \leqslant \frac{1}{4}(1-\lambda)^2 A^2. \tag{15}$$

For any $b \geqslant -1$, (4), (5) and (15) give

$$\chi + bp \geqslant \frac{1}{8\pi^2}\int_M \left\{(1-2b)|W^-|^2 + \left[\left(\frac{5}{2}\lambda^2 + \lambda - \frac{1}{2}\right) + 3b\lambda^2\right]A^2\right\}\Omega. \tag{16}$$

Taking $b = \frac{1}{2}$ in (16), we have

$$\chi + \frac{1}{2}p \geqslant \frac{1}{8\pi^2}\int_M \left(4\lambda^2 + \lambda - \frac{1}{2}\right)A^2\Omega.$$

Thus

$$\chi + \frac{1}{2}p \geqslant 0, \quad \text{for } \frac{1}{4} \leqslant \lambda \leqslant 1. \tag{17}$$

Taking $b = \frac{1}{6}(1 - 2\lambda - 5\lambda^2)/\lambda^2$ in (16), we have

$$\chi + \frac{1 - 2\lambda - 5\lambda^2}{6\lambda^2}p \geqslant 0, \tag{18}$$

when $\frac{1}{4} \leqslant \lambda \leqslant 1$.

If we denote

$$\begin{aligned}
\varepsilon_1^- &= e_1 \wedge Je_1 - e_2 \wedge Je_2,\\
\varepsilon_2^- &= e_1 \wedge e_2 - Je_2 \wedge Je_1,\\
\varepsilon_3^- &= e_1 \wedge Je_2 - Je_1 \wedge e_2,
\end{aligned}$$

then

$$\begin{aligned}
\langle R_{\varepsilon_1^-}, \varepsilon_1^- \rangle &= \langle R_{\varepsilon_1^-}(e_1), Je_1\rangle - \langle R_{\varepsilon_1^-}(e_2), Je_2\rangle\\
&= -\frac{S}{2} + 2\langle R_{e_1Je_1}(e_1), Je_1\rangle + 2\langle R_{e_2Je_2}(e_2), Je_2\rangle\\
&= -\frac{S}{2} + 2H(e_1) + 2H(e_2),
\end{aligned}$$

from which it follows that

$$|R^{--}|^2 = \left(H(e_1) + H(e_2) - \frac{S}{4}\right)^2 + L^2, \tag{19}$$

where $L^2$ is the sum of the squares of all the other entries of the matrix $(R^{--})$.

For any $b$ from (9) we have

$$(1-2b)\left(H(e_1) + H(e_2) - \frac{1}{4}S\right)^2 + \frac{1+2b}{16}S^2 - 2|R^{+-}|^2$$

$$= (1-2b)\left(H(e_1) + H(e_2) - \frac{1}{4}S\right)^2 + \frac{1+2b}{16}S^2 - \frac{1}{2}(H(e_1) - H(e_2))^2$$

$$\begin{aligned} &= \left(\tfrac{1}{2} - 2b\right)\left[H(e_1) + H(e_2) - \frac{1}{2}S\right]^2 + \frac{b}{2}\left[S - H(e_1) - H(e_2)\right]^2 \\ &\quad - \frac{b}{2}\left[H(e_1) + H(e_2)\right]^2 + 2H(e_1)H(e_2). \end{aligned} \tag{20}$$

For $0 \leqslant b \leqslant \frac{1}{4}$ it follows from (13), (14) and (20) that

$$\begin{aligned} &(1-2b)\left(H(e_1) + H(e_2) - \frac{1}{4}S\right)^2 + \frac{1+2b}{16}S^2 - 2|R^{+-}|^2 \\ &\geqslant (8\lambda^2 b - 2b + 2\lambda^2)A^2 = 2((4\lambda^2 - 1)b + 2\lambda^2)A^2. \end{aligned} \tag{21}$$

For $0 \leqslant b \leqslant \frac{1}{4}$, (4′), (5′), (19) and (21) give

$$\chi + bp \geqslant \frac{1}{8\pi^2}\int_M \{(1-2b)L^2 + 2[(4\lambda^2 - 1)b + \lambda^2]A^2\}\Omega. \tag{22}$$

Taking $b = \lambda^2/(1 - 4\lambda^2)$ in (22) yields

$$\chi + \frac{\lambda^2}{1 - 4\lambda^2}p \geqslant \frac{1}{8\pi^2}\int_M \frac{1 - 6\lambda^2}{1 - 4\lambda^2}L^2\Omega.$$

Note that $0 \leqslant b \leqslant 1/4$. Thus if $\lambda \leqslant \sqrt{2}/4$, then

$$\chi + \frac{\lambda^2}{1 - 4\lambda^2}p \geqslant 0. \tag{23}$$

We consider again the case $b \leqslant 0$ in (20). In this case

$$\begin{aligned} &(1-2b)\left(H(e_1) + H(e_2) - \frac{1}{4}S\right)^2 + \frac{(1+2b)}{16}S^2 - 2|R^{+-}|^2 \\ &\geqslant (8b - 2b\lambda^2 + 2\lambda^2)A^2 = \left[(8 - 2\lambda^2)b + 2\lambda^2\right]A^2, \end{aligned}$$

from which for any $\lambda \geqslant 0$ we obtain the inequality

$$\chi + \frac{\lambda^2}{\lambda^2 - 4}p \geqslant \frac{1}{8\pi^2}\int_M \frac{4 + \lambda^2}{4 - \lambda^2}L^2\Omega \geqslant 0. \tag{24}$$

Therefore inequalities (3) follow from (17), (18), (23) and (24).

**Remarks.** 1. A result similar to Theorem 2 holds also in the case of nonpositive holomorphic curvature, but the pinching condition $-A \leqslant H(X) \leqslant \lambda A$ with $\lambda \leqslant 0$ must be substituted for $\lambda A \leqslant H(X) \leqslant A$ with $\lambda \geqslant 0$. It is easy to see that the proof is similar.

2. From (4′), (19) and (20) it follows

$$(25)\quad \chi = \frac{1}{8\pi^2}\int_M \left[ L^2 + \tfrac{1}{2}\left(H(e_1) + H(e_2) - \tfrac{1}{2}S\right)^2 + 2H(e_1)H(e_2)\right]\Omega,$$

which is nonnegative when holomorphic curvature has the same sign everywhere. This is the theorem of R. Bishop and S. I. Goldberg, which is a special case of Theorem 2 in this paper. It is easy to see that $\chi = 0$ forces $M$ to be flat.

## References


[1] M Berger, *Sur les variétés d'Einstein compactes*, C. R. III$^e$ Réunion Math. Expression Latine, Namur (1965) 35–55.

[2] R. L. Bishop & S. I. Goldberg, *Some implications of the generalized Gauss-Bonnet theorem*, Trans. Amer. Math. Soc. **112** (1964) 508–535.

[3] S. S. Chern, *On curvature and characteristic class of a Riemannian manifold*, Abh. Math. Sem. Univ. Hamburg **20** (1955) 117–126.

[4] H. Donnelly, *Topology and Einstein Kaehler metrics*, J. Differential Geometry **11** (1976) 259–264.

[5] A. Gray, *Invariants of curvature operators of four-dimensional Riemannian manifolds*, Proc. 13th Biennal Sem. Canad. Math. Congress, Vol. 2, 1971, 42–65.

[6] N. Hitchin, *Compact four-dimensional Einstein manifolds*, J. Differential Geometry **9** (1974) 435–441.

[7] Albert Polombo, *Nombres caracteristiques d'une surface Kählérienne compacte*, C. R. Acad. Sc. Paris Sér. A **283** (1976) 1025–1028.

[8] ______, *Nombres caracteristiques d'une variété Riemannienne de-dimension* 4, J. Differential Geometry **13** (1978) 145–162.

[9] J. A. Thorpe, *Some remarks on the Gauss-Bonnet integral*, J. Math. Mech. **18** (1969) 779–786.

[10] Y. L. Xin, *Remarks on characteristic classes of 4-dimensional Einstein manifolds*, J. Mathematical Phys. **21** (1980) 343–346.

[11] S. T. Yau, *Curvature restrictions on four-manifolds*, preprint.



State University of New York, Stony Brook
Fudan University, Shanghai

*Chin. Ann. of Math.*
4 (3) 1982

# ON INFINITE GALOIS THEORY FOR DIVISION RINGS

XU YONGHUA
*(Institute of Mathematics, Fudan University)*

Dedicated to Professor Su Bu-chin on the Occasion of his 80th Birthday and his 50th Year of Educational Work

In [1] we have established the finite structure theorem between complete rings of linear transformations and have used it to study the finite Galois theory of division rings. In this paper, we shall generalize the structure theorem to infinite case and then we study the infinite Galois theory of division rings, which is hard to deal with as we know. In § 1 we shall establish the infinite structure theorem between complete rings of linear transformations. In § 3 we shall use it to study the fundamental theorem of infinite Galois theory of division rings. And in § 4 we shall indicate that our theory includes the well known finite Galois theory of division rings.

## § 1.

If $G$ is a group of automorphisms of division ring $F$, then we write

$$I(G)=\{f\in F \mid f^{\sigma}=f \text{ for all } \sigma\in G\}.$$

If $K$ is a division subring of $F$, then we write $A(K)=\{\sigma\in G \mid K^{\sigma}=K \text{ for all } k\in K\}$. And we denote the algebra[1)] of $G$ by $E'$ and the complete ring of $P$-linear transformations of $\mathfrak{M}$ by $\mathscr{L}(\tilde{P}, \mathfrak{M})$ just as in [1], where $\tilde{P}$ is any division subring of $F$.

Let $\mathfrak{M}=\sum\limits_{i\in\Gamma} Fu_i$ be a left vector space over a division ring $F$, $\mathbf{E}$ be the ring of all endomorphisms of the additive group $(\mathfrak{M}, +)$. A subset $\{x_i\}_{i\in I}$ of $\mathfrak{M}$ is called having $\nu$-order if the cardinal number of $\{x_i\}_{i\in I}$ is $\aleph_\nu$. We denote it by Card. $\{x_i\}_I=\aleph_\nu$. Let $D_\nu$ be the class of all sets with order$\leqslant\aleph_\nu$ in $\mathfrak{M}$. Then we can define a mapping [ ] of $D_\nu$ into $\mathfrak{M}$ as follows: for any element $\{x_i\}_{i\in I_1}$ of $D_\nu$ with Card. $\{x_i\}_{i\in I_1}\leqslant\aleph_\nu$, it corresponds an element $[\{x_i\}_{i\in I_1}]$ of $\mathfrak{M}$, i. e.,

Manuscript received January 1, 1981.

1) Let $R$ be a ring with an identity, $\Phi$ its center and let $G$ be a group of automorphisms in $R$. Then we call the subalgebra $E'$ over $\Phi$ generated by the (regular) elements $c$ such that $I_c\in G$ the algebra of the group $G$.

$$\{x_i\}_{i\in I_1} \xrightarrow{[\ ]} [\{x_i\}_{i\in I_1}]$$

such that [ ] satisfies the following conditions:

(i) If all $x_i=0$ for $\{x_i\}_{i\in I_1}$, then $[\{x_i\}_{i\in I_1}]=0$,

(ii) $(f[\{x_i\}_{i\in I_1}])\sigma=[\{(fx_i)\sigma\}_{i\in I_1}]$ for $f\in F$, $\sigma\in \mathbf{E}$,

(iii) Let $\{a_i^{(\alpha)}\}_{\substack{i\in I\\ \alpha\in I'}}$ be a set of $F$, Card. $I\leqslant\aleph_\nu$, Card. $I'\leqslant\aleph_\nu$ and let $\{y_\alpha\}_{\alpha\in I'}$ be a set of $F$-linearly independent elements of $\mathfrak{M}$. If the system of equations

$$[\{a_i^{(\alpha)}x_i\}_{i\in I}]=y_\alpha \quad \text{for } \alpha\in I' \tag{1.1}$$

is solvable for $\{x_i\}_{i\in I}$ in $\mathfrak{M}$, then it follows that

(1) Card. $I'\leqslant$Card. $I$,

(2) there exists a subset $\{a_i^{(\alpha)}\}_{i\in I'}$ of $\{a_i^{(\alpha)}\}_{i\in I}$ such that for any element $\{Y_\alpha\}_{\alpha\in I'}$ of $D_\nu$ the following system of equations

$$[\{a_i^{(\alpha)}X_i\}_{i\in I'}]=Y_\alpha, \quad \alpha\in I' \tag{1.2}$$

has one and only one solution for $\{X_i\}_{i\in I'}$, that is, if two subsets $\{x_i\}_{i\in I'}$ and $\{x_i^0\}_{i\in I'}$ satisfy the equation (1.2), then this forces $x_i=x_i^0$ for $i\in I'$.

**Definition 1. 1.** *The above mapping [ ] of $\mathfrak{M}$ is called a $\nu$-soluable function. A vector space is called a $\nu$-soluable space if it has a $\nu$-soluable function. If the function [ ] is defined in the class of all subsets of $\mathfrak{M}$, then we call [ ] simply a soluable function and $\mathfrak{M}$ a soluable space.*

Since $\mathfrak{M}=x\mathbf{E}$ for any non-zero element $x$ of $\mathfrak{M}$, we can extend the $\nu$-soluable function to $\mathbf{E}$. Similarly, a subset $\{\varepsilon_i\}_{i\in I}$ of $\mathbf{E}$ is called a set with $\nu$-order if Card. $I=\aleph_\nu$. Denote the class of all subsets with orders $\leqslant\aleph_\nu$ by $D_\nu^*$, then we define a function of $D_\nu^*$ into $\mathbf{E}$ which is still denoted by [ ], if it is not to be confused with the preceding one. Furthermore these two functions have the following relations: for any element $x$ of $\mathfrak{M}$ we define $x[\{\varepsilon_i\}_{i\in I}]=[\{x\varepsilon_i\}_{i\in I}]$, where Card. $I\leqslant\aleph_\nu$. It is easy to see that the element $[\{x_i\}_{i\in I}]$ of $\mathfrak{M}$ can be written as $[\{x_i\}_{i\in I}]=[\{x\varepsilon_i\}_{i\in I}]$, where $\{\varepsilon_i\}_{i\in I}\subset\mathbf{E}$. It is also easy to see that $[\{\varepsilon_i\}_{i\in I}]\sigma=[\{\varepsilon_i\sigma\}_{i\in I}]$ for any element $\sigma$ of $\mathbf{E}$.

Let $\mathfrak{M}=\sum_{i\in I'}Fu_i$ be a space, G a group of automorphisms of $F$, and let $\mathbf{E}$ be the complete ring of endomorphisms of $(\mathfrak{M}, +)$. Let $T$ and $\mathscr{L}$ be subsets of $\mathbf{E}$, and $I$ be a set. Then $M(T, \mathscr{L};\ I)$ of $\mathbf{E}$ is defined as follows

$$M(T, \mathscr{L};\ I)=\{[\{t_j\sigma_j\}_{j\in I_1}]\,|\,t_j\in T,\ \sigma_j\in\mathscr{L};\ I_1\subseteq I\}, \tag{1.3}$$

where $t_j\in T$, $\sigma_j\in\mathscr{L}$ and $t_j\sigma_j\in\mathbf{E}$ for $j\in I_1$.

**Definition 1. 2.** *Let $\mathfrak{M}=\sum_j Fu_j$ be a space. A set $Q$ of $\mathbf{E}$ is called $\aleph_\nu$-transitive if and only if for any subset $\{x_i\}_{i\in I}$ of $F$-linearly independent elements of $\mathfrak{M}$ and any subset $\{y_i\}_{i\in I}$ of $\mathfrak{M}$, where Card. $I\leqslant\aleph_\nu$, there exists an element $\sigma\in Q$ such that $x_i\sigma=y_i$ for $i\in I$.*

**Definition 1. 3.** *Let* $\mathfrak{M}=\sum_{i\in\Gamma} Fu_i$, $P'=C_F(E')=\{f\in F\,|\,fe'=e'f$ *for all* $e'\in E\}$[2]. *We say that* $\mathfrak{M}$ *is a* $(\aleph_\nu, 1)$*-type space over* $P'$ *if and only if* $\mathfrak{M}$ *is a soluable space and there exists a set* $M(E'_L, \mathscr{L}(F, \mathfrak{M}); I)$ *which is* $\aleph_\nu$*-transitive in* $\mathfrak{M}$ *as the vector space over* $P'$ *where Card.* $I=\aleph_\nu$.

**Definition 1. 4.** $M(T, \mathscr{L}; I)$ *is called free if and only if that if* $[\{t_j\sigma_j\}_{j\in I_1}]=0$ *for any element* $[\{t_j\sigma_j\}_{j\in I_1}]$ *of* $M(T, \mathscr{L}; I)$, *then all* $t_j\sigma_j=0$ *for* $j\in I_1$. *In this case we write* $M(T, \mathscr{L}; I)=\oplus M(T, \mathscr{L}; I)$,

we are now ready to establish the following theorem:

**Theorem 1. 1.** *Let* $\mathfrak{M}=\sum_{i\in\Gamma} Fu_i$ *be a space,* $G$ *a group of automorphisms of* $F$, *and let* $E'$ *be the algebra of* $G$, $P'=C_F(E')$. *Denote the dimension of the left vector space* $F$ *over a subring* $P'$ *by* $[F: P']_L$ *and let* $I$ *be an index set with Card.* $I=[F: P']_L=\aleph_\nu$. *If* $\mathfrak{M}$ *is a* $(\aleph_\nu, 1)$*-type space over* $P'$, *then there exists a subset* $\{r_j\}_{j\in I}$ *of* $E'$ *such that*

$$\mathscr{L}(P', \mathfrak{M})=\oplus M(\{r_{jL}\}_{j\in I}, \mathscr{L}(F, \mathfrak{M}); I).$$

*Conversely, if* $\mathscr{L}(P', \mathfrak{M})=\oplus M(\{r_{jL}\}_{j\in I^*}, \mathscr{L}(F, \mathfrak{M}); I^*)$, *where* $\{r_{jL}\}_{j\in I^*}\subset E'_L$, *then there exists a subset* $\{r_{jL}\}_{j\in I'}$ *of* $\{r_{jL}\}_{j\in I^*}$ *such that*

$$\mathscr{L}(P', \mathfrak{M})=\oplus M(\{r_{jL}\}_{j\in I'}, \mathscr{L}(F, \mathfrak{M}); I')$$

*and* Card. $I'=[F: P']_L$

*Proof* we prove the first assertion. Since $P'$ is a division subring of $F$, there exists a $P'$–basis $\{f^{(\alpha)}\}_{\alpha\in I}$ such that $F=\sum_{\alpha\in I} P'f^{(\alpha)}$, where Card. $I=[F: P']_L=\aleph_\nu$. Hence $\mathfrak{M}=\sum_{i\in\Gamma} Fu_i=\sum_{i\in\Gamma,\alpha\in I} P'f^{(\alpha)}u_i=\sum_{i\in\Gamma,\alpha\in I} P'v_i^{(\alpha)}$, where $v_i^{(\alpha)}=f^{(\alpha)}u_i$. Clearly $\{v_i^{(\alpha)}\}_{i\in\Gamma, \alpha\in I}$ is also a $P'$–basis of $\mathfrak{M}$. Let $\{y_\alpha\}_{\alpha\in I}$ be an arbitrary set of $F$–linear independent elements of $\mathfrak{M}$. By assumption $\mathfrak{M}$ is a $(\aleph_\nu, 1)$-type space over $P'$. In view of Definition 1.3 and (1.3) there exists an element $\sigma=[\{r_{jL}\sigma'_j\}_{j\in I_1}]\in M(E'_L, \mathscr{L}(F, \mathfrak{M}); I)$ for $r_{jL}\in E'_L$, $\sigma'_j\in\mathscr{L}(F, \mathfrak{M})$, $I_1\subseteq I$ such that $\sigma$ satisfies $v_i^{(\alpha)}\sigma=y_\alpha$, $\alpha\in I$. Hence

$$y_\alpha=v_i^{(\alpha)}\sigma=[\{(f^{(\alpha)}u_i)(r_{jL}\sigma'_j)\}_{j\in I_1}]=[\{r_jf^{(\alpha)}(u_i\sigma'_j)\}_{j\in I_1}],\quad \alpha\in I. \tag{1.4}$$

Let $a_{\alpha j}=r_jf^{(\alpha)}$, $x_j=u_i\sigma'_j$ for $\alpha\in I$, $j\in I$, then (1.4) becomes the following formula

$$[\{a_{\alpha j}x_j\}_{j\in I_1}]=y_\alpha,\quad \alpha\in I. \tag{1.5}$$

From the porperty of soluable function[ ] it follows that Card. $I\leqslant$Card. $I_1$, furthermore there exists a subset $\{a_{\alpha j}\}_{j\in I_1}$ of $\{a_{\alpha j}\}_{j\in I}$ such that the following system of functional equations

$$[\{a_{\alpha j}X_j\}_{j\in I}]=Y_\alpha,\quad \alpha\in I \tag{1.6}$$

is soluable for any set $[Y_\alpha]_{\alpha\in I}$ of $\mathfrak{M}$ and it has a unique solution. Now we want to show that $\mathscr{L}(P', \mathfrak{M})\subseteq M(\{r_{jL}\}_{j\in I}, \mathscr{L}(F, \mathfrak{M}); I)$.

In fact, let $\sigma^*\in\mathscr{L}(P', \mathfrak{M})$, $v_i^{(\alpha)}\sigma^*=Y_\alpha(i)$, $\alpha\in I$, $i\in\Gamma$. Consider the system of functional equations (1.6). Then from the soluable property of (1.6) for arbitrary

2) $E'$ is the algebra of the group $G$, where $G$ is a group of automorphisms in $F$.

subset $\{Y_\alpha\}_{\alpha\in I}$ of $\mathfrak{M}$ it follows that for any element $i\in\Gamma$ the following system of equations

$$[\{a_{\alpha j}X_j(i)\}_{j\in I}]=Y_\alpha(i),\quad \alpha\in I \tag{1.7}$$

has a solution. Since $\mathfrak{M}$ is a soluable space, (1.7) has a unique solution $X_j(i)=x_j(i)\in\mathfrak{M}$ for $j\in I$. On the other hand, since $\mathfrak{M}=\sum_{i\in\Gamma}Fu_i$, there exists an element $\sigma'_j\in\mathscr{L}(F,\mathfrak{M})$ such that

$$u_i\sigma'_j=x_j(i) \text{ for } j\in I,\ i\in\Gamma. \tag{1.8}$$

Let $\bar{\sigma}=[\{r_{jL}\sigma'_j\}_{j\in I}]$, then we have

$$v_i^{(\alpha)}\bar{\sigma}=[(f^{(\alpha)}u_i)\{r_{jL}\sigma'_j\}_{j\in I}]=[\{r_jf^{(\alpha)}(u_i\sigma'_j)\}_{j\in I}]=[\{a_{\alpha j}x_j(i)\}_{j\in I}]=v_i^{(\alpha)}\sigma^*.$$

Since $\{v_i^{(\alpha)}\}_{\substack{\alpha\in I\\ i\in\Gamma}}$ is $P'$-basis of $\mathfrak{M}$, it follows that $\sigma^*=\bar{\sigma}\in M(\{r_{jL}\}_{j\in I},\mathscr{L}(F,\mathfrak{M});I)$. But $\sigma^*$ is an arbitrary element of $\mathscr{L}(P',\mathfrak{M})$, we have

$$\mathscr{L}(P',\mathfrak{M})\subset M(\{r_{jL}\}_{j\in I},\mathscr{L}(F,\mathfrak{M});\ I).$$

Let $p'\in P'$, $v\in\mathfrak{M}$ and $\sigma=[\{r_{jL}\sigma'_j\}_{j\in I^*}]$ is arbitrary element of $M(\{r_{jL}\}_{j\in I},\mathscr{L}(F,\mathfrak{M});\ I)$, where $\sigma'_j\in\mathscr{L}(F,\mathfrak{M})$, $I^*\subset I_1$, then

$$(p'v)[\{r_{jL}\sigma'_j\}_{j\in I^*}]=[\{r_jp'v\sigma'_j\}_{j\in I^*}]=p'(v[\{r_{jL}\sigma'_j\}_{j\in I^*}]),$$

this is true because $p'\in P'=C_F(E')$ and $r_j\in E'$. Since $p'$ and $v$ are arbitrary, it follows that $[\{r_{jL}\sigma'_j\}_{j\in I^*}]\in\mathscr{L}(P',\mathfrak{M})$. Hence $M(\{r_{jL}\}_{j\in I},\mathscr{L}(F,\mathfrak{M});\ I)\subseteq\mathscr{L}(P',\mathfrak{M})$.

We need to prove that $M(\{r_{jL}\}_{j\in I},\mathscr{L}(F,\mathfrak{M});\ I)=\oplus M(\{r_{jL}\}_{j\in I},\mathscr{L}(F,\mathfrak{M});\ I)$. If $M(\{r_{jL}\}_{j\in I},\mathscr{L}(F,\mathfrak{M});\ I)$ has an element $[\{r_{jL}\sigma_j\}_{j\in I'}]=0$, where $I'\subseteq I$, then

$$(f^{(\alpha)}u_i)[\{r_{jL}\sigma_j\}_{j\in I'}]=[\{a_{\alpha j}(u_i\sigma_j)\}_{j\in I'}]=0,\quad \alpha\in I.$$

From the property of soluable space of $\mathfrak{M}$ it follows that there exists a subset $\{a_{\alpha j}\}_{j\in I}$ of $\{a_{\alpha j}\}_{j\in I'}$, where $I\subseteq I'$ such that $[\{a_{\alpha j}(u_i\sigma_j)\}_{j\in I}]=0$ for $\alpha\in I$. From the property of unique solution it follows that $u_i\sigma_j=0$ for all $i\in\Gamma$. Hence $\sigma_j=0$ for $j\in I=I'$. This proves $r_{jL}\sigma_j=0$ for $j\in I'$.

Now we prove the second assertion. If $\mathscr{L}(P',\mathfrak{M})=\oplus M(\{r_{jL}\}_{j\in I^*},\mathscr{L}(F,\mathfrak{M});\ I^*)$, where $\{r_{jL}\}_{j\in I^*}\subset E'_L$, then we want to prove $[F:P']_L\leqslant\text{Card. } I^*$. In fact, if we write $F=\sum_{\alpha\in I}P'f^{(\alpha)}$, Card. $I=[F:\ P']_L$, then $\mathfrak{M}=\sum_\Gamma Fu_i=\sum_{i,\alpha}P'v_i^{(\alpha)}$, $v_i^{(\alpha)}=f^{(\alpha)}u_i$. Since $\{u_i\}_{i\in\Gamma}$ is a basis of $\mathfrak{M}$, it is easy to see that $\{v_i^{(\alpha)}\}_{\substack{i\in\Gamma\\ \alpha\in I}}$ is a $P'$-basis of $\mathfrak{M}$. Let $\{y_\alpha\}_{\alpha\in I}$ be a given set of $F$-linearly independent elements of $\mathfrak{M}$, then there exists an element $\sigma\in\mathscr{L}(P',\mathfrak{M})$ such that $v_i^{(\alpha)}\sigma=y_\alpha$, $\alpha\in I$. By assumption $\sigma=[\{r_{jL}\omega_j\}_{j\in I'}]$, Card. $I'\leqslant$Card. $I^*$, $\omega_j\in\mathscr{L}(F,\mathfrak{M})$. Hence we have

$$y_\alpha=[\{r_jf^{(\alpha)}(u_i\omega_j)\}_{j\in I'}]=[\{a_{\alpha j}x_j\}_{j\in I'}],\quad \alpha\in I, \tag{1.9}$$

where $a_{\alpha j}=r_jf^{(\alpha)}$, $x_j=u_i\omega_j$. Therefore from the property of soluable function [ ] it follows that Card. $I\leqslant$Card. $I'$. But Card. $I'\leqslant$Card. $I^*$, hence $[F:P']_L\leqslant$Card. $I^*$.

On the other hand, in view of (1.9) and the property of soluable function we can choose a subset $\{a_{\alpha j}\}_{j\in I}$ from $\{a_{\alpha j}\}_{j\in I'}$ such that the system of functional equations

$$[\{a_{\alpha j}X_j\}_{j\in I}]=Y_\alpha,\quad \alpha\in I$$

has a unique solution for any subset $\{Y_\alpha\}_{\alpha\in I}$. Repeating the proof of the first assertion we can immediatly obtain $\mathscr{L}(P',\mathfrak{M})=\oplus M(\{r_{jL}\}_{j\in I},\mathscr{L}(F,\mathfrak{M});I)$ and Card. $I=[F:P']_L$, $I\subseteq I^*$. This completes the proof.

**Definition 1.5.** *We say that* $\mathscr{L}(P',\mathfrak{M})=\oplus M(\{r_{jL}\}_{j\in I},\mathscr{L}(F,\mathfrak{M});I)$ *is an expression of* $\mathscr{L}(P',\mathfrak{M})$ *about* $E'$ *and* $\mathscr{L}(F,\mathfrak{M})$, *where* $\{r_j\}_{j\in I}\subset E'$. *A expression* $\oplus M(\{r_{jL}\}_{j\in I},\mathscr{L}(F,\mathfrak{M});I)$ *is called minimal if and only if any expression* $\oplus M(\{r^*_{jL}\}_{j\in I^*},\mathscr{L}(F,\mathfrak{M});I^*)$ *of* $\mathscr{L}(P',\mathfrak{M})$ *about* $E'$ *and* $\mathscr{L}(F,\mathfrak{M})$ *has the property Card.* $I\leqslant$ *Card.* $I^*$. *In this case we call Card.* $I$ *the cardinal number of expression of* $\mathscr{L}(P',\mathfrak{M})$.

From the last part of Theorem 1.1 and Definition 1.5 follows the following theorem:

**Theorem 1.2.** *Under the above hypotheses,* $\mathscr{L}(P',\mathfrak{M})$ *has a minimal expression about* $E'$ *and* $\mathscr{L}(F,\mathfrak{M})$. *And the cardinal number of its expression is equal to* $[F:P']_L$.

Let $G$ be a group of automorphisms of $F$. It is easy to see that for any element $\psi$ of $G$ there exists an $F$-semi-linear isomorphism $S$, which we denote by $(S,\psi)$ if we need to emphasize $\psi$. Let $\Theta=\{S=(S,\psi)\mid\psi\in G\}$. As (1.3) we write

$$M(\Theta,\mathscr{L}(P',\mathfrak{M});I)=\{[\{S_j\sigma'_j\}_{j\in I_1}]\mid S_j\in\Theta,\ \sigma'_j\in\mathscr{L}(P',\mathfrak{M});\ I_1\subseteq I\}.$$

**Definition 1.6.** *Let* $\mathfrak{M}=\sum_{i\in\Gamma}Fu_i$, $P=I(G)=\{f\in F\mid f^\sigma=f$ *for all* $\sigma\in G\}$. $\mathfrak{M}$ *is called a* $(\aleph_\nu,2)$*-type space over* $P$ *if* $\mathfrak{M}$ *is a soluable space and there exists a set* $M(\Theta,\mathscr{L}(P',\mathfrak{M});I)$, *which is* $\aleph_\nu$*-transitive in space* $\mathfrak{M}$ *over* $P$.

**Theorem 1.3.** *Let* $\mathfrak{M}=\sum_{i\in\Gamma}Fu_i$, $P'=C_F(E')$, $I(G)=P$, $[P':P]_L=\aleph_\nu$. *Suppose that* $\mathfrak{M}$ *is a* $(\aleph_\nu,2)$*-type space over* $P$, *then there exists in* $\Theta$ *a subset* $\{v_j\}_{j\in I}$ *with cardinal number* $\aleph_\nu$ *such that* $\mathscr{L}(P,\mathfrak{M})=\oplus M(\{S_j)_{j\in I},\mathscr{L}(P',\mathfrak{M});I)$. *Conversely, if*

$$\mathscr{L}(P,\mathfrak{M})=\oplus M(\{S_j\}_{j\in I},\mathscr{L}(P',\mathfrak{M});I),$$

*then there exists in* $\{S_j\}_{j\in I}$ *a subset* $\{S_j\}_{j\in I'}$ *such that* $\mathscr{L}(P,\mathfrak{M})=\oplus M(\{S_j\}_{j\in I'},\mathscr{L}(P',\mathfrak{M});I')$ *and Card,* $I'=[P':P]_L$.

*Proof* Now we prove the first assertion. Since $E'$ is the algebra of $G$, $P'=C_F(E')$, by [1] $E'^\psi=E'$, $P'^\psi=P'$ for any element $\psi$ of $G$. Let

$$\mathfrak{M}=\sum_{j\in\Gamma'}Pw_j=\sum_{j\in I'',\alpha\in I'}Pv_j^{(\alpha)}$$

be $P'$-and $P$-spaces respectively, where $P'=\sum_{\alpha\in I'}Pf'^{(\alpha)}$, $v_j^{(\alpha)}=f'^{(\alpha)}w_j$. It is easy to see that $\{v_j^{(\alpha)}\}_{j\in\Gamma',\alpha\in I'}$ is a $P$-basis of $\mathfrak{M}$. Let $\{y_\alpha\}_{\alpha\in I'}$ be a set of $F$-linearly independent elements of $\mathfrak{M}$. By assumptions there exists an element $\sigma=[\{S_j\sigma'_j\}_{j\in I_1}]\in M(\Theta,\mathscr{L}(P',\mathfrak{M});I)$, $I_1\subseteq I$ such that

$$y_\alpha=v_i^{(\alpha)}\sigma=(f'^{(\alpha)}w_j)[\{S_j\sigma'_j\}_{j\in I_1}]=[(f'^{(\alpha)}w_j)\{S_j\sigma'_j\}_{j\in I_1}]$$
$$=[\{f'^{(\alpha)\psi_j}(w_jS_j\sigma'_j)\}_{j\in I_1}]=[\{a_{\alpha j}x_j(i)\}_{j\in I_1}],\ \alpha\in I',$$

where $a_{\alpha j}=f'^{(\alpha)\psi_j}$, $f'^{(\alpha)}\in P'$, $x_j(i)=w_j(S_j\sigma'_j)$. Then from the property of soluable function [ ] it follows that Card. $I' \leqslant$ Card. $I_1$, and there exists a subset $\{a_{\alpha j}\}_{j\in I'}$ of $\{a_{\alpha j}\}_{j\in I_1}$ such that the following system of functional equations

$$[\{a_{\alpha j}X_j(i)\}_{j\in I'}] = Y_\alpha(i), \quad \alpha\in I' \tag{1.10}$$

has a unique solution in $\mathfrak{M}$ for any subset $\{Y_\alpha(i)\}_{\alpha\in I'}$.

Now we let $\sigma^*\in \mathscr{L}(P, \mathfrak{M})$, $v_i^{(\alpha)}\sigma^* = Y_\alpha(i)$ for $\alpha\in I'$. Since $S_j = (S_j, \psi_j)$ is an $F$-semi-linear isomorphism, we know that $\{w_i S_j\}_{i\in \Gamma'}$ is also a $P'$-basis of $\mathfrak{M}$. Therefore for $j\in I'$ there exists an element $\sigma_j''\in \mathscr{L}(P', \mathfrak{M})$ such that $w_i S_j \sigma_j'' = X_j(i)$ for $i\in \Gamma'$. Let $\bar\sigma = [\{S_j\sigma_j''\}_{j\in I'}]$, then

$$v_i^{(\alpha)}\bar\sigma = [\{f'^{(\alpha)\psi_j}(w_i S_j \sigma_j'')\}_{j\in I'}] = [\{a_{\alpha j}X_j(i)\}_{j\in I'}] = Y_\alpha(i) = v_i^{(\alpha)}\sigma^*$$

for all $i\in \Gamma'$, $\alpha\in I'$. Hence $\sigma^* = \bar\sigma\in M(\{S_j\}_{j\in I'}, \mathscr{L}(P', \mathfrak{M});\ I')$. This proves that

$$\mathscr{L}(P, \mathfrak{M}) \sqsubseteq M(\{S_j\}_{j\in I'}, \mathscr{L}(P', \mathfrak{M});\ I').$$

On the other hand, let $p\in P$, $v\in \mathfrak{M}$, $[\{S_j\sigma_j'\}_{j\in J}]\in M(\{S_j\}_{j\in I'}, \mathscr{L}(P', \mathfrak{M});\ I')$, then $(pv)[\{S_j\sigma_j'\}_{j\in J}] = p(v[\{S_j\sigma_j'\}_{j\in J}])$, it follows that $[\{S_j\sigma_j'\}_J]\in \mathscr{L}(P, \mathfrak{M})$. This proves $M(\{S_j\}_{j\in I'}, \mathscr{L}(P', \mathfrak{M});\ I') \sqsubseteq \mathscr{L}(P, \mathfrak{M})$.

Similarly to Theorem 1.1 we can prove the remainder of this theorem.

**Theorem 1.4.** *Under the same assumptions as in Theorem* 1.3, $\mathscr{L}(P, \mathfrak{M})$ *has a minimal expression about* $\Theta$ *and* $\mathscr{L}(P', \mathfrak{M})$ *and the cardinal number of this expression is* $[P':P]_L$.

*Proof* It follows from Theorem 1.3.

Now let $F = \sum_{\alpha\in I_1} P'f^{(\alpha)}$, $P' = \sum_{\beta\in I_2} Pg'^{(\beta)}$, then $\mathfrak{M} = \sum_{\Gamma} Fu_i = \sum_{\alpha\in I_1, \beta\in I_2} P(g'^{(\beta)}f^{(\alpha)}u_i)$.

Let $v_i^{(\alpha,\beta)} = g'^{(\beta)}f^{(\alpha)}u_i$, $\alpha\in I_1$, $\beta\in I_2$, then $\{v_i^{(\alpha,\beta)}\}$ is a $P$-basis of $\mathfrak{M}$. Write

$$M(\Theta E_L', \mathscr{L}(F, \mathfrak{M});\ I_1\times I_2) = \{[\{S_k r_{jL} w_{kj}\}_{k\in I_1', j\in I_2'}\},$$

where $I_1'\sqsubseteq I_1$, $I_2'\sqsubseteq I_2$, $S_k\in\Theta$, $r_{jL}\in E_L'$, $\omega_{jk}\in \mathscr{L}(F, \mathfrak{M})$.

**Definition 1.7.** *Let* $\mathfrak{M} = \sum_{\Gamma} Fu_i$, $E'$ *be the algebra of* $G$, *and let* $\Theta$ *be the same as above. Then* $\mathfrak{M}$ *is called a* $(\aleph_\nu, 3)$*-type space over* $P$ *if* $\mathfrak{M}$ *is a soluable space and there exists a set* $M(\Theta E_L', \mathscr{L}(F, \mathfrak{M});\ I)$ *which is* $\aleph_\nu$*-transitive over* $P$.

**Theorem 1.5.** *Let* $\mathfrak{M} = \sum_{i\in\Gamma} Fu_i$, $G$ *be a group of automorphims of* $F$. *Let* $P = I(G)$ *as above,* $E'$ *be the algebra of* $G$, $P' = C_F(E')$, $\Theta = \{S_j = (S_j, \psi_j)\mid \psi_j\in G\}$, *and* $[F:P]_L = \aleph_\nu$. *Suppose that* $\mathfrak{M}$ *is a* $(\aleph_\nu, 3)$*-type space over* $P$, *then there exist subsets* $\{S_j\}_{j\in I_1}$ *of* $\Theta$ *and* $\{r_{jL}\}_{j\in I_2}$ *of* $E_L'$ *such that* $\mathscr{L}(P, \mathfrak{M}) = \oplus M(\{S_k r_{jL}\}_{k\in I_1, j\in I_2}, \mathscr{L}(F, \mathfrak{M});\ I_1\times I_2)$ *holds, where Card.* $I_1\times I_2 = \aleph_\nu$.

*Conversely, if* $\mathscr{L}(P, \mathfrak{M}) = \oplus M(\{S_k r_{jL}\}_{k\in I_1', j\in I_2'}, \mathscr{L}(F, \mathfrak{M});\ I_1'\times I_2')$, *where* $S_k\in\Theta$, $r_j\in E'$, *then there exist subsets* $I_1\subset I_1'$ *and* $I_2\subset I_2'$ *such that* $\mathscr{L}(P, \mathfrak{M}) = \oplus M(\{S_k r_{jL}\}_{k\in I_1, j\in I_2}, \mathscr{L}(F, \mathfrak{M});\ I_1\times I_2)$ *and Card.* $I_1\times I_2 = \aleph_\nu$.

*Proof* Let $\{y_{\beta,\alpha}\}_{\beta\in I_1, \alpha\in I_2}$ be a set of $F$-linearly independent elements of $\mathfrak{M}$, then by the assumptions, there exists an element $\sigma \in M(\Theta E'_L, \mathscr{L}(F, \mathfrak{M}); I_1 \times I_2)$ such that

$$y_{\beta,\alpha} = v_i^{(\alpha,\beta)}\sigma, \quad \beta \in I_1, \quad \alpha \in I_2.$$

Since $\sigma = [\{S_k r_{jL} w_{kj}\}_{\substack{k \in I_1 \\ j \in I_2}}]$, where $S_k \in \Theta$, $r_{jL} \in E'_L$, $w_{kj} \in \mathscr{L}(F, \mathfrak{M})$, we have

$$y_{\beta,\alpha} = [\{g'^{(\beta)\psi_k I_{r_j^{-1}}} f^{(\alpha)\psi_k I_{r_j^{-1}}}(u_i S_k r_{jL} w_{kj})\}_{\substack{k \in I_1 \\ j \in I_2}}] = [\{a_{k,j}^{(\beta,\alpha)} x_{kj}(i)\}_{\substack{k \in I_1 \\ j \in I_2}}],$$

where $a_{k,j}^{(\beta,\alpha)} = g'(\beta)\psi_k I_{r_j^{-1}} f(\alpha)\psi_k I_{r_j^{-1}}$, $x_{kj}(i) = u_i S_k r_{jL} w_{kj}$, $I_{r_j^{-1}}$ is an inner isomorphism. Analogous to the proofs of Theorem 1.1 and 1.3 we can prove all assertions of our theorem.

**Theorem 1. 6.** *Let the assumptions be as in Theorem* 1.5, *then* $\mathscr{L}(P, \mathfrak{M})$ *has a minimal expression about* $\Theta E'_L$ *and* $\mathscr{L}(F, \mathfrak{M})$. *And the cadinal number of the minimal expression is equal to* $[F: P]_L$.

**Remark.** Let the assumptions and symbols be as in Theorem 1.5. Then from Theorem 1.5 it follows that

$$\mathscr{L}(P, \mathfrak{M}) = \oplus M(\{S_k r_{jL}\}_{\substack{k \in I_1 \\ j \in I_2}}, \mathscr{L}(F, \mathfrak{M}); I_1 \times I_2). \tag{1.11}$$

Putting $S^*_{kj} = S_k r_{jL}$, it is easy to see that $S^*_{kj}$ is a semilinear transformation of $\mathfrak{M} = \sum_{i \in \Gamma} F u_i$, and $\psi^*_{kj} = \psi_k I_{r_j^{-1}}$ is the associated isomorphism, where $S_k = (S_k, \psi_k)$, $I_{r_j^{-1}}$ is an inner isomorphism, $r_j \in E'$.

## § 2.

We take an example to explain the preceding theory.

We assume that $\mathfrak{M} = \sum F u_i$, $G$ is a group of automorphisms of $F$, $P = I(G)$, $E'$ is the algebra of $G$, $P' = C_F(E')$ and $[F: P]_L = n < \infty$. Denote the class of all finite subsets $\{x_i\}_{i \in I}$ of $\mathfrak{M}$ by $D$, where $I$ always denotes a finite set. Then we can define a function [ ] of $D$ into $\mathfrak{M}$, i. e., $[\{x_i\}_{i \in I}] = \sum_{i \in I} x_i$. Clearly [ ] satisfies all conditions of Definition 1.1. Now we check it. We have $(f \sum_{i \in I} x_i)\sigma = (\sum_{i \in I} f x_i)\sigma$ for $f \in F$, $\sigma \in \mathbf{E}$, hence condition (ii) is satisfied. It is easy to see from [1] and [2] that the condition (iii) of Definition 1.1 is satisfied. Therefore the above defined function is a soluable one with finite order.

On the other hand, we can introduce a similar function to $\mathbf{E}$, the complete ring of endomorphisms of $(\mathfrak{M}, +)$. Let $D^*$ be the class of all finite subsets of $\mathbf{E}$. Then we define a function [ ] of $D^*$ into $\mathbf{E}$, i. e. $[\{\varepsilon_i\}_{i \in I}] = \sum_{i \in I} \varepsilon_i$. Clearly we have $x \sum_{i \in I} \varepsilon_i = \sum_{i \in I} x\varepsilon_i$ for any $x \in \mathfrak{M}$.

Now let $E'$ be the algebra of $G$, $P' = C_F(E')$, $E'_L \mathscr{L}(F, \mathfrak{M}) = \{\sum_{j < \infty} r_{jL}\omega_j \mid r_j \in E', \omega_j \in \mathscr{L}(F, \mathfrak{M})\}$. Then $E'_L \mathscr{L}(F, \mathfrak{M})$ is finite transitive in $\mathfrak{M}$ over $P'$, i. e., let $\{x_i\}_{i \in I}$ be a finite set of $P'$-linearly independent elements, $\{y_i\}_{i \in I}$ be a subset of $\mathfrak{M}$, then there exists an element $\sigma \in E'_L \mathscr{L}(F, \mathfrak{M})$ such that $x_i\sigma = y_i$ for $i \in I$. Hence from the proof of Theorem 1.1 it follows that $\mathfrak{M}$ is a $([F:P']_L, 1)$-type space with the above finite soluable function[ ]. Without going into the matter in detail it is clear that Theorem 1.1 coincides with the Theorem 1 in [1]. In this case we have

$$\oplus M(\{r_{jL}\}_{j\in I}, \mathscr{L}(F, \mathfrak{M}); I) = \sum_{j\in I} \oplus r_{jL}\mathscr{L}(F, \mathfrak{M}),$$

Card. $I = [F:P']_L$.

In the same way we write $\Theta\mathscr{L}(P', \mathfrak{M}) = \sum_{S_j\in\Theta} S_j\mathscr{L}(P', \mathfrak{M})$. It is clear that $\Theta\mathscr{L}(P', \mathfrak{M})$ is finite transitive in $\mathfrak{M}$ over $P$. Owing to the soluable function [ ] with finite order mentioned above, it is easy to see that $\mathfrak{M}$ is a $([P': P]_L, 2)$-type space over $P$. In this case Theorem 1.1 coincides again with the Theorem 1 in [1].

Let $\Theta E'_L\mathscr{L}(F, \mathfrak{M}) = \sum S_k r_{jL}\mathscr{L}(F, \mathfrak{M})$, then it is clear that $\Theta E'_L\mathscr{L}(F, \mathfrak{M})$ is finite transiive in $\mathfrak{M}$ over $P$. Similarly we know that $\mathfrak{M}$ is a $([F: P]_L, 3)$-type space over $P$. Theorem 1.1 coincides with the Theorem 1 in [1] once more.

## § 3.

In this section, we are going to investigate infinite Galois theory of division ring $F$.

Let $\tilde{\mathbf{E}}$ denote the set of all endomorphisms of the additive group $(F, +)$. Let $D_\nu$ be the class of all subsets $\{f_i\}_I$ of $F$ whose cardinal number $\leqslant \aleph_\nu$ and $\tilde{D}_\nu$ be the class of all subsets $\{\tilde{\varepsilon}_i\}_I$ of $\mathbf{E}$ whose cardinal number $\leqslant \aleph_\nu$. As above, let [ ] denote a function from $D_\nu$ to $F$ and at the same time let [ ] also denote a function from $\tilde{D}_\nu$ to $\mathbf{E}$ such that $a[\{\tilde{\varepsilon}_i\}_I] = [\{a\tilde{\varepsilon}_i\}_I]$ holds for all $a\in F$. Write $F_R = \{f_R | f\in F, af_R = af, a\in F\}$. It is clear that $F_R \subseteq \tilde{\mathbf{E}}$ and $F$ is a ring isomorphic to $F_R$. If $\{f_{iR}\}_I$ is any subset of $F_R$ and Card. $I \leqslant \aleph_\nu$, then $\{f_{iR}\}\in\tilde{D}_\nu$ and $[\{f_{iR}\}_I]\in\mathbf{E}$. For the convenience of future, we demand that $[\{f_{iR}\}_I]\in F_R$. And we can easily prove that $[\{f_{iR}\}_I] = [\{g_{iR}\}_I]$ if and only if $[\{f_i\}_I] = [\{g_i\}_I]$.

Let $[\{a_jv_j\}_I, [\{b_jv_j\}_I]$ be two elements of $\mathfrak{M}$ and Card. $I \leqslant \aleph_\nu$, where $a_j, b_j\in F$, $v_j\in V$. If $\{v_j\}_{I_1}$ is any maximal set of $F$-linearly independent elements of $\{v_j\}_I$ and we write $\{v_j)_{I_2} = \{v_j\}_I - \{v_j\}_{I_1}$, then for any element $v_j\in\{v_j\}_{I_2}$, we have $v_j = \sum_{i\in I_1} g_i^{(j)} v_i$, where all but a finite number of $g_i^{(j)}$ are zero. Hence we have

$$\begin{aligned} [\{a_jv_j\}_I] &= [\{a_iv_i\}_{I_1} \cup \{a_j \sum_{i\in I_1} g_i^{(j)} v_i\}_{j\in I_2}], \\ [\{b_jv_j\}_I] &= [\{b_iv_i\}_{I_1} \cup \{b_j \sum_{i\in I_1} g_i^{(j)} v_i\}_{j\in I_2}], \end{aligned} \tag{3.1}$$

where $\cup$ represents the union of sets.

**Definition 3. 1.** *Let $\mathfrak{M} = \sum_{i\in\Gamma} Fu_i$ be a $\nu$-soluable space (see §1), $\tilde{\mathbf{E}}$ be the complete ring of endomorphisms of $(F, +)$, and let $\tilde{D}_\nu$ be the class of all subsets $\{\tilde{\varepsilon}_i\}_{i\in I}$ with Card. $I \leqslant \aleph_\nu$. The function[ ]: $\tilde{D}_\nu \to \tilde{\mathbf{E}}$ will be called a $\nu$-function about $F$ if it satisfies the following conditions:*

(i) *If* $\{f_{iR}\}_{i\in I}\in\tilde{D}_v$, $f_{iR}\in F_R$, *then* $[\{f_{iR}\}_{i\in I}]\in F_R$.

(ii) $[\{a_jv_j\}_{j\in I}]=[\{b_jv_j\}_{j\in I}]$ *if and only if*

$$[a_{iR}\cup\{a_{jR}g_{iR}^{(j)}\}_{j\in I_2}]=[b_{iR}\cup\{b_{jR}g_{iR}^{(j)}\}_{j\in I_2}]$$

*for all* $i\in I_1$ *(see* (3.1)).

Now we consider the element $[\{\tilde{\varepsilon}_i\}_{i\in I}]$ of $\tilde{\mathbf{E}}$. It is easy to see that for any element $\tilde{\sigma}\in\tilde{\mathbf{E}}$ we have $[\{\tilde{\varepsilon}_i\}_{i\in I}]\tilde{\sigma}=[\{\tilde{\varepsilon}_i\tilde{\sigma}\}_{i\in I}]$. Let $[\{\tilde{\varepsilon}_i\}_{i\in I}]$ and $[\{\tilde{\delta}_i\}_{i\in I'}]$ be elements of $\tilde{\mathbf{E}}$, and $K$ is a subring of $F$, then we say that $[\{\tilde{\varepsilon}_i\}_{i\in I}]$ is equal to $[\{\tilde{\delta}_i\}_{i\in I'}]$ over $K$ denoted by $[\{\tilde{\varepsilon}_i\}_{i\in I}]\underset{K}{=\!=}[\{\tilde{\delta}_i\}_{i\in I'}]$ if and only if $k[\{\tilde{\varepsilon}_i\}_{i\in I}]=k[\{\tilde{\delta}_i\}_{i\in I'}]$ for all $k\in K$.

Let $K$ be a subring of $F$, $\tilde{\varepsilon}$ be an element of $\tilde{\mathbf{E}}$. We say that $\tilde{\varepsilon}F_R$ is (right) $F_R$-cyclic module relative to $K_R$ if there exists an automorphism $\sigma$ of $F_R$ such that $k_R\tilde{\varepsilon}=\tilde{\varepsilon}k_R^\sigma$ for any $k_R\in K_R$. When $K_R=F_R$, then $\tilde{\varepsilon}F_R$ is shortly said to be $F_R$-cyclic module.

For the sake of simplicity we use the element $1_K$ of $\tilde{\mathbf{E}}$ to denote the identity of $K$, i. e. $k1_K=k$ for all $k\in K$. It is easy to see that $1_KF_R$ is an $F_R$-cyclic module relative $K_R$.

**Definition 3. 2.** *Let* $\lambda,\lambda'\in\tilde{\mathbf{E}}$. *Then two* $F_R$*-cyclic modules* $\lambda F_R$ *and* $\lambda'F_R$ *are said to be* $(K_R, F_R)$*-bimodule (i. e. left* $K_R$*-module and also right* $F_R$*-module) isomorphic relative to* $K$ *denoted by* $\lambda F_R\underset{K}{\cong}\lambda'F_R$ *if there exists an element* $\delta_R\in F_R$ *for the element* $\lambda$ *such that the relation* $\sum k_{iR}\lambda f_{iR}\underset{K}{=\!=}\sum k^*_{iR}\lambda f^*_{iR}$ *holds if and only if the relation* $\sum k_{iR}\lambda'\delta_Rf_{iR}\underset{K}{=\!=}\sum_i k^*_{iR}\lambda'\delta_Rf^*_{iR}$ *holds. Conversely, if there exists an element* $\delta'_R\in F_R$ *for the element* $\lambda'$ *such that the relation* $\sum_j\tilde{k}_{jR}\lambda'\tilde{f}_{jR}\underset{K}{=\!=}\sum_j\tilde{k}^*_{jR}\lambda'\tilde{f}^*_{jR}$ if and only if

$$\sum_j\tilde{k}_{jR}\lambda\delta'_R\tilde{f}_{jR}\underset{K}{=\!=}\sum_j\tilde{k}^*_{jR}\lambda\delta'_R\tilde{f}^*_{jR}.$$

Clearly we have (i) $\lambda F_R\underset{K}{\cong}\lambda F_R$ (ii) if $\lambda F_R\underset{K}{\cong}\lambda'F_R$ then $\lambda'F_R\underset{K}{\cong}\lambda F_R$, and (iii) if $\lambda F_R\underset{K}{\cong}\lambda'F_R$, $\lambda'F_R\underset{K}{\cong}\lambda''E_R$, then $\lambda F_R\underset{K}{\cong}\lambda''F_R$.

Now we let $\{\tilde{\varepsilon}_i\}_{i\in I}$ be a subset of $\tilde{\mathbf{E}}$, $F_R\subset\tilde{\mathbf{E}}$, and

$$M(\{\tilde{\varepsilon}_i\}_{i\in I},F_R;I)=\{[\{\tilde{\varepsilon}_if_{iR}\}_{i\in I}]\,|\,f_{iR}\in F_R\}.$$

Then we can introduce the following definition.

**Definition 3. 3.** *Let* $\{\tilde{\varepsilon}_i\}_{i\in I}\in\tilde{D}_v$, $H=M(\{\tilde{\varepsilon}_i\}_{i\in I},F_R;I)$. $H$ *is said to be a homogeneous Galois* $(K_R,F_R)$*-bimodule if and only if there exists a subset* $\{\lambda_j\}_{j\in J}$ *satisfying the following conditions:*

(i) $\lambda_jF_R$ *is* $F_R$*-cyclic module and* $H=\sum_{j\in J}\lambda_jF_R$[3],

(ii) *For any* $F_R$*-module* $\lambda_jF_R$ *we have* $\lambda_jF_R\underset{K}{\cong}1_KF_R$.

**Lemma 3. 1.** *Let* $H$ *be a homogeneous Galois* $(K_R,F_R)$*-bimodule,*

$$A=\{\varphi\in H\,|\,\varphi k_R\underset{K}{=\!=}k_R\varphi \textit{ for all } k_R\in K_R\},$$

3) $\sum_{j\in J}\lambda_jF_R=\{\sum_j\lambda_jf_{jR}\,|\,f_{jR}\in F_R\}$ and $\sum_j\lambda_jf_{jR}$ express a sum of elements $\lambda_j$ and $f_{jR}$ in ring $\tilde{\mathbf{E}}$.

*then* $AF_R=H$.

*Proof* Since $H=\sum_{j\in J}\lambda_jF_R$, it is enough to show that $\lambda_j\in AF_R$. In fact, $\lambda_jF_R\underset{K}{\cong}1_KF_R$ by the assumption, hence from the relation $k_R1_K\underset{K}{\cong}1_Kk_R$ for any $k_R\in K_R$ it follows that $k_R\lambda_j\delta_R\underset{K}{=\!=}\lambda_j\delta_Rk_R$, where $\delta_R$ is an element of $F_R$. Hence $\lambda_j\delta_R\in A$. Therefore $\lambda_j\in AF_R$.

**Lemma 3. 2.** *Let $H$ be a homogeneous Galois $(K_R, F_R)$-bimodule, $M_R=C_{F_R}(K_R)$. Let $\varphi_1, \cdots, \varphi_n$ be elements of $A$. Suppose we have a non-trivial relation $\sum_{i=1}^n\varphi_if_{iR}=0$, $f_{iR}\in F_R$, then there exist $m_{1R}, \cdots, m_{nR}\in M_R$ such that $\sum_{i=1}^n\varphi_im_{iR}=0$ is a non-trivial relation too.*

*Proof* Without loss of generality we may assume that $\sum_{i=1}^n\varphi_if_{iR}=0$ and the relation is a shortest one in the sense that the number of non-zero elements of $\{f_{iR}\}$ is least. If $\varphi_1=0$, then our lemma is clear. Hence we assume $\varphi_1\neq0$, $\varphi_1+\varphi_2f_{2R}+\cdots+\varphi_nf_{nR}=0$ and $f_{iR}\neq0$ for $i=2, \cdots, n$. Then we have

$$k_R(\varphi_1+\varphi_2f_{2R}+\cdots+\varphi_nf_{nR})-(\varphi_1+\varphi_2f_{2R}+\cdots+\varphi_nf_{nR})k_R=0$$

for all $k_R\in K_R$. Hence $\sum_{i=2}^n\varphi_i(k_Rf_{iR}-f_{iR}k_R,)=0$. Therefore $k_Rf_{iR}-f_{iR}k_R=0$ for $i=2,\cdots, n$. This proves that $f_{iR}\in M_R$ for $i=2, \cdots, n$.

**Lemma 3. 3.** *Let $H=M(\{\tilde{s}_i\}_{i\in I}, F_R; I)$ is a homogeneous Galois $(K_R, F_R)$-bimodule, $\tilde{K}=I(A(K))$*[4]*, then $H$ is a homogeneous Galois $(\tilde{K}_R, F_R)$-bimodule.*

*Proof* By Definition 3.3 it is enough to show that $\lambda_jF_R\underset{\tilde{K}}{\cong}1_{\tilde{K}}F_R$ for any $\lambda_jF_R$. We attend first to the following. Let $\sigma$ be an automorphism of $F$. Owing to the isomorphism of $F$ onto $F_R$, it is clear that $\sigma$ can be regarded as an automophism of $F_R$ because we can define $f_R^\sigma$: $f^*f_R^\sigma=f^*f^\sigma$ for $f^*\in F$. Clearly $\sigma$ is an automorphism of $F_R$ and we have $(f_R)^\sigma=(f^\sigma)_R$. This proves that $A(K)=A(K_R)$.

Since $k_R1_K=1_Kk_R$ for $k_R\in K_R$, by assumptions it follows that $k_R\lambda\delta_R\underset{K}{=\!=}\lambda\delta_Rk_R$ for $\delta\in F$, where $\lambda=\lambda_j$. Since by assumption $\lambda F_R$ is an $F_R$-cyclic module, we have $k_R\lambda\delta_R=\lambda\delta_Rk_R^{\sigma I_{\delta_R^{-1}}}\underset{K}{=\!=}\lambda\delta_Rk_R$. By the definition of $F_R$-cyclic module we know $k_R\lambda=\lambda k_R^\sigma$ for any automorphism $\sigma$ of $F_R$. Operating the identity $1\in K$ on two sides of above equation we get $(1\lambda)\delta k_R=(1\lambda)\delta k_R^{\sigma I_{\delta^{-1}}}$. Hence $fk_R^{\sigma I_{\delta_R^{-1}}}=fk_R$ for any $f\in F$. From this follows $k_R^{\sigma I_{\delta_R^{-1}}}=k_R$ for $k_R\in K_R$. Therefore $\sigma\ I_{\delta^{-1}}\in A(K)$. This proves $\tilde{k}_R^{\sigma I_{\delta_R^{-1}}}=\tilde{k}_R$, $\tilde{k}_R\in\tilde{K}_R$, Since $\lambda F_R$ is $F_R$-cyclic module, it follows that

$$\sum_l\tilde{k}_{lR}1_{\tilde{K}}f_{lR}\underset{\tilde{K}}{=\!=}\sum_l\tilde{k}_{lR}f_{lR}\underset{\tilde{K}}{=\!=}\sum_l\tilde{k}_{lR}^{\sigma I_{\delta_R^{-1}}}f_{lR}=\sum_l\delta_R^{-1}\tilde{k}_{lR}^\sigma\delta_Rf_{lR}. \quad (3.2)$$

Now we make a correspondence $\sigma'$: $\sum_l\tilde{k}_{lR}1_{\tilde{K}}f_{lR}\to\sum_l\tilde{k}_{lR}\lambda\delta_Rf_{lR}$. We want to show

4) $A(K)=\{\sigma\in G|k^\sigma=k$ for all $k\in K\}$, $I(A(K))=\{f\in F|f^\sigma=f$ for all $\sigma\in A(K)\}$.

that if $0 \underset{\widetilde{K}}{=\!=} \sum_l \tilde{k}_{lR} 1_{\widetilde{K}} f_{lR}$, then $0 \underset{\widetilde{K}}{=\!=} \sum_l \tilde{k}_{lR} \lambda \delta_R f_{lR}$. If $\tilde{k}^* \sum_l \tilde{k}_{lR} 1_{\widetilde{K}} f_{lR} = 0$ for any element $\tilde{k}^* \in \widetilde{K}$, then by (3.2) we have $0 = \sum_l \tilde{k}^* \delta^{-1} \tilde{k}^{\sigma}_{lR} \delta_R f_{lR} = \tilde{k}^* \delta^{-1} \sum_l \tilde{k}^{\sigma}_{lR} \delta_R f_{lR}$. Hence

$$\sum_l \tilde{k}^{\sigma}_{lR} \delta_R f_{lR} = 0.$$

On the other hand we have

$$\sum_l \tilde{k}^* \tilde{k}_{lR} \lambda \delta_R f_{lR} = \sum_l \tilde{k}^* \lambda \tilde{k}^{\sigma}_{lR} \delta_R f_{lR} = (\tilde{k}^* \lambda) \sum_l \tilde{k}^{\sigma}_{lR} \delta_R f_{lR} = 0.$$

It follows that $\sum_l \tilde{k}_{lR} \lambda \delta_R f_{lR} \underset{\widetilde{K}}{=\!=} 0$. Conversely, if $0 \underset{\widetilde{K}}{=\!=} \sum_l \tilde{k}_{lR} \lambda \delta_R f_{lR}$, then according to the inverse course of above proof we can obtain $0 \underset{\widetilde{K}}{\cong} \sum_l \tilde{k}_{lR} 1_{\widetilde{K}} f_{lR}$. Therefore under the above correspondence $\sigma' \colon \sum_l \tilde{k}_{lR} 1_{\widetilde{K}} f_{lR} \rightarrow \sum_l \tilde{k}_{lR} \lambda \delta_R f_{lR}$ the $(\widetilde{K}, F_R)$-bimodules $\lambda F_R$ and $1_K F_R$ are $(\widetilde{K}_R, F_R)$-bimodule isomorphic relative to $\widetilde{K}$. Therefore we complete our proof.

Having above preparations we can turn back to the Galois theory of division ring $F$.

Let $K$ be a division subring of $F$, $G$ a group of automorphisms of $F$ and $K \supset P = I(G)$. From Theorem 1.5 it follows that if $\mathfrak{M}$ is a $(\aleph_\nu, 3)$-type space over $P$, then $\mathscr{L}(P, \mathfrak{M}) = \oplus M(\{S_j\}_{j \in I}, \mathscr{L}(F, \mathfrak{M}); I)$, where $S_j$ are $F$-semi-linear isomorphisms for $j \in I$ and Card. $I = \aleph_\nu$. Hence $\sigma = [\{S_j \omega_j\}_{j \in \bar{I}}]$ for any element $\sigma \in \mathscr{L}(P, \mathfrak{M})$, where $\omega_j \in \mathscr{L}(F, \mathfrak{M})$. In order to indicate that $\sigma$ is relative to the set $\bar{I}$, we write specifically $\sigma = \sigma_{\bar{I}}$. Let $I'$ be a subset of $I$, $\sigma_{I'} = [\{S_j \omega_j\}_{j \in I'}]$, we call $\sigma_{I'}$ an associative element of $\sigma$. If $\sigma_{I'} = [\{S_j \omega_j\}_{j \in I'}]$ is an arbitrary associative element of $\sigma$, then we can form a set $H(\sigma_{I'}) = M(\{\psi_j\}_{j \in I'}, F_R; I') = \{[\{\psi_j f_{jR}\}_{j \in I'}] \mid f_{jR} \in F_R\}$, where $S_j = (S_j, \psi_j)$, $j \in I'$.

We are now ready to give the following definition.

**Definition 3. 4.** *Let $K$ be a division subring of $F$ and $K \supset P$, $\mathscr{L}(K, \mathfrak{M})$ be the complete ring of $K$-linear transformations of $\mathfrak{M}$, $\mathscr{L}(P, \mathfrak{M}) = \oplus M(\{S_j\}_{j \in I}, \mathscr{L}(F, \mathfrak{M}); I)$ by* §1. *Let $\sigma_{I'}$ be an associative element of $\sigma \in \mathscr{L}(P, \mathfrak{M})$, and let $H(\sigma_{I'}) = M(\{\psi_j\}_{j \in I'}, F_R; I')$, where $\psi_j$ is associative isomorphism of $S_j$, i. e., $S_j = (S_j, \psi_j)$. We say that $K$ is a homogenize division subring if and only if $H(\sigma_{I'})$ is a homogeneous Calois$(K_R, F_R)$-bimodule, where $H(\sigma_{I'})$ contains $1_K$.*

**Lemma 3. 4.** *Let $\mathfrak{M} \underset{i \in \Gamma}{=\!=} F u_i$ be a $(\aleph_\nu, 3)$-type space over $P$ with $\nu$-function about $F$ (see Definition 3.1) and let $K$ be a homogenize division subring of $F$ and $K \supset P$. Write $\widetilde{K} = I(A(K))$. Then $\mathscr{L}(K, \mathfrak{M}) = \mathscr{L}(\widetilde{K}, \mathfrak{M})$, $K = \widetilde{K}$.*

*Proof* Let $H$ be a homogeneous Galois $(K_R, F_R)$-bimodule. As Lemma 3.1. $A = \{\varphi \in H \mid \varphi k_R \underset{K}{=\!=} k_R \varphi$ for all $k_R \in K_R\}$. Let $\widetilde{A} = \{\tilde{\varphi} \in H \mid \tilde{\varphi} \tilde{k}_R \underset{\widetilde{K}}{=\!=} \tilde{k}_R \tilde{\varphi}$ for all $\tilde{k}_R \in \widetilde{K}_R\}$. It is clear that $\widetilde{A} \subseteq A$. By Lemmas 3.1 and 3.3 $H = \widetilde{A} F_R = A F_R$. Write $\widetilde{M}_R = C_{F_R}(\widetilde{K}_R)$. Clearly $\widetilde{M}_R = M_R = C_{F_R}(K_R)$. Hence $A$ and $\widetilde{A}$ are both right vector spaces over $M_R$ and $\widetilde{A}$ is a subspace of $A$. If $\varphi \in A$, then from $H=\widetilde{A}F_R \supset A$ it follows that there exist $\tilde{\varphi}_1, \cdots, \tilde{\varphi}_n \in \widetilde{A}$ such that $\varphi=\sum_{i=1}^{n} \tilde{\varphi}_i f_{iR}$, $f_{iR} \in F_R$. By Lemma 3.2 there exist $m_{1R}, \cdots, m_{nR} \in M_R$ such that $\varphi=\sum_{i=1}^{n} \tilde{\varphi}_i m_{iR} \in \widetilde{A}$. Hence $\widetilde{A}=A$.

We want to show $\mathscr{L}(K, \mathfrak{M}) \sqsubseteq \mathscr{L}(\widetilde{K}, \mathfrak{M})$. Since $\mathscr{L}(P, \mathfrak{M})=\oplus M(\{S_j\}_{j\in I}, \mathscr{L}(F, \mathfrak{M}); I)$, Card. $I=\aleph_\nu=[F: P]_L$, for any element $\sigma \in \mathscr{L}(K, \mathfrak{M})$ it has the form $\sigma=[\{S_j\omega_j\}_{j\in \bar{I}}]$, where $S_j=(S_j, \psi_j)$, $\omega_j \in \mathscr{L}(F, \mathfrak{M})$. Take an element $u \in \mathfrak{M}$ such that $u\sigma \neq 0$, then for any $k \in K$ we have

$$(ku)\sigma=[\{(ku)S_j\omega_j\}_{j\in \bar{I}}]=[\{k^{\psi_i}v_i\}_{i\in I_1} \cup \{k^{\psi_j} \sum_{i\in I_1} g_i^{(j)} v_i\}_{j\in I_2}],$$

$$k(u\sigma)=[\{k(uS_j\omega_j)\}_{j\in \bar{I}}]=[\{kv_i\}_{i\in I_1} \cup \{k \sum_{i\in I_1} g_i^{(j)} v_i\}_{j\in I_2}],$$

where $v_j=uS_j\omega_j$, $\{v_i\}_{i\in I_1}$ is the set of maximal $F$-linearly independent elements of $\{v_j\}_{j\in \bar{I}}$ and $\{v_j\}_{j\in I_2}=\{v_j\}_{j\in \bar{I}_2}-\{v_i\}_{i\in I_1}$. Since $\mathfrak{M}$ has $\nu$-function about $F$, we obtain the following equation

$$[k^{\psi_i} \cup \{k^{\psi_j} g_i^{(j)}\}_{j\in I_2}]=[k \cup \{kg_i^{(j)}\}_{j\in I_2}],\ i \in I. \tag{3.3}$$

That is

$$[\psi_i \cup \{\psi_j g_{iR}^{(j)}\}_{j\in I_2}] \underset{K}{=\!=} [1_R \cup \{g_{iR}^{(j)}\}_{j\in I_2}] \in F_R \tag{3.4}$$

for any $i \in I_1$. If $[k \cup \{kg_i^{(j)}\}_{j\in I_2}]=0$ for all $i \in I_1$, then $u\sigma=0$, this is impossible. Hence we may assume $[1_R \cup \{g_{iR}^{(j)}\}_{j\in I_2}]=b_R^{-1} \neq 0$. Now we let $H=\{[\psi_1 f_R \cup \{\psi_j f'_R\}_{j\in \bar{I}_2}] \mid f_R, f'_R \in F_R;\ \bar{I}_2 \sqsubseteq I_2\}$ It is easy to see that $H$ contains $1_R$. It is clear that $k1_R=k=k1_K$ for any $k \in K$. Therefore $H$ is a homogeneous Galois $(K_R, F_R)$-bimodule by the assumptions. From the property of homogenize subring $K$ and Definition 3.3 it follows that $\psi_1 F_R$ and $\psi_j F_R$ are all $(K_R, F_R)$-bimodule isomorphis with $1_K F_R$ relative to $K$, $j \in I_2$.

By Lemma 3.3 $H$ is a homogeneous Galois $(\widetilde{K}_R, F_R)$-bimodule. From (3.4) it follows that

$$k[\psi_i b_R \cup \{\psi_j g_{iR}^{(j)} b_R\}_{j\in I_2}]=k \in K, \tag{3.5}$$

where $b_R^{-1}=[\{1_R \cup \{g_{iR}^{(j)}\}_{j\in I_2}]$. Since $[\psi_1 b_R \cup \{\psi_j g_{iR}^{(j)} b_R\}_{j\in I_2}] \underset{K}{=\!=} 1_R$, we have

$$k_R[\psi_1 b_R \cup \{\psi_j g_{1R}^{(j)} b_R\}_{j\in I_2}] \underset{K}{=\!=} [\psi_1 b_R \cup \{\psi_j g_{1R}^{(j)} b_R\}_{j\in I_2}] k_R.$$

Therefore $[\psi_1 b_R \cup \{\psi_j g_{1R}^{(j)} b_R\}_{j\in I_2}] \in A$. By Lemma 3.2, $A=\widetilde{A}$. And we have

$$\tilde{k}_R[\psi_1 b_R \cup \{\psi_j g_{1R}^{(j)} b_R\}_{j\in I_2}] \underset{\widetilde{K}}{=\!=} [\psi_1 b_R \cup \{\psi_j g_{1R}^{(j)} b_R\}_{j\in I_2}] \tilde{k}_R,$$

for any $\tilde{k}_R \in \widetilde{K}_R$. Hence

$$\tilde{k}[\psi_1 b_R \cup \{\psi_j g_{1R}^{(j)} b_R\}_{j\in I_2}]=\tilde{k},$$

$$\tilde{k}[\psi_1 \cup \{\psi_j g_{1R}^{(j)}\}_{j\in I_2}]=\tilde{k} b_R^{-1}=\tilde{k}[1_R \cup \{g_{1R}^{(j)}\}_{j\in I_2}], \tag{3.6}$$

for all $\tilde{k} \in \widetilde{K}$.

If $[1_R \cup \{g_{iR}^{(j)}\}_{j\in I_2}]=0$, by (3.4) it follows that

$$k_R[\psi_i \cup \{\psi_j g_{iR}^{(j)}\}_{j\in I_2}] \underset{K}{=\!=} 0 \underset{K}{=\!=} [\psi_i \cup \{\psi_j g_{iR}^{(j)}\}_{j\in I_2}] k_R.$$

Since $A=\widetilde{A}$, it follows that

$$\tilde{k}[\psi_i \cup \{\psi_j g_{iR}^{(j)}\}_{j\in I_2}] = 0 = \tilde{k}[1_R \cup \{g_{iR}^{(j)}\}_{j\in I_2}] \tag{3.7}$$

for all $\tilde{k}\in\tilde{K}$. In view of the property of $\nu$-function [ ] of $\mathfrak{M}$ about $F$ and (3.6), (3.7) we have

$$(\tilde{k}u)\sigma = [\{\tilde{k}^{\psi_j}v_j\}_{j\in\bar{I}}] = [\{\tilde{k}^{\psi_i}v_i\}_{i\in I_1} \cup \{\tilde{k}^{\psi_j}v_j\}_{j\in I_2}] = [\{\tilde{k}v_i\}_{i\in I_1} \cup \{\tilde{k}v_j\}_{j\in I_2}] = \tilde{k}(u\sigma).$$

If $u\sigma=0$, then $v\sigma\neq 0$ for $\sigma\neq 0$. Hence $(u+v)\sigma\neq 0$. From the preceding results it follows that $(\tilde{k}(u+v))\sigma = \tilde{k}((u+v)\sigma)$, therefore $(\tilde{k}u)\sigma + (\tilde{k}v)\sigma = \tilde{k}(u\sigma) + \tilde{k}(v\sigma)$. But $v\sigma\neq 0$, it follows that $(\tilde{k}v)\sigma = \tilde{k}(v\sigma)$, and $(\tilde{k}u)\sigma = \tilde{k}(u\sigma)$. This proves $\sigma\in\mathscr{L}(\tilde{K}, \mathfrak{M})$, hence $\mathscr{L}(K, \mathfrak{M}) \sqsubseteq \mathscr{L}(\tilde{K}, \mathfrak{M}) \sqsubseteq \mathscr{L}(K, \mathfrak{M})$. This completes the proof.

**Definition 3. 5.** *Let $\tilde{G}$ be a subgroup of $G$, $\tilde{K}=I(\tilde{G})$. Let $E'$ be the algebra of $G$. Then $\tilde{G}$ is called a homogenize group if and only if for any element $\sigma=[\{S_j\omega_j\}_{j\in I}]$ of* $\mathscr{L}(\tilde{K}, \mathfrak{M})$, *where $S_j=(S_j, \psi_j)$, we have*

(i) *there exists $\delta_j$ in $E'$ such that $\psi_j I_{\delta_j}\in\tilde{G}$,*

(ii) *if $1_{\tilde{K}}\in H(\sigma_{I'}) = M(\{\psi_j\}_{I'}, F_R; I')$, then $H(\sigma_{I'})$ is a homogenize Galois $(K_R, F_R)$-bimodule, where $I'\subset I$.*

(iii) *if $S=(S, \psi)\in\mathscr{L}(\tilde{K}, \mathfrak{M})$, $S=[\{S_j\omega_j\}_{j\in I}]$, $S_j=(S_j, \psi_j)$, $\omega_j\in\mathscr{L}(F, \mathfrak{M})$ and $\psi_j F_R \underset{\tilde{K}}{\cong} 1_K F_R$ for $j\in I$, then $\psi\in\tilde{G}$.*

**Lemma 3. 5.** *Let $\mathfrak{M}$ be a $(\aleph_\nu, 3)$-type space over $P$ (see § 1) with $\nu$-function about $F$. Let $K$ be a homogenize subring of $F$, $\tilde{G}=A(K)$, then $\tilde{G}$ is a homogenize subgroup.*

*Proof* By Lemma 3.4 $K=\tilde{K}=I(\tilde{G})$. Let $\sigma\in\mathscr{L}(K, \mathfrak{M})$, then $\sigma=[\{S_j\omega_j\}_{j\in I}]$, where $S_j=(S_j, \psi_j)$, $\omega_j\in\mathscr{L}(F, \mathfrak{M})$. We first check (i) of Definition 3.5. In fact, we choose an arbitrary element $S_j=(S_j, \psi_j)$ and an element $u$ such that $v_j=uS_j\omega_j\neq 0$. Let $v_i=uS_i\omega_i$ and $\{v_i\}_{i\in I_1}$ be a set of maximal $F$-linearly independent elements and $v_j\notin\{v_i\}_{i\in I_1}$. Write $\{v_i\}_{i\in I_2}=\{v_i\}_{i\in I}-\{v_i\}_{i\in I_1}$. Then by (3.4) there exists an element $1\in I_1$ such that

$$[\psi_1 \cup \{\psi_j g_{1R}^{(j)}\}_{j\in I_2}] \underset{K}{=\!=} [1_R \cup \{g_{1R}^{(j)}\}_{j\in I_2}] \neq 0.$$

Set $H=\{[\psi_1 f_R \cup \{\psi_j f'_R\}_{j\in\bar{I}_2}] \mid f_R, f'_R\in F_R;\ \bar{I}_2\sqsubseteq I_2\}$, then $1_K\in H$. From the property of homogenize subring of $K$ it follows that $\psi_1 F_R$ and $\psi_j F_R$ are all $(K_R, F_R)$-bimodule isomorphic with $1_K F_R$ relative to $K$ for $j\in I_2$. As the proof of Lemma 3.3 we have $\psi_j I_{\delta_{jR}} \underset{K}{=\!=} 1_K$, hence $\psi_j I_{\delta_j}\in\tilde{G}$. Since $P^{\psi_j}=p$ for any element $p$ of $P$, we have $p^{I_{\delta_j}}=p$, and $\delta_j\in E'$. This proves that (i) of Definition 3.5 is satisfied. From the property of homogenize subring of $K$ it follows that (ii) of Definition 3.5 is also satisfied. Now we check (iii) of Definition 3.5. If $S=(S, \psi)\in\mathscr{L}(K, \mathfrak{M})$, then $k^\psi=k$ for any element $k\in K$. Hence $\psi\in A(K)=\tilde{G}$. This proves $\tilde{G}$ is a homogenize subgroup.

**Lemma 3. 6.** *Let $\tilde{G}$ be a homogenize subgroup of $G$, then $\tilde{K}=I(\tilde{G})$ is a homogenize subring and $A(\tilde{K})=\tilde{G}$.*

*Proof* From the property of homogenize subgroup and Definition 3.5. it follows that $\widetilde{K}$ is a homogenize subring. Now we need to prove the last assertion. Let $\widetilde{\widetilde{G}}=A(\widetilde{K})$, then $\widetilde{\widetilde{G}}$ is a homogenize subgroup by Lemma 3.5. Let $\widetilde{\widetilde{K}}=I(\widetilde{\widetilde{G}})$, then $\widetilde{\widetilde{K}}$ is a homogenize subring. Hence $\mathscr{L}(\widetilde{K}, \mathfrak{M})=\mathscr{L}(\widetilde{\widetilde{K}}, \mathfrak{M})$ by Lemma 3.4. Let $\psi\in\widetilde{\widetilde{G}}$, then $\psi\in\widetilde{G}$ by the property of homogenize subgroup $\widetilde{G}$ (see Definition 3.5). Therefore $\widetilde{\widetilde{G}}=\widetilde{G}$.

We are now ready to establish the following fundamental theorem of infinite Galois theory of division ring $F$.

**Theorem 3.1.** *(Fundamental theorem). Let $F$ be a division ring, $G$ be a group of automorphisms of $F$, $P=I(G)$ and $[F: P]_L=\aleph_\nu$. Let $G'$ be an arbitrary homogenize subgroup of $G$, $K$ be an arbitrary homogenize subring of $F$ and $K\supset P$. Then the correspondences $G'\to I(G')$ and $K\to A(K)$ are inverses of each other.*

## § 4.

We will show that the theory established in § 3 implies the usual finite Galois theory of division ring $F$. Let $\mathbf{E}$ be the complete set of endomorphisms of $(F, +)$, $\{f_i\}_{i\in I}$ be a finite subset of $F$, then the function $[\ \ ]$ is defined as follows: $[\{f_i\}_{i\in I}]=\sum_{i\in I} f_i$, where $\Sigma$ expresses the finite sum of $F$. Clearly $[\{f_i\}_{i\in I}]\in F$. Let $\mathfrak{M}=\Sigma F u_i$ and $\{v_i\}_{i\in I}$ be a finite subset of $\mathfrak{M}$. Let $\{v_i\}_{i\in I_1}$ be a set of maximal $F$-linearly independent elements of $\{v_i\}_{i\in I}$, $\{v_i\}_{i\in I_2}=\{v_i\}_I-\{v_i\}_{I_1}$, then any element $v_j$ of $\{v_j\}_{I_2}$ can be expressed as $v_j=\sum_{i\in I_1} g_i^{(j)} v_i$. It is clear that $\sum_{j\in I} a_j v_j=\sum_{j\in I} b_j v_j$ holds if and only if $b_i+\sum_{j\in I_2} b_j g_i^{(j)}=a_i+\sum_{j\in I_2} a_j g_i^{(j)}$. This proves that the above defined function $[\ \ ]$ satisfies all conditions of Definition 3.1. Therefore $\mathfrak{M}$ has a 0-function (i. e. $\nu=0$) about $F$.

**Theorem 4.1.** *Let $\mathfrak{M}=\Sigma F u_i$, $G$ be a group of automorphisms of $F$, $P=I(G)$. Let $[F: P]_L<\infty$, then*

(i) *if $K$ is a division subring of $F$ and $K\supset P$, then $K$ is a homogenize subring (see § 3),*

(ii) *$\widetilde{G}$ is a homogenize subgroup of $G$ if and only if $\widetilde{G}$ is an $N$-subgroup*[5)]*.*

*Proof* From [1] it follows that $\mathscr{L}(P, \mathfrak{M})=\sum_{j\in I}\oplus S_j\mathscr{L}(F, \mathfrak{M})$, where $I=\{1, \cdots, n\}$ and $\mathscr{L}(K, \mathfrak{M})=\sum_{j\in I_1}\oplus S_j\mathscr{L}(F, \mathfrak{M})$, $I_1\subset I$. Moreover, $A(K)=\widetilde{G}=\{\psi_j \mid S_j=(S_j, \psi_j)$ for $j\in I_1\}$. Now we want to prove that $K$ is a homogenize subring in the meaning of definition 3.4. In fact, if $\sigma\in\mathscr{L}(K, \mathfrak{M})$, then $\sigma=\sum_{j\in I_1} S_j\omega_j$. We still denote the associative element of $\sigma$ by $\sigma_{I'}$ and write $H(\sigma_{I'_1})=\sum_{j\in I'_1}\psi_j F_R$. It is enough to show

1) Let $R$ be a ring with an identity, $\Phi$ its center and let $G$ be a group of automorphisms in $R$. Then we call the subalgebra $E'$ over $\Phi$ generated by the (regular) elements $c$ such that $I_c\in G$ the algebra of $G$. We say that $G$ is an $N$-subgroup if and only if for every regular $c\in E'$, $I_{c'}\in G$.

that $H(\sigma_{I'_1})$ is a homogeneous Galois $(K_R, F_R)$-bimodule. Since $\psi_j$ is an automorphism of $F$, it is easy to see that $\psi_j F_R$ is an $F_R$-cyclic module. Hence the condition (i) of Definition 3.3 is satisfied. It remainds to show that $\psi_j F_R \underset{K}{\cong} 1_K F_R$. But it is clear because $k^{\psi_j}=k$ holds for every $\psi_j$, $j\in I'_1$. Hence $\psi_j \underset{K}{=\!=} 1_K$ for $j\in I'_1$. This proves $I(\sigma_{I'_1})$ is a homogeneous Galois $(K_R, F_R)$-bimodule. Therefore $K$ is a homogenize division subring.

Now we want to show the assertion (i) of Theorem 4.1. Let $\tilde{G}$ be a homogenize subgroup and $E'$ be the algebra of $\tilde{G}$, $\delta'\in E'$, then $\delta'_L\in \mathscr{L}(\tilde{K}, \mathfrak{M})$, where $\tilde{K}=I(\tilde{G})$. By (iii) of Definition 3.5, the associative isomorphism $I_{\delta'}$ of $\delta'_L$ belongs to $\tilde{G}$, hence $\tilde{G}$ is an $N$-subgroup. Conversely, let $\tilde{G}$ be an $N$-subgroup, $\tilde{K}=I(\tilde{G})$, then from [1] it follows that every $S_j=(S_j, \psi_j)$ in the form of the element $\sigma=\sum_{j\in I} S_j\omega_j$ of $\mathscr{L}(\tilde{K}, \mathfrak{M})$ mplies $\psi_j\in\tilde{G}$. Let $\tilde{G}_0$ be a subgroup of inner automorphisms of $\tilde{G}$, then $\psi_j I_\delta\in\tilde{G}$ for any $I_\delta\in\tilde{G}_0$. Hence the condition (i) of Definition 3.5 is satisfied. Let $I'\subset I$, $\sigma_{I'}=\sum_{j\in I'} S_j\omega_j$, $H(\sigma_{I'})=\sum_{j\in I'}\psi_j F_R$, then clearly we have $\psi_j F_R\underset{\tilde{K}}{\cong} 1_{\tilde{K}} F_R$, where $\tilde{K}=I(\tilde{G})$ since $\tilde{k}^{\psi_j}=\tilde{k}$ for $\psi_j\in\tilde{G}$. This implies that the condition (ii) of Definition 3.5 is satisfied. On the other hand, if $S=(S, \psi)\in\mathscr{L}(\tilde{K}, \mathfrak{M})$, then from the well known proprety that every $N$-subgroup is a Galois subgroup, it follows that $\psi\in\tilde{G}$. This proves $\tilde{G}$ is a homogenize subgroup. This completes the proof.

From Theorems 4.1 and 3.1 we can now obtain again the finite Galois theory for division rings.

**Theorem 4. 2.** *Let $F$ be a division ring, $G$ be a group of automorphisms of $F$, $P=I(G)$ and $[F: P]_L<\infty$. Let $G'$ be any $N$-subgroup of $G$, and $K$ be any division subring of $F$ containing $P$. Then the correspondences $G'\to I(G')$ and $K\to A(K)$ are inverses of each other.*

## References

[1] Xu Yonghua, A finite structure theorem between primitive rings and its application to Galois theory, *Chin. Ann. of Math.* **1**: 2 (1980), 183—197.

[2] Xu Yonghua, A theory of ring that are isomorphic to the complete rings of linear transformations (VI), *Acta Math. Sinica*, **23** (1980), 646—657.

J. DIFFERENTIAL GEOMETRY
16 (1981) 577–593

# DIVISION ALGEBRAS AND FIBRATIONS OF SPHERES BY GREAT SPHERES

C. T. YANG

*Dedicated to Professor Buchin Su on his* 80*th birthday*

Smooth fibrations of spheres by *great spheres* occur naturally in the study of the Blaschke conjecture. In fact, if $M$ is a Blaschke manifold, $m$ is a point of $M$, $T_mM$ is the tangent space of $M$ at $m$, $\exp_m: T_mM \to M$ is the exponential map at $m$, and $\mathrm{Cut}(m)$ is the cut locus of $m$ in $M$, then $\exp_m^{-1}(\mathrm{Cut}(m))$ is a sphere $S_m$ in $T_mM$ of center 0, and $\exp_m: S_m \to \mathrm{Cut}(m)$ is a smooth *great sphere fibration* of the sphere $S_m$. For general information of the Blaschke conjecture, see [2].

If $\mathbf{K}$ is the real, complex, quaternionic or Cayley algebra, $n$ is the dimension of $\mathbf{K}$ as a euclidean space, which is 1, 2, 4 or 8, and $S^{2n-1}$ is the unit $(2n-1)$-sphere in the euclidean $2n$-space $\mathbf{K} \times \mathbf{K}$, then there is a natural smooth great $(n-1)$-sphere fibration of $S^{2n-1}$ such that any $(u, w), (u', w') \in S^{2n-1}$ belong to the same fibre iff either $w = w' = 0$ or $uw^{-1} = u'w'^{-1}$. When $n > 1$, this fibration, as well as isomorphic ones, is often referred as the *Hopf fibration*. Related to this result, Adams' theorem [1] says that a smooth fibration of $S^{2n-1}$ by $(n-1)$-spheres can occur only when $n = 1, 2, 4$ or 8, and a classical theorem of Hurwitz [4] says that any division algebra $\mathbf{K}$, which possesses a norm such that for any $v, w \in \mathbf{K}$, $|vw| = |v||w|$, must be the real, complex, quaternionic or Cayley algebra. If $n = 1$ or 2, then any $n$-dimensional division algebra is the real or complex algebra, and any fibration of $S^{2n-1}$ by $(n-1)$-spheres is unique up to an isomorphism. Hence in these cases, the correspondence between $n$-dimensional division algebras and smooth great $(n-1)$-sphere fibrations of $S^{2n-1}$ is trivial.

In this paper, we show that for $n = 4$ or 8, each $n$-dimensional division algebra $\mathbf{K}$ determines a smooth great $(n-1)$-sphere fibration of $S^{2n-1}$, and every smooth great $(n-1)$-sphere fibration of $S^{2n-1}$, up to an isomorphism, is determined by an $n$-dimensional division algebra $\mathbf{K}$. However, it is possible

Received December 23, 1981. The author is supported in part by the National Science Foundation.

that two division algebras, not isomorphic to each other, may determine isomorphic smooth great $(n-1)$-sphere fibrations of $S^{2n-1}$. Such an example can be found using division algebras constructed in Bruck [3].

We also show that any division algebra of dimension $> 1$ contains the complex algebra as a subalgebra. Results of a subsequent paper of the author's joint work with Herman Gluck and Frank Warner will be used to show that any smooth great 3-sphere fibration of $S^7$ is isomorphic to the Hopf fibration, and hence any Blaschke manifold which has the integral cohomology ring of the quaternionic projective 2-space $\mathbf{H}P^2$ is homeomorphic to $\mathbf{H}P^2$.

The author wishes to express his gratitude to many colleagues of his for numerous dicussions, and especially to McKenzie Y. Wang for bringing Bruck's paper to his attention, and to Stephen S. Shatz for showing him an algebraic proof of the result that any division algebra of dimension $> 1$ contains the complex algebra.

Throughout this paper, $\mathbf{R}$ denotes the real algebra, and $\mathbf{C}$ the complex algebra. Let $\mathbf{K}$ be the euclidean $n$-space, $n \geqslant 1$, which is often regarded as a vector space over $\mathbf{R}$. By a *regular multiplication* on $\mathbf{K}$, we mean a bilinear function

$$m: \mathbf{K} \times \mathbf{K} \to \mathbf{K}$$

such that for any $a, b \in \mathbf{K}$ with $a \neq 0$, each of

$$m(v, a) = b, \quad m(a, w) = b$$

has a unique solution in $\mathbf{K}$. $\mathbf{K}$ together with a regular multiplication on $\mathbf{K}$ is called a *regular algebra* which we also denote by $\mathbf{K}$. If $m$ is the only regular multiplication on $\mathbf{K}$ under our consideration, we often write $vw$ in place of $m(v, w)$. We note that a regular multiplication may not be associative, and a regular algebra may have no identity, and that a regular algebra may not have a norm such that the norm of a product is equal to the product of the norms. On the other hand, it can be shown that any 1-dimensional regular algebra must be $\mathbf{R}$, and that the dimension of any regular algebra is $1, 2, 4$ or $8$. A *division algebra* is defined to be a regular algebra having an identity. Notice that the real, complex, quaternionic and Cayley algebras are division algebras.

Let $\{e_1, \cdots, e_n\}$ be a basis of $\mathbf{K}$ as a vector space over $\mathbf{R}$. Then for any bilinear function $m: \mathbf{K} \times \mathbf{K} \to \mathbf{K}$, there are $n^3$ real numbers $a_{ijk}$, $i, j, k = 1, \cdots, n$, such that

$$m\left(\sum_{i=1}^{n} v_i e_i, \sum_{k=1}^{n} w_k e_k\right) = \sum_{j=1}^{n}\left(\sum_{i,k=1}^{n} v_i a_{ijk} w_k\right) e_j.$$

Hence regular multiplications are always smooth.

**Proposition 1.** *Any 1-dimensional regular algebra is the real algebra.*

*Proof.* Let **K** be a 1-dimensional regular algebra, and let $a$ be an element of **K** different from the zero of **K**. By definition, $ae = a$ for some $e \in \mathbf{K}$. $e$ is different from the zero of **K**; otherwise, $a = ae = a(0e) = 0(ae) = 0e = e$, contradicting to our assumption.

Let $a = te$, $t \in \mathbf{R}$. Then $t \neq 0$, and $te = (te)e = te^2$ so that $e^2 = e$. Hence $e$ is the identity of **K**, and **K** can be naturally identified with **R** by setting $re = r$ for all $r \in \mathbf{R}$.

**Theorem 1.** *Any division algebra of dimension* $> 1$ *contains a subalgebra isomorphic to the complex algebra.*

**Corollary 1.** *Any 2-dimensional division algebra is the complex algebra.*

Let **K** be a division algebra of dimension $n > 1$, and let $S^{2n-1}$ be the unit $(n-1)$-sphere in **K**. We may assume that the identity $e$ of **K** is contained in $S^{n-1}$; otherwise all we have to do is to use a new norm on **K** which is equal to $|e|^{-1}$ times the old one.

**Lemma 1.** *The map $f: S^{n-1} \to S^{n-1}$ defined by $f(x) = x^2/|x^2|$ is of degree* 2.

*Proof.* Let

$$\phi: S^{n-1} \times S^{n-1} \to S^{n-1}$$

be the map defined by

$$\phi(x, y) = xy/|xy|.$$

Notice that $\phi$ is well-defined and continuous, since $xy \in \mathbf{K} - \{0\}$ for any $x, y \in \mathbf{K} - \{0\}$.

Let $\Delta$ be the diagonal set in $S^{n-1} \times S^{n-1}$. Let $S^{n-1}$ be oriented, and let $S^{n-1} \times \{e\}$, $\{e\} \times S^{n-1}$ and $\Delta$ be so oriented that the natural projection of each of them onto $S^{n-1}$ is orientation-preserving. Let

$$\phi_*: H_{n-1}(S^{n-1} \times S^{n-1}) \to H_{n-1}(S^{n-1})$$

be the induced homomorphism of integral homology groups by $\phi$. Then

$$\phi_*[S^{n-1} \times \{e\}] = [S^{n-1}] = \phi_*[\{e\} \times S^{n-1}],$$

$$[\Delta] = [S^{n-1} \times \{e\}] + [\{e\} \times S^{n-1}],$$

so that

$$\phi_*[\Delta] = 2[S^{n-1}].$$

Since $\phi(x, x) = f(x)$ for any $x \in S^{n-1}$, our assertion follows.

*Proof of Theorem* 1. By Lemma 1, the map

$$g: \mathbf{K} \to \mathbf{K}$$

defined by $g(x) = x^2$ is onto. Therefore there is an element $i$ of $\mathbf{K} - \{0\}$ such that

$$i^2 = g(i) = -e.$$

The linear 2-subspace of **K** having $\{e, i\}$ as a basis is clearly a subalgebra of **K** isomorphic to **C**.

As mentioned earlier, Stephen S. Shatz has an algebraic proof of Lemma 1, and hence Theorem 1 can be proved algebraically.

**Theorem 2.** *Let* **K** *be a regular algebra of dimension* $n > 1$, *and let* $S^{2n-1}$ *be the unit* $(2n-1)$*-sphere in the euclidean* $2n$*-space* $\mathbf{K} \times \mathbf{K}$. *Then* **K** *determines a smooth great* $(n-1)$*-sphere fibration of* $S^{2n-1}$ *such that any* $(u, w), (u', w') \in S^{2n-1}$ *belong to the same fibre iff either* $w = w' = 0$ *or* $u = vw$ *and* $u' = vw'$ *for some* $v \in \mathbf{K}$. *Moreover, the fibrations determined by two isomorphic regular algebras are smoothly isomorphic.*

Notice that if **K** is the complex, quaternionic or Cayley algebra, then the fibration determined by **K** is the Hopf fibration.

*Proof*. Let $\Sigma^n = \mathbf{K} \cup \{\infty\}$ be the one-point compactification of **K**. Then $\Sigma^n$ can be made a smooth manifold as follows. For any $u \in \mathbf{K} - \{0\}$, we let

$$\lambda_u: \Sigma^n - \{0\} \to \mathbf{K}$$

be the homeomorphism such that $\lambda_u(\infty) = 0$ and $v\lambda_u(v) = u$ for any $v \in \mathbf{K} - \{0\} = \Sigma^n - \{0, \infty\}$. Since $\lambda_u: \mathbf{K} - \{0\} \to \mathbf{K} - \{0\}$ is a diffeomorphism, there is a smooth structure on $\Sigma^n$ such that the inclusion map of **K** into $\Sigma^n$ is a smooth imbedding, and $\lambda_u$ is a diffeomorphism for some $u \in \mathbf{K} - \{0\}$. The smooth structure on $\Sigma^n$ is independent of the choice of $u$. In fact, for any $u, u' \in \mathbf{K} - \{0\}$, $u$ and $u'$ can be joined by a smooth path in $\mathbf{K} - \{0\}$, and hence $\lambda_u, \lambda_{u'}: \mathbf{K} - \{0\} \to \mathbf{K} - \{0\}$ are isotopic.

Let

$$\pi: S^{2n-1} \to \Sigma^n$$

be the map such that $\pi(u, 0) = \infty$ for any $u \in S^{n-1}$, and $\pi(u, w)w = u$ for any $(u, w) \in S^{2n-1}$ with $w \neq 0$. Since the multiplication on **K** is bilinear, it follows that $\pi^{-1}v$ is a great $(n-1)$-sphere in $S^{2n-1}$ for any $v \in \Sigma^n$.

There is a smooth imbedding

$$g_0: \mathbf{K} \times S^{n-1} \to S^{2n-1}$$

given by

$$g_0(v, w) = \left( vw/\sqrt{|vw|^2 + 1}\,,\, w/\sqrt{|vw|^2 + 1} \right),$$

and for any $v \in \mathbf{K}$, $\pi g_0(\{v\} \times S^{n-1}) = v$. Also there is a smooth imbedding

$$g_1: S^{n-1} \times (\Sigma^n - \{0\}) \to S^{2n-1}$$

given by

$$g_1(u, v) = \left( u/\sqrt{1 + |\lambda_u(v)|^2}, \lambda_u(v)/\sqrt{1 + |\lambda_u(v)|^2} \right),$$

and for any $v \in \Sigma^n - \{0\}$, $\pi g_1(S^{n-1} \times \{v\}) = v$. Hence

$$\pi \colon S^{2n-1} \to \Sigma^n$$

is a smooth great $(n - 1)$-sphere fibration.

Let $\mathbf{K}_1$ be a regular algebra isomorphic to $\mathbf{K}$, and let

$$\pi_1 \colon S_1^{2n-1} \to \Sigma^n$$

be the smooth great $(n - 1)$-sphere fibration determined by $\mathbf{K}_1$, where $S_1^{2n-1}$ is the unit $(2n - 1)$-sphere in $\mathbf{K}_1 \times \mathbf{K}_1$. Then $\pi_1 \colon S_1^{2n-1} \to \Sigma_1^n$ is smoothly isomorphic to $\pi \colon S^{2n-1} \to \Sigma^n$. In fact, if $f \colon \mathbf{K}_1 \to \mathbf{K}$ is an isomorphism, then

$$f \times f \colon \mathbf{K}_1 \times \mathbf{K}_1 \to \mathbf{K} \times \mathbf{K}$$

defined by $(f \times f)(u_1, w_1) = (fu_1, fw_1)$ is a nonsingular linear map so that

$$h \colon S_1^{2n-1} \to S^{2n-1}$$

defined by $h(u_1, w_1) = (fu_1, fw_1)/|(fu_1, fw_1)|$ is a diffeomorphism. It is easy to see that $h$ maps fibres of $\pi_1 \colon S_1^{2n-1} \to \Sigma_1^n$ into fibres of $\pi \colon S^{2n-1} \to \Sigma^n$. Hence the proof is completed.

As a consequence of Theorem 2 and Adams' theorem, we have

**Corollary 2.** *The dimension of any regular algebra is* 1, 2, 4 *or* 8.

Let $GL(\mathbf{K})$ be the group of nonsingular linear maps of $\mathbf{K}$ into $\mathbf{K}$. Two regular multiplications $m$ and $m_1$ on $\mathbf{K}$ are said to be *equivalent* if there exist $\mu, \nu, \omega \in GL(\mathbf{K})$ such that $m_1(\nu \times \omega) = \mu m$, that means, the diagram

$$\begin{array}{ccc} \mathbf{K} \times \mathbf{K} & \xrightarrow{m_1} & \mathbf{K} \\ \uparrow{\scriptstyle \nu \times \omega} & & \uparrow{\scriptstyle \mu} \\ \mathbf{K} \times \mathbf{K} & \xrightarrow{m} & \mathbf{K} \end{array}$$

is commutative.

**Proposition 2.** *Let $m$ and $m_1$ be equivalent regular multiplications on the euclidean $n$-space $\mathbf{K}$. Then the smooth great $(n - 1)$-sphere fibrations of $S^{2n-1}$ determined by the regular algebras $(\mathbf{K}, m)$ and $(\mathbf{K}, m_1)$ are smoothly isomorphic.*

*Proof.* Let

$$\pi \colon S^{2n-1} \to \Sigma^n, \quad \pi_1 \colon S^{2n-1} \to \Sigma_1^n$$

be the smooth great $(n-1)$-sphere fibrations determined by $(\mathbf{K}, m)$ and $(\mathbf{K}, m_1)$. Since $m$ and $m_1$ are equivalent, there are $\mu, \nu, \omega \in GL(\mathbf{K})$ such that

$m_1(\nu \times \omega) = \mu m$. Then $\mu \times \omega$: $\mathbf{K} \times \mathbf{K} \to \mathbf{K} \times \mathbf{K}$ is a nonsingular linear map so that $h$: $S^{2n-1} \to S^{2n-1}$ defined by $h(u, w) = (\mu u, \omega w)/|(\mu u, \omega w)|$ is a diffeomorphism. It is easily seen that $h$ maps fibers of $\pi$: $S^{2n-1} \to \Sigma^n$ into fibres of $\pi_1$: $S^{2n-1} \to \Sigma_1^n$.

**Proposition 3.** *On the euclidean n-space* **K**, *any regular multiplication is equivalent to one having an identity.*

*Proof.* Let $m$ be a regular multiplication on **K**, and let

$$\Phi, \Psi: \mathbf{K} - \{0\} \to GL(\mathbf{K})$$

be the smooth maps such that

$$\Phi(v)w = m(v, w), \quad \Psi(w)v = m(v, w).$$

Let $e \in \mathbf{K} - \{0\}$, let

$$\mu, \nu, \omega: \mathbf{K} \to \mathbf{K}$$

be the elements of $GL(\mathbf{K})$ given by

$$\mu(u) = \Psi(e)^{-1}u, \quad \nu(v) = v, \quad \omega(w) = \Psi(e)^{-1}\Phi(e)w,$$

and let $m'$ be the regular multiplication on **K** such that

$$m'(v \times \omega) = \mu m.$$

Then for any $v', w' \in \mathbf{K} - \{0\}$,

$$m'(v', w') = \Psi(e)^{-1}m\left(v', \Phi(e)^{-1}\Psi(e)w'\right) = \begin{cases} \Psi(e)^{-1}\Phi(v')\Phi(e)^{-1}\Psi(e)w', \\ \Psi(e)^{-1}\Psi\left(\Phi(e)^{-1}\Psi(e)w'\right)v'. \end{cases}$$

Therefore

$$m'(e, w') = \Psi(e)^{-1}\Phi(e)\Phi(e)^{-1}\Psi(e)w' = w',$$

so that

$$e = m'(e, e) = \Psi(e)^{-1}\Psi\left(\Phi(e)^{-1}\Psi(e)e\right)e.$$

From the last equality, we infer that $\Phi(e)^{-1}\Psi(e)e = e$ and hence

$$m'(v', e) = \Psi(e)^{-1}\Psi\left(\Phi(e)^{-1}\Psi(e)e\right)v' = v'.$$

As a consequence of Propositions 2 and 3, we have

**Corollary 3.** *Any smooth great* $(n-1)$*-sphere fibration of* $S^{2n-1}$ *determined by a regular algebra is smoothly isomorphic to one determined by a division algebra.*

Now we are in a position to construct, from a given smooth great $(n-1)$-sphere fibration of $S^{2n-1}$, an $n$-dimensional division algebra **K** such that the smooth great $(n-1)$-sphere fibration of $S^{2n-1}$ determined by **K** is smoothly isomorphic to the given one. Since it is trivial for $n = 1$ or 2, in the following we assume that

$$n = 4 \text{ or } 8.$$

Let **K** be the euclidean $n$-space, and $S^{n-1}$ the unit $(n-1)$-sphere in **K**. Let $GL(\mathbf{K})$ be the group of all nonsingular linear maps of **K** into **K**, and $SL(\mathbf{K})$ the subgroup of $GL(\mathbf{K})$ consisting of all the $g \in GL(\mathbf{K})$ with det $g = 1$.

Let $L_i$ be a normed real vector $n$-space, and $S_i^{n-1}$ the unit $(n-1)$-sphere in $L_i$, $i = 1, 2$. A diffeomorphism $f\colon S_1^{n-1} \to S_2^{n-1}$ is called a *linear* diffeomorphism if there is a nonsingular linear map $g\colon L_1 \to L_2$ such that for any $x \in S_1^{n-1}, f(x) = g(x)/|g(x)|$.

**Lemma 2.** *Whenever $g \in GL(\mathbf{K})$, we have a linear diffeomorphism*

$$\bar{g}\colon S^{n-1} \to S^{n-1}$$

*defined by $g(x) = g(x)/|g(x)|$. Conversely, whenever $f\colon S^{n-1} \to S^{n-1}$ is a linear diffeomorphism, there is a unique $g \in GL(\mathbf{R})$ such that $\bar{g} = f$ and* det $g = \pm 1$, *and $g'g^{-1}$ is in the center of $GL(\mathbf{K})$ for any $g' \in GL(\mathbf{K})$ with $\bar{g}' = f$. Hence*

$$\overline{SL}(\mathbf{K}) = \{\bar{g} \mid g \in SL(\mathbf{K})\}$$

*acts on $S^{n-1}$ as a smooth transformation group.*

For any map $\alpha\colon S^{n-1} \to SL(\mathbf{K})$, we have a map $\bar{\alpha}\colon S^{n-1} \to \overline{SL}(\mathbf{K})$ defined by $\bar{\alpha}(v) = \overline{\alpha(v)}$, called the *associated map* of $\alpha$.

**Lemma 3.** *Let $S_i^{n-1}$ be $S^{n-1}$ or a great $(n-1)$-sphere in $S^{2n-1}$, $i = 1, 2$. Then any linear diffeomorphism $f\colon S_1^{n-1} \to S_2^{n-1}$ maps great circles into great circles, and any map $f\colon S_1^{n-1} \to S_2^{n-1}$ which maps great circles into great circles is a linear diffeomorphism.*

Lemma 2 is quite obvious and Lemma 3 is a consequence of the well-known theorem in projective geometry that any map of a projective space of dimension $> 1$ into itself which maps projective lines into projective lines is a projective transformation.

Let

$$\pi\colon S^{2n-1} \to \Sigma^n$$

be a given smooth great $(n-1)$-sphere fibration of $S^{2n-1}$. We first observe that $\Sigma^n$ is homeomorphic to the $n$-sphere. In fact, if $S^n$ is a great $n$-sphere in $S^{2n-1}$ containing a fibre $F$, then $F$ is a great $(n-1)$-sphere in $S^n$, and $\Sigma^n$ is obtained from a closed hemisphere in $S^n$ with boundary $F$ by identifying $F$ to a single point.

Let $F_0$ and $F_1$ be two distinct fibres. Whenever $x$ is a point of $S^{2n-1} - F_i$, $F_i$ and $x$ determine a great $n$-sphere in $S^{2n-1}$. The closed hemisphere in this great $n$-sphere of boundary $F_i$ containing $x$ will be denoted by $F_i x$.

Let

$$h_0: S^{2n-1} - F_1 \to F_0, h_1: S^{2n-1} - F_0 \to F_1$$

be the smooth maps such that for any $x \in S^{2n-1} - F_{1-i}$, $h_i(x)$ is the point of intersection of $F_{1-i}x$ with $F_i$, $i = 0, 1$. Let

$$x_0 = \pi F_0, \quad x_1 = \pi F_1.$$

Then

$$\pi \times h_0: S^{2n-1} - F_1 \to (\Sigma^n - \{x_1\}) \times F_0,$$
$$h_1 \times \pi: S^{2n-1} - F_0 \to F_1 \times (\Sigma^n - \{x_0\})$$

are diffeomorphisms, which are local trivializations of the fibration over $\Sigma^n - \{x_1\}$ and $\Sigma^n - \{x_0\}$ respectively.

Let $S$ be the $(n-1)$-sphere of unit tangent vectors of $\Sigma^n$ at $x_0$ with respect to any preassigned Riemannian metric on $\Sigma^n$. Then for any $(v, w) \in S \times F_0$, there is a tangent vector $\tau(v, w)$ of $F_1 w$ at $w$ such that

$$d\pi(\tau(v, w)) = v.$$

Now we define a smooth map

$$\xi: S \times F_0 \to F_1$$

as follows. Let $(v, w) \in S \times F_0$. Then there is a smooth map $f: [0, 1] \to F_1 w$ such that $f(t) = w$ iff $t = 0$, and $f'(0) = \tau(v, w)$. It is not hard to see that $\lim_{t\to 0} F_0 f(t)$ exists and is a closed hemisphere of boundary $F_0$ with $\tau(v, w)$ as a tangent vector at $w$. $\xi(v, w)$ is defined to be the point of intersection of $\lim_{t\to 0} F_0 f(t)$ with $F_1$.

The following lemma plays a key role in our paper.

**Lemma 4.** *For any $v \in S$, $w \to \xi(v, w)$ is a linear diffeomorphism of $F_0$ onto $F_1$, and for any $w \in F_0$, $v \to \xi(v, w)$ is a linear diffeomorphism of $S$ onto $F_1$.*

*Proof.* Let $v \in S$ and let $f: [0, 1] \to \Sigma^n - \{x_1\}$ be a smooth map such that $f(t) = x_0$ iff $t = 0$, and $f'(0) = v$. Then for any $w \in F_0$, we have a smooth map $f_w: [0, 1] \to F_1 w$ such that $\pi f_w = f$. Clearly $f_w(t) = w$ iff $t = 0$, and $f_w'(0) = \tau(v, w)$. Moreover,

$$\xi(v, w) = \lim_{t\to 0} h_1 f_w(t).$$

Let $C$ be a great circle in $F_0$. Then for any $t \in (0, 1]$, $C_t = \{f_w(t) \mid w \in C\}$ is the intersection of $\pi^{-1}f(t)$ with the great $(n + 1)$-sphere in $S^{2n-1}$ determined by $F_1$ and $C$, so that it is a great circle in $\pi^{-1}f(t)$. Therefore $h_1(C_t)$, which is the intersection of $F_1$ with the great $(n + 1)$-sphere in $S^{2n-1}$ determined by $F_0$ and $C_t$, is a great circle in $F_1$. Hence $\xi(v, C) = \lim_{t\to 0} h_1(C_t)$ is a great circle in $F_1$. From this result and Lemma 3 we conclude that $w \to \xi(v, w)$ is a linear diffeomorphism of $F_0$ onto $F_1$.

Let $w \in F_0$. For any great circle $C$ in $S$ we have a great $(n+1)$-sphere $S^{n+1}$ in $S^{2n-1}$ containing $F_0$ such that for any $v \in C$, $\tau(v, w)$ is a tangent vector of $S^{n+1}$ at $w$. It can be seen that $\xi(C, w)$ is the intersection of $F_1$ and $S^{n+1}$ so that it is a great circle in $F_1$. Hence by Lemma 3, $v \to \xi(v, w)$ is a linear diffeomorphism of $S$ onto $F_1$.

Since $\Sigma^n$ is 1-connected, we may assume that $\pi: S^{2n-1} \to \Sigma^n$ is oriented. Then for any $v \in S$, $w \to \xi(v, w)$ is an orientation-preserving linear diffeomorphism of $F_0$ onto $F_1$. We let $S$ be so oriented that for any $w \in F_0$, $v \to \xi(v, w)$ is also an orientation-preserving linear diffeomorphism of $S$ onto $F_1$.

Let $S^{n-1}$ be naturally oriented, and let us identify $F_0$, $F_1$ and $S$ with $S^{n-1}$ by orientation-preserving linear diffeomorphisms. Then $\xi: S \times F_0 \to F_1$ becomes a smooth map

$$\xi: S^{n-1} \times S^{n-1} \to S^{n-1}$$

such that for some smooth maps

$$\phi, \psi: S^{n-1} \to SL(\mathbf{K}),$$

we have

$$\xi(v, w) = \bar{\phi}(v)w = \bar{\psi}(w)v,$$

where $\bar{\phi}, \bar{\psi}: S^{n-1} \to \overline{SL}(\mathbf{K})$ are the associated maps of $\phi$ and $\psi$.

The following result can be proved in the same way as Proposition 3.

**Lemma 5.** *For any $e \in S^{n-1}$, we let*

$$\mu_e = \psi(e)^{-1}, \quad \nu_e = identity, \quad \omega_e = \psi(e)^{-1}\phi(e),$$

*let*

$$\phi_e, \psi_e: S^n \to SL(\mathbf{K})$$

*be the smooth maps defined by*

$$\phi_e(v) = \mu_e \phi(\nu_e^{-1} v)\omega_e^{-1},$$
$$\psi_e(w) = \mu_e \psi(\omega_e^{-1} w)\nu_e^{-1},$$

*and let*

$$\xi_e: S^{n-1} \times S^{n-1} \to S^{n-1}$$

*be the smooth map defined by*

$$\xi_e(v, w) = \bar{\mu}_e \xi(\bar{\nu}_e^{-1} v, \bar{\omega}_e^{-1} w).$$

*Then*

$$\phi_e(e) = \psi_e(e) = identity,$$
$$\xi_e(v, w) = \bar{\phi}_e(v)w = \bar{\psi}_e(w)v,$$

*where $\bar{\phi}_e, \bar{\psi}_e: S^{n-1} \to \overline{SL}(\mathbf{K})$ are the associated maps of $\phi_e$ and $\psi_e$.*

**Lemma 6.** **K** *can be made a division algebra with identity e such that for any* $v, w \in S^{n-1}$,

$$\xi_e(v, w) = vw/|vw|\,.$$

The following results are needed in the proof of Lemma 6.

**Sublemma 1.** *Let U be a nonnull open subset of* **R**, *and let* $\nu, \omega\colon U \to \mathbf{R}$ *and* $\alpha\colon \mathbf{R} \to \mathbf{R}$ *be smooth maps such that*

$$\alpha(r) > 0$$

*for any* $r \in \mathbf{R}$, *and*

$$\alpha(r) = \frac{1 + \nu(s)r}{1 + \omega(s)r}$$

*for any* $r \in \mathbf{R}$ *and* $s \in U$. *Then*

$$\nu = \omega, \quad \alpha = 1.$$

*Proof.* By hypothesis,

$$\alpha(r)(1 + \omega(s)r) = 1 + \nu(s)r.$$

Partially differentiating the equality with respect to $s$, we obtain

$$\alpha(r)\omega'(s)r = \nu'(s)r.$$

Therefore

$$\alpha(r)\omega'(s) = \nu'(s).$$

If $\omega'(s) \not\equiv 0$, then $\alpha(r) = \nu'(s)/\omega'(s)$ which is independent of the choice of $r$. Therefore $\alpha(r) = \alpha(0) = 1$ and hence

$$\alpha = 1.$$

If $\omega'(s) \equiv 0$, then $\nu'(s) \equiv 0$. Therefore there are $\nu, \omega \in \mathbf{R}$ such that

$$\alpha(r) = \frac{1 + \nu r}{1 + \omega r}.$$

Since $\alpha(r) > 0$ for all $r \in \mathbf{R}$, it follows that $\nu = \omega$. Hence $\alpha(r) = 1$ for all $r \in \mathbf{R}$ or

$$\alpha = 1.$$

**Sublemma 2.** *Let* $\lambda_1, \lambda_2, \mu_1, \mu_2, \alpha\colon \mathbf{R} \to \mathbf{R}$ *be smooth maps such that*

$$\alpha(r) > 0$$

*for any* $r \in \mathbf{R}$, *and*

$$\alpha(r) = \frac{1 + \lambda_1(s)r + \lambda_2(s)r^2}{1 + \mu_1(s)r + \mu_2(s)r^2}$$

*for any* $r, s \in \mathbf{R}$. *Then either* $\lambda_1, \lambda_2, \mu_1, \mu_2$ *are constant maps or* $\alpha = 1$.

*Proof*. By hypothesis,

$$\alpha(r)\big(1+\mu_1(s)r+\mu_2(s)r^2\big)=1+\lambda_1(s)r+\lambda_2(s)r^2.$$

Partially differentiating the equality with respect to $s$, we obtain

$$\alpha(r)\big(\mu_1'(s)r+\mu_2'(s)r^2\big)=\lambda_1'(s)r+\lambda_2'(s)r^2.$$

Therefore

$$\alpha(r)\big(\mu_1'(s)+\mu_2'(s)r\big)=\lambda_1'(s)+\lambda_2'(s)r.$$

Assume first that $\mu_1'(s)\equiv 0$. Then

$$\lambda_1'(s)=\alpha(0)\mu_2'(s)0\equiv 0,$$

so that

$$\alpha(r)\mu_2'(s)=\lambda_2'(s).$$

If $\mu_2'(s)\equiv 0$, then $\lambda_2'(s)\equiv 0$. Hence $\lambda_1,\lambda_2,\mu_1,\mu_2$ are constant maps. If $\mu_2'(s)\not\equiv 0$, then there is a nonnull open subset $U$ of $\mathbf{R}$ such that for any $s\in U$, $\mu_2'(s)\neq 0$. Therefore for any $r\in\mathbf{R}$ and $s\in U$, $\alpha(r)=\lambda_2'(s)/\mu_2'(s)$ which is independent of the choice of $r$. Hence $\alpha(r)=\alpha(0)=1$ or $\alpha=1$.

Assume next that $\mu_1'(s)\not\equiv 0$. Then there is a nonnull open subset $U$ of $\mathbf{R}$ such that $\mu_1'(s)\neq 0$ and

$$\lambda_1'(s)=\alpha(0)\mu_1'(s)=\mu_1'(s)$$

for any $s\in U$. Therefore for any $r\in\mathbf{R}$ and $s\in U$,

$$\alpha(r)=\frac{1(\lambda_2'(s)/\mu_1'(s))r}{1+(\mu_2'(s)/\mu_1'(s))r}.$$

Hence by Sublemma 1,

$$\alpha=1.$$

*Proof of Lemma* 6. In this proof, we drop the subscript $e$ from $\xi_e,\phi_e,\psi_e$ so that $\xi,\phi,\psi$ are actually $\xi_e,\phi_e,\psi_e$ of Lemma 5.

Let

$$\Phi,\Psi:\mathbf{K}-\{0\}\to GL(\mathbf{K})$$

be the maps such that for any $v,w\in\mathbf{K}-\{0\}$,

$$\Phi(v)=\frac{|v|}{|\phi(v/|v|)e|}\phi(v/|v|),\quad \Psi(w)=\frac{|w|}{|\psi(w/|w|)e|}\psi(w/|w|).$$

Then for any $v,w\in\mathbf{K}-\{0\}$,

$$\Phi(v)e=v=\Psi(e)v,\quad \Psi(w)e=w=\Phi(e)w,$$

$$\Phi(v)w/|\Phi(v)w|=\Psi(w)v/|\Psi(w)v|.$$

If we are able to show that for any $v, w \in \mathbf{K} - \{0\}$,

$$\Phi(v)w = \Psi(w)v,$$

then $\mathbf{K}$ can be made a division algebra such that for any $v, w \in \mathbf{K} - \{0\}$, $vw = \Phi(v)w = \Psi(w)v$ so that for any $v, w \in S^{n-1}$, $\xi(v, w) = vw/|vw|$.

In the following, we let $v$ and $w$ be two fixed elements of $\mathbf{K} - \{0\}$. If $v = re$ for some $r \in \mathbf{R}$, then

$$\Phi(v)w = r\Phi(e)w = r\Psi(w)e = \Psi(w)v.$$

If $w = re$ for some $r \in \mathbf{R}$, then $\Phi(v)w = r\Phi(v)e = r\Psi(e)v = \Psi(w)v$. Hence we may assume that $v, w \notin \mathbf{R}e$. Let $\gamma$ be the real number such that

$$\Phi(v)w = \gamma\Psi(w)v.$$

We claim that $\gamma = 1$.

Assume first that $e, v, w$ are not linearly independent. Then for some $t, t' \in \mathbf{R}$,

$$w = te + t'v, \ t' \neq 0$$

Let $\{e_1, \cdots, e_n\}$ be a basis of $\mathbf{K}$ such that

$$e_1 = e, e_2 = v,$$

and let $\gamma_1, \cdots, \gamma_n \in \mathbf{R}$ be such that

$$\Psi(e_2)e_2 = \gamma_1 e_1 + \cdots + \gamma_n e_n.$$

If $\gamma_1 = \gamma_3 = \gamma_4 = \cdots = \gamma_n = 0$, then

$$\Psi(e_2)(e_2 - \gamma_2 e_1) = \Psi(e_2)e_2 - \gamma_2\Psi(e_2)e_1 = \gamma_2 e_2 - \gamma_2 e_2 = 0,$$

which is impossible. Therefore $\gamma_k \neq 0$ for some $k \neq 2$. We may assume that

$$\gamma_1 \neq 0.$$

In fact, if $\gamma_1 = 0$, then $\gamma_k \neq 0$ for some $k > 2$, so that by replacing $e_k$ by $e_k + e_1$ we obtain a new $\gamma_1$ different from 0.

For any $r, s \in \mathbf{R}$, there are smooth real valued functions

$$\alpha = \alpha(r), \quad \beta = \beta(s)$$

such that

$$\Phi(e_1 + re_2)e_2 = \alpha\Psi(e_2)(e_1 + re_2), \quad \Psi(e_1 + se_2)e_2 = \beta\Phi(e_2)(e_1 + se_2).$$

Clearly $\alpha(0) = 1$ and $\alpha(r) \neq 0$ for all $r \in \mathbf{R}$. Hence $\alpha(r) > 0$ for all $r \in \mathbf{R}$. Similarly $\beta(0) = 1$ and $\beta(s) > 0$ for all $s \in \mathbf{R}$. Now

$$\begin{aligned}\Phi(e_1 + re_2)(e_1 + se_2) &= \Phi(e_1 + re_2)e_1 + s\Phi(e_1 + re_2)e_2 \\ &= e_1 + re_2 + s\alpha\Psi(e_2)(e_1 + re_2) \\ &= e_1 + re_2 + s\alpha e_2 + rs\alpha(\gamma_1 e_1 + \cdots + \gamma_n e_n).\end{aligned}$$

$$\Psi(e_1 + se_2)(e_1 + re_2) = e_1 + r\beta e_2 + se_2 + rs\beta\gamma(\gamma_1 e_1 + \cdots + \gamma_n e_n).$$

Since the coefficients of $e_1, \cdots, e_n$ in $\Phi(e_1 + re_2)(e_1 + se_2)$ and those in $\Psi(e_1 + se_2)(e_1 + re_2)$ are proportional, we infer that

$$\frac{1 + rs\alpha\gamma_1}{1 + rs\beta\gamma\gamma_1} = \frac{r + s\alpha + rs\alpha\gamma_2}{r\beta + s + rs\beta\gamma\gamma_2} = \frac{\alpha\gamma_k}{\beta\gamma\gamma_k}, \quad k > 2.$$

Therefore

$$\alpha = \frac{1 + ((\beta - 1)/s + \beta\gamma\gamma_2)r - (\beta\gamma\gamma_1)r^2}{1 + (\gamma_2 + s\gamma_1 - s\beta\gamma\gamma_1)r - (\beta\gamma_1)r^2}.$$

By Sublemma 2, either $\alpha = 1$ or

$$((\beta - 1)/s + \beta\gamma\gamma_2)' = (\beta\gamma\gamma_1)' = (\gamma_2 + s\gamma_1 - s\beta\gamma\gamma_1)' = (\beta\gamma_1)' = 0.$$

In the first case,

$$(\beta - 1)/s + \beta\gamma\gamma_2 = \gamma_2 + s\gamma_1 - s\beta\gamma\gamma_1, \quad \beta\gamma\gamma_1 = \beta\gamma_1.$$

Since $\beta\gamma_1 \neq 0$, it follows from the second equality that $\gamma = 1$. Then the first equality becomes

$$(\beta - 1)(1/s + \gamma_2 + s\gamma_1) = 0.$$

Therefore $\beta - 1 = 0$ or $\beta = 1$. In the second case, $\beta'(s) = 0$ so that $\beta(s) = \beta(0) = 1$. Then

$$0 = (\gamma_2 + s\gamma_1 - s\beta\gamma\gamma_1)' = \gamma_1(1 - \gamma),$$

so that $\gamma = 1$. Therefore $\alpha = 1$. Hence we always have

$$\alpha = 1, \beta = 1, \gamma = 1.$$

Since $v = e_2$ and $w = te_1 + t'e_2$, it follows that when $t = 0$,

$$\Phi(v)w = t'\Phi(e_2)e_2 = t'\Psi(e_2)e_2 = \Psi(w)v,$$

and when $t \neq 0$,

$$\Phi(v)w = t\Phi(e_2)(e_1 + (t'/t)e_2) = t\Psi(e_1 + (t'/t)e_2)e_2 = \Psi(w)v.$$

Assume now that $e, v, w$ are linearly independent. Then there is a basis $\{e_1, \cdots, e_n\}$ of $\mathbf{K}$ such that

$$e_1 = e, e_2 = v, e_3 = w.$$

Let $\gamma_1, \cdots, \gamma_n \in \mathbf{R}$ be such that

$$\Psi(e_3)e_2 = \gamma_1 e_1 + \cdots + \gamma_n e_n.$$

Then

$$\Phi(e_1 + re_2)(e_1 + se_3) = e_1 + re_2 + s\alpha e_3 + rs\alpha(\gamma_1 e_1 + \cdots + \gamma_n e_n),$$

$$\Psi(e_1 + se_3)(e_1 + re_2) = e_1 + r\beta e_2 + se_3 + rs\beta\gamma(\gamma_1 e_1 + \cdots + \gamma_n e_n).$$

Therefore

$$\frac{1+rs\alpha\gamma_1}{1+rs\beta\gamma\gamma_1}=\frac{r+rs\alpha\gamma_2}{r\beta+rs\beta\gamma\gamma_2}=\frac{s\alpha+rs\alpha\gamma_3}{s+rs\beta\gamma\gamma_3}=\frac{\alpha\gamma_k}{\beta\gamma\gamma_k},\quad k>3.$$

We may assume that one of $\gamma_1, \gamma_2, \gamma_3$ is not 0. In fact, if $\gamma_1=\gamma_2=\gamma_3=0$, then for some $k>0$, $\gamma_k\neq 0$. By replacing $e_k$ by $e_k+e_1$ we obtain a new $\gamma_1$ different from 0.

If either $\gamma_1$ or $\gamma_3$ is not 0, from

$$\frac{1+rs\alpha\gamma_1}{1+rs\beta\gamma\gamma_1}=\frac{s\alpha+rs\alpha\gamma_3}{s+rs\beta\gamma\gamma_3}$$

we obtain that

$$\alpha=\frac{1+(\beta\gamma\gamma_3)r}{1+(\gamma_3+s\beta\gamma\gamma_1-s\gamma_1)r}.$$

By Sublemma 1, $\alpha=1$, so that

$$\beta\gamma\gamma_3=\gamma_3+s\beta\gamma\gamma_1-s\gamma_1,$$

or

$$(\beta\gamma-1)(\gamma_3-s\gamma_1)=0.$$

Since either $\gamma_1$ and $\gamma_3$ is not 0, it follows that $\beta\gamma-1=0$. Hence

$$\gamma=1,\ \beta=1,\ \alpha=1.$$

If $\gamma_2$ is not 0, we have

$$\beta=\frac{1+(\alpha\gamma_2)s}{1+(\gamma\gamma_2+r\alpha\gamma_1-r\gamma\gamma_1)s}.$$

Similarly,

$$\gamma=1,\ \alpha=1,\ \beta=1.$$

Since $v=e_2$ and $w=e_3$, it follows that

$$\Phi(v)w=\Phi(e_2)e_3=\Psi(e_3)e_2=\Psi(w)v.$$

Hence the proof of Lemma 6 is completed.

**Theorem 3.** *Let* **K** *be a division algebra of dimension $n$, $n=4$ or 8, and let*

$$\pi\colon S^{2n-1}\to\Sigma^n$$

*be the smooth great $(n-1)$-sphere fibration determined by* **K** *as seen in Theorem 2. Then* **K** *can be recovered from the fibration by the construction given above.*

*Proof.* Let

$$F_0=\{0\}\times S^{n-1},\quad F_1=S^{n-1}\times\{0\},$$

and let $\Sigma^n$ be assigned a Riemannian metric such that the smooth imbedding $\rho$ of $D^n = \{x \in \mathbf{K} \mid |x| \leq 1\}$ into $\Sigma^n$ given by

$$\rho(v) = \pi\left(vw/\sqrt{|vw|^2 + 1}, w/\sqrt{|vw|^2 + 1}\right)$$

is isometric. Then we have natural linear diffeomorphisms of $F_0$, $F_1$ and $S$ onto $S^{n-1}$, of which the first two are projections and the last is $(d\rho)^{-1}$.

Let us use these diffeomorphisms to identify $F_0$, $F_1$ and $S$ with $S^{n-1}$. Then $\xi: S \times F_0 \to F_1$ becomes

$$\xi: S^{n-1} \times S^{n-1} \to S^{n-1}$$

defined by

$$\xi(v, w) = vw/|vw|.$$

Hence the regular multiplication constructed in Lemma 6 is the same as that in **K**.

**Theorem 4.** *Let*

$$\pi: S^{2n-1} \to \Sigma^n$$

*be a given smooth great $(n-1)$-sphere fibration, $n = 4$ or $8$, and let* **K** *be the $n$-dimensional division algebra constructed from the fibration as seen earlier. Then the fibration is smoothly isomorphic to that determined by* **K**.

*Proof.* With respect to a preassigned Riemannian metric on $\Sigma^n$, there is a $\delta > 0$ such that if $D_\delta$ is the closed $n$-disk in the tangent space of $\Sigma^n$ at $x_0$ of center 0 and radius $\delta$, then the exponential map exp imbeds $D_\delta$ smoothly into $\Sigma^n - \{x_1\}$. Let $D$ be the compact smooth $n$-manifold obtained from the disjoint union of $\Sigma^n - \{x_0\}$ and $S \times [0, \delta)$ by identifying every $(v, t) \in S \times (0, \delta)$ with $\exp tv \in \Sigma^n \times \{x_0\}$. It is clear that $D$ is a smooth closed $n$-disk, and its boundary is $S \times \{0\} = S$.

Let

$$\lambda: D_\delta \times F_0 \to S^{2n-1}$$

be the smooth imbedding such that $\lambda(v, w) \in F_1 w$ and $\pi\lambda(v, w) = \exp v$ for any $(v, w) \in D_\delta \times F_0$. Then we have a compact smooth $(2n-1)$-manifold $W$ obtained from the disjoint union of $S^{2n-1} - F_0$ and $S \times [0, \delta) \times F_0$ by identifying every $(v, t, w) \in S \times (0, \delta) \times F_0$ with $\lambda(tv, w) \in S^{2n-1} - F_0$. It is clear that the boundary of $W$ is $S \times \{0\} \times F_0 = S \times F_0$, and that $\pi: S^{2n-1} - F_0 \to \Sigma^n - \{x_0\}$ can be naturally extended to a smooth fibration

$$\pi: W \to D.$$

From the construction of $\xi: S \times F_0 \to F_1$, it can be shown that $\xi$ can be naturally extended to a smooth fibration

$$\xi: W \to F_1$$

such that for any $x \in W - (S \times F_0)$, $\xi(x)$ is the point of intersection of $F_0 x$ with $F_1$. Hence

$$h_1 = (\xi \times \pi)^{-1}: F_1 \times D \to W$$

is a diffeomorphism.

The inclusion map of $S^{2n-1} - F_0$ into $S^{2n-1}$ can be extended to a smooth map

$$h_2: W \to S^{2n-1}$$

such that $h_2(v, w) = w$ for any $(v, w) \in S \times F_0 = \partial W$. Therefore we have a smooth map

$$h = h_2 h_1: F_1 \times D \to S^{2n-1}$$

such that the fibration $\pi: S^{2n-1} \to \Sigma^n$ is induced by the projection fibration $F_1 \times D \to D$. Moreover, whenever $(u, v), (u', v') \in F_1 \times D$, $h(u, v) = h(u', v')$ iff either $(u, v) = (u', v')$ or $u = u'$, $v, v' \in S = \partial D$ and for some $w, w' \in F_0$. $u = \xi(v, w) = \xi(v', w') = u'$.

In the construction of the division algebra **K**, we identify $F_0$, $F_1$ and $S$ with $S^{n-1} \subset \mathbf{K}$. Then we have a smooth map

$$h': F_1 \times D \to S^{2n-1}$$

given as follows. Let us regard $D - \{x_1\}$ as $\{v \in \mathbf{K} \mid 0 < |v| \leqslant 1\}$. Then for any $(u, v) \in F_1 \times (D - \{x_1\})$ there is a unique $w(u, v) \in \mathbf{K}$ with $vw(u, v) = u$. The map $h'$ is given by

$$h'(u, v) = \begin{cases} (u, 0) & \text{if } v = x_1, \\ \left( \dfrac{u}{\sqrt{1 + |w(u, v)|^2}}, \dfrac{w(u, v)}{\sqrt{1 + |w(u, v)|^2}} \right) & \text{otherwise.} \end{cases}$$

Now it is not hard to see that the identity map of $F_1 \times D$ induces a smooth isomorphism between the fibration $\pi: S^{2n-1} \to \Sigma^n$ and that determined by the division algebra **K**.

**Corollary 4.** *Up to a smooth isomorphism, every smooth great $(n-1)$-sphere fibration of $S^{2n-1}$ is determined by an $n$-dimensional division algebra.*

**Remark.** It is possible to have many $n$-dimensional division algebras, not isomorphic to one another but determining isomorphic smooth great $(n-1)$-sphere fibrations of $S^{2n-1}$. In fact, whenever $\alpha, \beta, \gamma, \alpha', \beta', \gamma'$ are positive real numbers satisfying

$$\alpha + \beta + \gamma - \alpha\beta\gamma = \alpha' + \beta' + \gamma' - \alpha'\beta'\gamma',$$

there is a 4-dimensional division algebra which, as the quaternionic algebra, has $\{e, i, j, k\}$ as a basis, but in which the multiplication is given by:

| | $e$ | $i$ | $j$ | $k$ |
|---|---|---|---|---|
| $e$ | $e$ | $i$ | $j$ | $k$ |
| $i$ | $i$ | $-e$ | $\gamma k$ | $-\beta' k$ |
| $j$ | $j$ | $-\gamma' k$ | $-e$ | $\alpha i$ |
| $k$ | $k$ | $\beta j$ | $-\alpha' i$ | $-e$ |

Also for any $\theta \in [0, \pi/2]$, there is a 4-dimensional division algebra which has $\{e, i, j, k\}$ as a basis and in which the multiplication is given by:

| | $e$ | $i$ | $j$ | $k$ |
|---|---|---|---|---|
| $e$ | $e$ | $i$ | $j$ | $k$ |
| $i$ | $i$ | $-e$ | $k$ | $-j$ |
| $j$ | $j$ | $-k$ | $-e\cos\theta + i\sin\theta$ | $i\cos\theta + e\sin\theta$ |
| $k$ | $k$ | $j$ | $-i\cos\theta - e\sin\theta$ | $-e\cos\theta + i\sin\theta$ |

For details, see Bruck [4]. Since all these division algebras are homotopic to the quaternionic algebra, the smooth great 3-sphere fibrations of $S^7$ determined by them are smoothly isomorphic to the Hopf fibration.

## References

[1] J. F. Adams, *On the non-existence of elements of Hopf invariant one*, Ann. of Math. **72** (1960) 20–104.

[2] A. Beese, *Manifolds all of whose geodesics are closed*, Ergebnisse Math. und ihrer Grenzgebiete, Vol. 93, Springer, Berlin, 1978.

[3] A. Hurwitz, *Über der Komposition der Quadratischen Form vom beliebig vielen Variablen*, Nachr. Gesellsch. Wiss. Göttingen, 1898, 309–306; Math. Werke, Bd. II, 565–571.

[4] R. H. Bruck, *Some results in the theory of linear non-associative algebras*, Trans. Amer. Math. Soc. **56** (1944) 141–199.

UNIVERSITY OF PENNSYLVANIA

*Chin. Ann. of Math.*
**3** (4) 1982

# QUALITATIVE THEORY OF THE QUADRATIC SYSTEMS IN THE COMPLEX SPACE

YE YANQIAN (YEH YEN-CHIEN)
(*Nanjing University*)

Dedicated to Professor Su Bu-chin on the Occasion of his 80th Birthday and his 50th Year of Educational Work

In the past 25 years many works have been done by Chinese and Soviet mathematicians on the qualitative theory of the real quadratic differential equation

$$\frac{dy}{dx}=\frac{Q_2(x,\ y)}{P_2(x,\ y)},\tag{1}$$

where $x$ and $y$ are real variables, $P_2$ and $Q_2$ are polynomials of degree $\leqslant 2$ with real coefficients. (See [1, 2]). Meanwhile, in the Soviet school, researches about global and local aspects of the general polynomial system

$$\dot{Z}=P(Z)\tag{2}$$

in complex variables have also been carried out gradually (see [3]), where $P(Z)$ is polynomial in $Z=(Z_1,\ \cdots,\ Z_n)$ of degree $\leqslant N$, and $Z\in C^n$. Up to now, one can see very little affinity between these two areas, although they have obviously a common objective: to solve the second part of the Hilbert's 16th problem.

As an intermediate work, the purpose of this paper is to carry out a rudimentary investigation of the qualitative property of solutions or integral surfaces of equation (1), where, instead, we will assume throughout this paper that $x$, $y$ are complex variables, while $P_2$ and $Q_2$ still remain to be polynomials of degree $\leqslant 2$ with real coefficients.

## § 1. Some properties of solutions of a complex differential equation with real coefficients

Consider the equation

$$\frac{dy}{dx}=\frac{Q(x,\ y)}{P(x,\ y)},\tag{3}$$

where $P$ and $Q$ are power series of the complex variables $x$ and $y$ with real coefficients. Let $x=x_1+ix_2$, $y=y_1+iy_2$, then on separating real and imaginary parts, any integral $f(x,\ y)=0$ of (3) can be thought as a 2-dimensional surface

Manuscript received April 8, 1981.

$$S:\ f_1(x_1, x_2, y_1, y_2)=0,\quad f_2(x_1, x_2, y_1, y_2)=0 \tag{4}$$

in the 4-dimensional $(x_1, x_2, y_1, y_2)$ real space.

Let

$$F(x, y, C)=0 \tag{5}$$

be the general integral of (3), then, generally speaking, (4) can be obtained from (5) by giving $C$ a definite value, real or complex. Furthermore, a solution of (3) may be a multiple-valued function, which can have even infinitely many branches. Hence in the sequel we must fix a single-valued branch of this solution, otherwise, even the uniqueness of solutions of a boundary value problem will not be ensured.

*Example 1.* Consider a very simple special case of (3)

$$\frac{dy}{dx}=-\mu\frac{y}{x}, \tag{6}$$

where $\mu>0$ is real and irrational. The general solution of (6) is

$$x^{\mu}y=C. \tag{7}$$

Let $x=re^{i\theta}=x_1+ix_2$, $y=y_1+iy_2$, $C=C_1+iC_2$, then one can get from (7) a family of integral surfaces

$$\begin{cases} r^{\mu}(y_1\cos\mu(\theta+2k\pi)-y_2\sin\mu(\theta+2k\pi))=C_1, \\ r^{\mu}(y_1\sin\mu(\theta+2k\pi)+y_2\cos\mu(\theta+2k\pi))=C_2 \end{cases} \tag{8}$$
$$(k=0,\ \pm1,\ \pm2,\ \cdots),$$

for any fixed value of $C$.

In order to find the intersection of (8) with the real plane $x_2=y_2=0$, one needs only to put in (8) $x_2=y_2=\theta=0$, $r=x_1$. Now, there are two different cases.

*Case A.* $k=0$. Then we get from (8)

$$x_1^{\mu}y_1=C_1,\quad 0=C_2,$$

Which means that the intersection curve of (8) for $C=C_1$ and $k=0$ with the real plane is just the integral curve of (6) which corresponds to the same constant $C_1$, if (6) is considered to have real variables. Moreover, we see that if $C_2\neq0$, then (8) for $k=0$ does not intersect the real plane.

*Case B.* $k\neq0$. Here the intersection curve must satisfy both the two equations

$$x_1^{\mu}y_1\cos 2k\pi\mu=C_1,\quad x_1^{\mu}y_1\sin 2k\pi\mu=C_2. \tag{9}$$

For $C_2\neq0$, (9) yields

$$C_1=C_2\cot 2k\pi\mu, \tag{10}$$

from which $C_1$ is determined by $k$, $\mu$ and $C_2$. The intersection curve (9) is then

$$x_1^{\mu}y_1=\frac{C_1}{\cos 2k\pi\mu}=C_1'=|C|. \tag{11}$$

That is to say, for $C_2\neq0$, $k\neq0$, only when $C_1$ satisfies (10), the integral surface (8) will intersect the real plane at a curve (11), but here $C_1'\neq C_1$, different from that of Case A.

It is easily seen that for $C_2=0$ and $k\neq 0$, (8) will intersect the real plane only when $C_1=0$, and thus the intersection will always be $x_1=0$ and $y_1=0$ for all $k\neq 0$.

To sum up, we see that for a fixed real number $C_1'\neq 0$, one can determine infinitely many pairs of $C_1$ and $C_2$, say, $(C_1^k, C_2^k)$ by (11) and (10), corresponding to $k=\pm 1, \pm 2, \cdots$, such that all the surfaces (8) with $C=C_1^k+iC_2^k$ will intersect the real plane at the same curve (11). In other words, if we take

$$x_1^{\mu}y_1=C_1' \quad \text{when} \quad x_2=y_2=0$$

as a boundary condition for the equation (6), then the solution is not unique. Since in this paper we are only interested in the intersection curve of integral surfaces of a complex equation (3) with the real plane, we will agree here after to take only the fundamental single-valued branch of any multi-valued solution.

It is interesting to note if we put $\mu=\sigma+i\tau$ in (6), such that $\sigma$, $\tau$ and $\sigma/\tau$ are all positive irrational, then parallel to (8) and (9) we will have now

$$\begin{cases} C_1=r^{\sigma}e^{-\tau(\theta+2k\pi)}[y_1\cos(\tau\ln r+\sigma(\theta+2k\pi))-y_2\sin(\tau\ln r+\sigma(\theta+2k\pi))], \\ C_2=r^{\sigma}e^{-\tau(\theta+2k\pi)}[y_1\sin(\tau\ln r+\sigma(\theta+2k\pi))+y_2\cos(\tau\ln r+\sigma(\theta+2k\pi))] \end{cases} \tag{8}$$

and (for $x_2=y_2=0$, $\theta=0$, $\gamma=x_1$)

$$\begin{cases} C_1=x_1^{\sigma}y_1e^{-2k\pi\sigma}\cos(\tau\ln x_1+2k\pi\sigma), \\ C_2=x_1^{\sigma}y_1e^{-2k\pi\sigma}\sin(\tau\ln x_1+2k\pi\sigma). \end{cases} \tag{9}$$

Thus for any fixed $C=C_1+iC_2$, (9) implies

$$x_1^{\sigma}y_1=|C|e^{2k\pi\tau}, \qquad |C|=\sqrt{C_1^2+C_2^2} \tag{$\overline{11}$}$$

For $k=0, \pm 1, \pm 2, \cdots$ the family of intersection curves ($\overline{11}$) will be dense in the real plane, which reveals the density property of equation (6).

It is easy to prove the following fundamental.

**Theorem 1.** *If an integral surface $S$ of* (3) *intersects the real plane at a curve $l$ (it may consist of more than one connected components), then $l$ is an integral curve of the corresponding real equation*

$$\frac{dy_1}{dx_1}=\frac{Q(x_1, y_1)}{P(x_1, y_1)}. \tag{12}$$

*Proof* Rewrite (3) in the form of a dynamic system ($t$ real)

$$\frac{dx}{dt}=P(x, y), \quad \frac{dy}{dt}=Q(x, y). \tag{13}$$

Separating real and imaginary parts gives

$$\begin{aligned} \frac{dx_1}{dt}&=P_r(x_1, x_2, y_1, y_2), & \frac{dx_2}{dt}&=P_j(x_1, x_2, y_1, y_2), \\ \frac{dy_1}{dt}&=Q_r(x_1, x_2, y_1, y_2), & \frac{dy_2}{dt}&=Q_j(x_1, x_2, y_1, y_2). \end{aligned} \tag{14}$$

Notice that for any positive integer $n$, we have

$$(x_1+ix_2)^n=[x_1^n+x_2(\cdots)]+ix_2[\cdots], \text{ etc.,}$$

(14) gives when $x_2=y_2=0$

$$\frac{dx_1}{dt}=P_r(x_1,\ 0,\ y_1,\ 0)=P(x_1,\ y_1),\qquad \frac{dx_2}{dt}=P_i(x_1,\ 0,\ y_1,\ 0)\equiv 0,$$

$$\frac{dy_1}{dt}=Q_r(x_1,\ 0,\ y_1,\ 0)=Q(x_1,\ y_1),\qquad \frac{dy_2}{dt}=Q_i(x_1,\ 0,\ y_1,\ 0)\equiv 0,$$

which means that any trajectory of (14), starting from a point of $l$ will always remain in the real plane, travelling the whole or a connected component of $l$, and hence is also a trajectory of the dynamical system obtained from (12).

**Remark**. If we put in (14) $x_1=y_1=0$ (or $x_1=y_2=0$, or $x_2=y_1=0$), in general we will not get $\frac{dx_1}{dt}=\frac{dy_1}{dt}\equiv 0$ $\left(\text{or } \frac{dx_1}{dt}=\frac{dy_2}{dt}\equiv 0,\text{ or } \frac{dx_2}{dt}=\frac{dy_1}{dt}\equiv 0\right)$. So, even if the plane $x_1=y_1=0$, $x_1=y_2=0$ or $x_2=y_1=0$ intersects $S$ at a curve $l'$, it need not be a trajectory of (14).

Theorem 1 implies that, in order to investigate the property of intersection curves of the general solution surfaces $F(x,\ y,\ C)=0$ of (3) with the real plane, it is sufficient to consider $C$ to be a real constant only. Because at this time $F(x_1,\ y_1,\ C)=0$ will represent a general solution of (12), and the locus of this family of curves will already fill up the whole real plane.

## § 2. Properties of Solutions of a Complex Quadratic System with Real Coefficients

For the complex quadratic differential system with real coefficients corresponding to equation (1), i. e. (with t real)

$$\frac{dx}{dt}=P_2(x,\ y),\qquad \frac{dy}{dt}=Q_2(x,\ y),\tag{1*}$$

the definition of a critical point is just the same as that for a real quadratic system. But now a critical point may have complex coordinates. Hereafter we will call it a real (complex) critical point, if it lies (does not lie) on the real plane $x_2=y_2=0$.

We can find in [4] many interesting properties of the trajectories of a real quadratic system, and now we are going to do the same investigation with regard to (1*).

The general equations of a 2-dimensional plane in the real 4-dimensional space are

$$\alpha_1x_1+\alpha_2x_2+\alpha_3y_1+\alpha_4y_2+\alpha_5=0,\quad \beta_1x_1+\beta_2x_2+\beta_3y_1+\beta_4y_2+\beta_5=0,\tag{15}$$

Of course, (15) can not always be written as an equation of a complex line

$$ax+by+c=0,\tag{16}$$

since (16) is equivalent to

$$a_1x_1-a_2x_2+b_1y_1-b_2y_2+c_1=0,\quad a_1x_2+a_2x_1+b_1y_2+b_2y_1+c_2=0,\tag{17}$$

where $a=a_1+ia_2$, $b=b_1+ib_2$, $c=c_1+ic_2$. Hence we call (15) a general (2-dim.) plane, and (16) a special (2-dim.) plane.

The proof of the following theorem is just the same as that of the corresponding theorem for a real quadratic system.

**Theorem 2.** *The special plane* (16) *has at most 2 tangent points*[1] *with integral surfaces of* (1*), *otherwise,* (16) *is itself a solution of* (1).

**Theorem 3.** *If* (1) *has no special plane solution, then any three critical points of* (1*) *can not be collinear.*

*Proof* Suppose on the contrary, there are 3 critical points $P_1$, $P_2$ and $P_3$ lying on a straight line $L$. Obriously, we can find a special plane (16), passing through $L$. Since any $P_i$ can be taken as a tangent point of (16) with some integral surface of (1), Theorem 2 implies that (16) is a solution of (1), contrary to the hypothesis.

**Remark.** If (1) has special plane solution, then (1*) may even have 4 collinear critical points. For example, if $P_2(x, y) = x^2 - y^2$, $Q(x, y) = xy$, then (1) has a solution $y = 0$, while the origin is a critical point of multiplicity 4 of (1*).

The following theorem is a generalization of the well-known property; "Any closed orbit of a real quadratic differential system is a convex oval" in [4]:

**Theorem 4.** *If a plane $\Pi_1$ intersects an integral surface $S$ of* (1) *at a closed orbit $l$ of* (1*), *then $l$ must be convex.*

*Proof* Let (15) be the equations of $\Pi_1$. Introduce the following coordinate transformation with real coefficients

$$\begin{aligned} x_1' &= \alpha_2 x_1 - \alpha_1 x_2 + \alpha_4 y_1 - \alpha_3 y_2, \\ x_2' &= \alpha_1 x_1 + \alpha_2 x_2 + \alpha_3 y_1 + \alpha_4 y_2 + \alpha_5, \\ y_1' &= \beta_2 x_1 - \beta_1 x_2 + \beta_4 y_1 - \beta_3 y_2, \\ y_2' &= \beta_1 x_1 + \beta_2 x_2 + \beta_3 y_1 + \beta_4 y_2 + \beta_5. \end{aligned} \tag{18}$$

(18) can also be written as

$$x' = ax + by + c, \quad y' = \lambda x + \mu y + \nu, \tag{19}$$

where $x' = x_1' + ix_2'$, $y' = y_1' + iy_2'$, $x = x_1 + ix_2$, $y = y_1 + iy_2$, $a = \alpha_2 + i\alpha_1$, $b = \alpha_4 + i\alpha_3$, $c = i\alpha_5$, $\lambda = \beta_2 + i\beta_1$, $\mu = \beta_4 + i\beta_3$, $\nu = i\beta_5$. Suppose in the new coordinates (1*) becomes

$$\frac{dx_1'}{dt} = R_1(x_1', y_1', x_2', y_2'), \quad \frac{dy_1'}{dt} = R_2(x_1', y_1', x_2', y_2'),$$

$$\frac{dx_2'}{dt} = R_3(x_1', y_1', x_2', y_2'), \quad \frac{dy_2'}{dt} = R_4(x_1', y_1', x_2', y_2').$$

Now $\Pi_1$ has equation $x_2' = y_2' = 0$, i. e., it is the new real plane, and $l$ is a closed orbit of the new real quadratic system

$$\frac{dx_1'}{dt} = R_1(x_1', y_1', 0, 0), \quad \frac{dy_1'}{dt} = R_2(x_1', y_1', 0, 0).$$

Hence from [4], $l$ must be convex.

---

1) i. e., a point at which $\frac{Q_2(x, y)}{P_2(x, y)} = -\frac{a}{b}$, and hence any critical point on (16) can be considered as a tangent point of this plane.

**Remark.** If the closed curve $l$ in Theorem 4 is not an orbit of (1*), then we can not prove its convexity.

Similar to [4], we have also the following two theorems, the proof is omitted.

**Theorem 5.** *No* (2-dim.) *plane can intersect any integral surface of* (1) *at three disjoint closed orbits of* (1*) *with relative position*

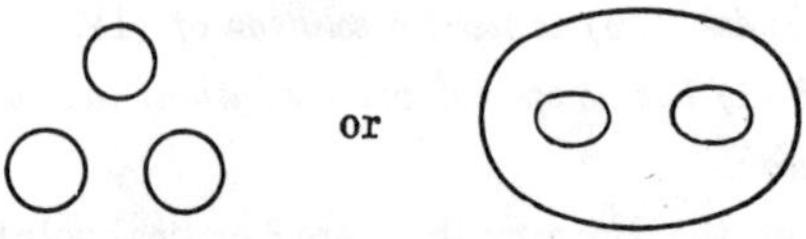

**Theorem 6.** *If an integral surface $S$ of* (1) *intersects the real plane at a closed curve (it must be an orbit of* (1*)), *then $S$ necessarily has common points with the plane*

$$D: \frac{\partial}{\partial x} P_2 + \frac{\partial}{\partial y} Q_2 = 0. \tag{20}$$

**Remark 1.** The common part of $S$ and D need not be a 1-dimensional curve.

*Example 2.* The equation (a, b, c real)

$$\frac{dy}{dx} = \frac{x(ax+by+c)}{-y(ax+by+c)+1-x^2-y^2}$$

has $x^2+y^2=1$ as one of its integral surface, which intersects the real plane $x_2=y_2=0$ at a circle $x_1^2+y_1^2=1$. It can easily be proved that now

$$\frac{\partial P_2}{\partial x} + \frac{\partial Q_2}{\partial y} = (b-2)x - ay = 0$$

has only two points in common with $x^2+y^2=1$.

**Remark 2.** Suppose the condition of Theorem 4 is satisfied, since (20) is invariant under the coordinate transformation (19), $D$ will still have common points with $S$. However, if the (closed) intersection curve of $S$ and $\Pi_1$ is not an orbit of (1*), then $S$ may have no point in common with the plane $D$.

*Example 3.* The equation

$$\frac{dy}{dx} = \frac{x+(x^2-y^2-1)}{y+(x^2-y^2-1)} \tag{21}$$

has an integral surface

$$S: \quad x^2-y^2=1 \quad \text{or} \quad x_1^2-y_1^2-x_2^2+y_2^2=1, \quad x_1x_2-y_1y_2=0,$$

it intersects the plane $x_2=y_1=0$ at the circle $x_1^2+y_2^2=1$, but the plane

$$\frac{\partial P_2}{\partial x} + \frac{\partial Q_2}{\partial y} = 2(x-y) = 0$$

has no common point with $S$. It is easily seen that the circle $x_2=y_1=0$, $x_1^2+y_2^2=1$ is not an orbit of the corresponding dynamical system.

## § 3. Variation of periodic orbits with respect to the parameter.

In this section we will give three examples in order to show that the investigation of a polynomial system with real coefficients and complex variables $x$ and $y$ will make clearer the property of periodic orbits of the corresponding real system.

*Example 4*. The system

$$\frac{dx}{dt}=-y+x(x^2+y^2-\lambda), \quad \frac{dy}{dt}=x+y(x^2+y^2-\lambda) \tag{22}$$

has an integral surface

$$x^2+y^2=\lambda, \tag{23}$$

which in the $(x_1, x_2, y_1, y_2)$ space can be written as

$$x_1^2+y_1^2-x_2^2-y_2^2=\lambda, \quad x_1x_2+y_1y_2=0. \tag{24}$$

When $\lambda>0$ (23) intersects the real plane at a circle $x_1^2+y_1^2=\lambda$, which is the unique (unstable) limit cycle of the corresponding real system. This circle tends to the origin as $\lambda\to 0$. If $\lambda<0$, then (23) has no common point with the real plane. The origin in the real plane is a stable focus when $\lambda>0$, an unstable focus when $\lambda\leqslant 0$.

Rewrite (22) as

$$\begin{cases} \dfrac{dx_1}{dt}=-y_1+x_1(x_1^2+y_1^2-x_2^2-y_2^2-\lambda)-2x_2(x_1x_2+y_1y_2), \\ \dfrac{dy_1}{dt}=x_1+y_1(x_1^2+y_1^2-x_2^2-y_2^2-\lambda)-2y_2(x_1x_2+y_1y_2), \\ \dfrac{dx_2}{dt}=-y_2+x_2(x_1^2+y_1^2-x_2^2-y_2^2-\lambda)+2x_1(x_1x_2+y_1y_2), \\ \dfrac{dy_2}{dt}=x_2+y_2(x_1^2+y_1^2-x_2^2-y_2^2-\lambda)+2y_1(x_1x_2+y_1y_2). \end{cases} \tag{25}$$

Putting in (25) $x_1=y_1=0$ gives

$$\frac{dx_1}{dt}=\frac{dy_1}{dt}\equiv 0, \quad \frac{dx_2}{dt}=-y_2+x_2(-x_2^2-y_2^2-\lambda), \quad \frac{dy_2}{dt}=x_2+y_2(-x_2^2-y_2^2-\lambda). \tag{26}$$

From this we see that all intersection curves of integral surfaces of (22) with the imaginary plane $x_1=y_1=0$ are trajectories of (25), and it is easily seen that when $\lambda>0$ the imaginary plane has no common point with (24), when $\lambda=0$ they have a unique common point (0, 0, 0, 0), when $\lambda<0$, they intersect at the circle $x_2^2+y_2^2=-\lambda$, which is the unique (stable) limit cycle of (26) in the imaginary plane.

It is well known that $\lambda=0$ is a bifurcation value of the so-called Hopf bifurcation of (22), when (22) is considered to be a real system. But for a complex system (22), $\lambda=0$ is only a bifurcation value of the integral surface (24). When $\lambda$ varies from positive to negative, the shape of (24) undergoes a sudden change, so that the intersection circle of (24) with $x_2=y_2=0$ shrinks to the origin, and then another intersection circle of (24) with $x_1=y_1=0$ appears from the origin and grows gradually. Moreover, all these intersection circles are trajectories of the corresponding 4-dimensional systems (25).

A somewhat different example is the following:

*Example 5*. Consider the system

$$\frac{dx}{dt}=-y(ax+by+c)+\lambda-x^2-y^2,\quad \frac{dy}{dt}=x(ax+by+c). \tag{27}$$

Now, (23) or (24) is again an integral surface. It has been proved in [5] that if $\lambda>0$ and

$$c^2>\lambda(a^2+b^2),$$

then (27) has a unique limit cycle $x_1^2+y_1^2=\lambda$ in the real plane, which will shrink to the origin when $\lambda\to 0$. Let us consider the case when $\lambda<0$. Put $\lambda=-\mu^2$ and rewrite (27) as

$$\begin{cases}\dfrac{dx_1}{dt}=-a(x_1y_1-x_2y_2)-b(y_1^2-y_2^2)-cy_1+x_2^2+y_2^2-x_1^2-y_1^2-\mu^2,\\ \dfrac{dx_2}{dt}=-a(x_1y_2+x_2y_1)-2by_1y_2-cy_2-2(x_1x_2+y_1y_2),\\ \dfrac{dy_1}{dt}=a(x_1^2-x_2^2)+b(x_1y_1-x_2y_2)+cx_1,\\ \dfrac{dy_2}{dt}=2ax_1x_2+b(x_1y_2+x_2y_1)+cx_2.\end{cases} \tag{28}$$

Let $x_1=y_1=0$ in (28), we get

$$\frac{dx_1}{dt}=y_2(ax_2+by_2)+x_2^2+y_2^2-\mu^2,\quad \frac{dy_1}{dt}=-x_2(ax_2+by_2),$$
$$\frac{dx_2}{dt}=-cy_2,\quad \frac{dy_2}{dt}=cx_2, \tag{29}$$

so that we can not get the desired closed orbit in $x_1=y_1=0$, contrary to that happened in Example 4. Now let us examine the equation satisfied by the projection on $(x_2, y_2)$ plane of the family of trajectories on (23). We solve from (24)

$$y_1=\pm\frac{x_2\sqrt{x_2^2+y_2^2-\mu^2}}{\sqrt{x_2^2+y_2^2}},\quad x_1=\frac{\mp y_2\sqrt{x_2^2+y_2^2-\mu^2}}{\sqrt{x_2^2+y_2^2}}, \tag{30}$$

substituting in the second and fourth equations of (28) gives

$$\begin{cases}\dfrac{dx_2}{dt}=\pm\dfrac{[a(y_2^2-x_2^2)\sqrt{x_2^2+y_2^2-\mu^2}-2bx_2y_2\sqrt{x_2^2+y_2^2-\mu^2}]}{\sqrt{x_2^2+y_2^2}}-cy_2,\\ \dfrac{dy_2}{dt}=\mp\dfrac{\sqrt{x_2^2+y_2^2-\mu^2}}{\sqrt{x_2^2+y_2^2}}[2ax_2y_2+b(y_2^2-x_2^2)]+cx_2.\end{cases} \tag{31}$$

From (31) we see that the intersection circle $x_2^2+y_2^2=\mu^2$ of (24) and the $(x_2, y_2)$ plane is a closed orbit. Introduce in (31) the curvilinear coordinates

$$x_2=\mu(1-n)\cos\theta,\quad y_2=\mu(1-n)\sin\theta, \tag{32}$$

under which $n=0$ corresponds to $x_2^2+y_2^2=\mu^2$. In order that (31) can be a real system, assume $n\leqslant 0$. Noticing that under (32) we have

$$x_2^2+y_2^2-\mu^2=\mu^2(n^2-2n),\quad y_2^2-x_2^2=-\mu^2(1-n)^2\cos 2\theta,$$
$$2x_2y_2=\mu^2(1-n)^2\sin 2\theta$$

and

$$\frac{dx_2}{dt}=-\mu(1-n)\sin\theta\frac{d\theta}{dt}-\mu\cos\theta\frac{dn}{dt},\quad \frac{dy_2}{dt}=\mu(1-n)\cos\theta\frac{d\theta}{dt}-\mu\sin\theta\frac{dn}{dt} \quad (33)$$

and putting $\tau=\mu(1-n)t$, we can solve from (31) and (33)

$$\begin{cases}\dfrac{d\theta}{d\tau}=c\pm\mu\sqrt{n(n-2)}\,(-a\sin\theta+b\cos\theta),\\ \dfrac{dn}{d\tau}=\pm(1-n)\mu\sqrt{n(n-2)}\,(a\cos\theta+b\sin\theta).\end{cases} \quad (34)$$

Since $\dfrac{d\theta}{d\tau}$ has the same sign as $C$ when $|n|$ is sufficiently small, we see from (34) that $x_2^2+y_2^2=\mu^2$ is a limit cycle of (31) from outside. For $\lambda=-\mu^2$ there is no point on (24) satisfying $x_2^2+y_2^2<\mu^2$, the projecting system (31) is thus only defined for $x_2^2+y_2^2\geqslant\mu^2$. However, we have in (30), (31) and (34) both plus and minus signs, so $x_2^2+y_2^2=\mu^2$ is still a twosided limit cycle on (24).

**Remark.** For $\lambda=\mu^2>0$, if we solve $x_2$, $y_2$ from (24), substitute in the first and third equations in (28) and introduce in $(x_1, y_1)$ plane the same curvilinear coordinates as (32), and also the change of time scale: $\tau=\mu(1-n)t$, we will finally get

$$\begin{cases}\dfrac{d\theta}{d\tau}=\dfrac{c}{\mu(1-n)}+a\cos\theta+b\sin\theta,\\ \dfrac{dn}{d\tau}=\dfrac{n(2-n)}{1-n}(b\cos\theta-a\sin\theta).\end{cases}$$

From the condition $c^2>\mu^2(a^2+b^2)$ we see that $x_1^2+y_1^2=\mu^2$ is also a limit cycle on (24). But different from (31), the projection of trajectories of this integral surface on the $(x_1, y_1)$ plane satisfies the equations

$$\begin{cases}\dfrac{dx_1}{dt}=-y_1(ax_1+by_1+c)-\dfrac{x_1(x_1^2+y_1^2-\mu^2)(ay_1-bx_1)}{x_1^2+y_1^2},\\ \dfrac{dy_1}{dt}=x_1(ax_1+by_1+c)-\dfrac{y_1(x_1^2+y_1^2-\mu^2)(ay_1-bx_1)}{x_1^2+y_1^2},\end{cases}$$

which are defined for all $(x_1, y_1)$.

It is also easily seen that in Example 4, the integral surface (24) is filled with closed orbits, and there is no limit cycle for the dynamical sub-system on (24).

In the above two examples we see that with the variation of the parameter limit cycle in the real plane shrinks to the origin and then reappears in the imaginary plane or in a certain integral surface. In the following we will give a third example, which shows when two limit cycles in the real plane approach to each other, coincide and then disappear, how we can find them out again.

*Example 6*. Consider the system

$$\begin{cases}\dfrac{dx}{dt}=[-y+x(x^2+y^2-1)^2]\cos\lambda-[x+y(x^2+y^2-1)^2]\sin\lambda,\\ \dfrac{dy}{dt}=[x+y(x^2+y^2-1)^2]\cos\lambda+[-y+x(x^2+y^2-1)^2]\sin\lambda.\end{cases} \quad (35)$$

As a real system, (35) has been discussed somehow in [5]. One finds that (35) can be obtained from

$$\frac{dx}{dt}=-y+x(x^2+y^2-1)^2,\quad \frac{dy}{dt}=x+y(x^2+y^2-1)^2 \tag{36}$$

by rotating its vector field through an angle $\lambda$. For $0<\lambda<\frac{\pi}{4}$, (35) has two limit cycles

$$x^2+y^2=1\pm\sqrt{\tan\lambda}. \tag{37}$$

They coincide and become a semi-stable cycle $x^2+y^2=1$ as $\lambda\to 0$, and then disappear as $\lambda$ becomes negative. Let us examine (35) as a complex system for any real $\lambda$, by using the method of the former two examples. The integral surface of (35)

$$(x^2+y^2-1)^2=\tan\lambda \tag{38}$$

can be written as

$$\begin{cases}(x_1^2-x_2^2+y_1^2-y_2^2-1)^2-4(x_1x_2+y_1y_2)^2=\tan\lambda,\\ (x_1^2-x_2^2+y_1^2-y_2^2-1)(x_1x_2+y_1y_2)=0.\end{cases} \tag{39}$$

If $\tan\lambda\geqslant 0$, then (39) is equivalent to

$$x_1^2-x_2^2+y_1^2-y_2^2-1=\pm\sqrt{\tan\lambda},\quad x_1x_2+y_1y_2=0, \tag{40}$$

if $\tan\lambda\leqslant 0$, (39) is equivalent to

$$x_1^2-x_2^2+y_1^2-y_2^2-1=0,\quad x_1x_2+y_1y_2=\pm\frac{1}{2}\sqrt{-\tan\lambda}. \tag{41}$$

In the following we discuss only the latter case. First, write out the equivalent 4-dimensional system of (35)

$$\begin{cases}\frac{dx_1}{dt}=\{-y_1+x_1(x_1^2-x_2^2+y_1^2-y_2^2-1)^2-4x_1(x_1x_2+y_1y_2)^2\\ \qquad -4x_2(x_1^2-x_2^2+y_1^2-y_2^2-1)(x_1x_2+y_1y_2)\}\cos\lambda\\ \qquad -\{x_1+y_1(x_1^2-x_2^2+y_1^2-y_2^2-1)^2-4y_1(x_1x_2+y_1y_2)^2\\ \qquad -4y_2(x_1^2-x_2^2+y_1^2-y_2^2-1)(x_1x_2+y_1y_2)\}\sin\lambda\\ \qquad =R_1(x,y)\cos\lambda-R_2(x,y)\sin\lambda,\\ \frac{dy_1}{dt}=R_2(x,y)\cos\lambda+R_1(x,y)\sin\lambda,\\ \frac{dx_2}{dt}=\{-y_2+x_2(x_1^2-x_2^2+y_1^2-y_2^2-1)^2-4x_2(x_1x_2+y_1y_2)^2\\ \qquad +4x_1(x_1^2-x_2^2+y_1^2-y_2^2-1)(x_1x_2+y_1y_2)\}\cos\lambda\\ \qquad -\{x_2+y_2(x_1^2-x_2^2+y_1^2-y_2^2-1)^2-4y_2(x_1x_2+y_1y_2)^2\\ \qquad +4y_1(x_1^2-x_2^2+y_1^2-y_2^2-1)(x_1x_2+y_1y_2)\}\sin\lambda\\ \qquad =R_3(x,y)\cos\lambda-R_4(x,y)\sin\lambda,\\ \frac{dy_2}{dt}=R_4(x,y)\cos\lambda+R_3(x,y)\sin\lambda.\end{cases} \tag{42}$$

Solve $x_1$, $y_1$ from (41), we get

$$x_1=\frac{1}{2(x_2^2+y_2^2)}\left[\sqrt{-\tan\lambda}\,x_2-y_2\sqrt{4y_2^2(x_2^2+y_2^2+1)+\tan\lambda}\,\right],$$

$$y_1=\frac{1}{2(x_2^2+y_2^2)}\left[\sqrt{-\tan\lambda}\,y_2+x_2\sqrt{4y_2^2(x_2^2+y_2^2+1)+\tan\lambda}\right].$$

Substituting in (42) gives

$$\begin{cases} \dfrac{dx_2}{dt}=[-y_2+x_2\tan\lambda]\cos\lambda-[x_2+y_2\tan\lambda]\sin\lambda=\dfrac{-y_2}{\cos\lambda}, \\ \dfrac{dy_2}{dt}=[x_2+y_2\tan\lambda]\cos\lambda+[-y_2+x_2\tan\lambda]\sin\lambda=\dfrac{x_2}{\cos\lambda}. \end{cases} \tag{43}$$

Its phase-portrait, after projecting on the $(x_2, y_2)$ plane, is a family of circles. Similarly, if we replace (41) by (40), solve from it $x_2$, $y_2$, or $x_1$, $y_1$, and then substitute in (42), the result will be the same as (43). So the situation is just the same as in Example 4, i. e., we can not find in the integral surface (38) limit cycles which disappear in the real plane as $\lambda$ varies from positive to negative.

It is worth noting that if we put in (42) $x_1=y_1=0$, then we get

$$\begin{cases} \dfrac{dx_2}{dt}=[-y_2+x_2(x_2^2+y_2^2+1)^2]\cos\lambda-[x_2+y_2(x_2^2+y_2^2+1)^2]\sin\lambda \\ \dfrac{dy_2}{dt}=[x_2+y_2(x_2^2+y_2^2+1)^2]\cos\lambda+[-y_2+x_2(x_2^2+y_2^2+1)^2]\sin\lambda, \\ \dfrac{dx_1}{dt}\equiv\dfrac{dy_1}{dt}\equiv 0. \end{cases} \tag{44}$$

Transform the first two in polar coordinates

$$\frac{dr}{dt}=r[(r^2+1)^2\cos\lambda-\sin\lambda], \quad \frac{d\theta}{dt}=\cos\lambda+(r^2+1)^2\sin\lambda. \tag{45}$$

From (45) we see that when $\lambda$ increases from $\frac{\pi}{4}$, a limit cycle

$$r=\sqrt{\sqrt{\tan\lambda}-1}$$

bifurcates from the origin in the $(x_2, y_2)$ plane, which is just come out of the one shrinking to the origin in the $(x_1, y_1)$ plane when $\lambda=\frac{\pi}{4}$. But the other limit cycle still remains in $(x_1, y_1)$ plane.

In order to find the missing semi-stable cycle of the real plane when $\lambda$ varies from positive to negative, let us look for a plane

$$x_1=kx_2, \quad y_1=ky_2, \tag{46}$$

where $k$ is real, such that it will intersect (41) at a circle. For this purpose, we substitute (46) in (41), and demand that

$$(k^2-1)(x_2^2+y_2^2)=1 \quad \text{and} \quad k(x_2^2+y_2^2)=\pm\frac{1}{2}\sqrt{-\tan\lambda} \tag{47}$$

will be the same equation. This gives

$$\sqrt{-\tan\lambda}\,k^2\pm 2k-\sqrt{-\tan\lambda}=0. \tag{48}$$

Obviously, we must take $|k|>1$, and then (48) will give only two values of $k$

$$k=\pm\frac{1+\sqrt{1-\tan\lambda}}{\sqrt{-\tan\lambda}} \quad (\lambda<0). \tag{49}$$

Substituting (49) in (46), and (46) in the latter two equations of (42) gives

$$\begin{cases} \dfrac{dx_2}{dt} = \{-y_2 + x_2[(5k^4-10k^2+1)(x_2^2+y_2^2)^2 - (6k^2-2)(x_2^2+y_2^2)+1]\}\cos\lambda \\ \qquad - \{x_2 + y_2[(5k^4-10k^2+1)(x_2^2+y_2^2)^2 - (6k^2-2)(x_2^2+y_2^2)+1]\}\sin\lambda \\ \qquad = M_1(x_2, y_2)\cos\lambda - M_2(x_2, y_2)\sin\lambda, \\ \dfrac{dy_2}{dt} = M_2(x_2, y_2)\cos\lambda + M_1(x_2, y_2)\sin\lambda. \end{cases} \tag{50}$$

Now, let us solve the equation

$$(5k^4-10k^2+1)(x_2^2+y_2^2)^2 - (6k^2-2)(x_2^2+y_2^2) + 1 - \tan\lambda = 0$$

in $x_2^2+y_2^2$, this gives two solutions

$$x_2^2+y_2^2 = \frac{1}{k^2-1} \quad \text{and} \quad x_2^2+y_2^2 = \frac{(k^2+1)^2}{(5k^4-10k^2+1)(k^2-1)}. \tag{51}$$

The first represents a circle in $(x_2, y_2)$ plane when $|k|>1$, while the second represents a circle when $|k|>\frac{1}{5}\sqrt{5+2\sqrt{5}}\sim 1.37$. Substitute (51) in (50), notice that the value in the [ ] equals $\tan\lambda$, so that (50) becomes (43). The circle[1)]

$$x_1 = kx_2, \quad y_1 = ky_2, \quad x_2^2+y_2^2 = \frac{1}{k^2-1} \tag{52}$$

lies in (41), along which also stands

$$\frac{d}{dt}(x_1 - kx_2) \equiv \frac{d}{dt}(y_1 - ky_2) \equiv 0. \tag{53}$$

This shows that (52) is actually a closed orbit of (35). But (53) does not hold along the circle

$$x_1 = kx_2, \quad y_1 = ky_2, \quad x_2^2+y_2^2 = \frac{(k^2+1)^2}{(5k^4-10k^2+1)(k^2-1)} \tag{54}$$

(except when $|k|=\sqrt{3}$, for this value (54) coincides with (52)), since $x_1^2 - x_2^2 + y_1^2 - y_2^2 - 1 \neq 0$ on (54). Therefore, (54) is not a closed orbit of (35).

In polar coordinates, (50) has the form

$$\begin{cases} \dfrac{dr}{dt} = r(5k^4-10k^2+1)\left(r^2 - \dfrac{1}{k^2-1}\right)\left(r^2 - \dfrac{(k^2+1)^2}{(5k^4-10k^2+1)(k^2-1)}\right)\cos\lambda, \\ \dfrac{d\theta}{dt} = 1 + (5k^4-10k^2+1)\left(r^2 - \dfrac{1}{k^2-1}\right)\left(r^2 - \dfrac{(k^2+1)^2}{(5k^4-10k^2+1)(k^2-1)}\right), \end{cases}$$

hence (52) is really a limit cycle in the plane (46). Since $k$ has two values in (49), (52) actually represents two limit cycles, each in a plane (46). They are just the limit cycles which disappear in the real plane when $\lambda$ changes from positive to negative. This is because as $\lambda \to 0$, we have $|k| \to \infty$, and in (52) $x_2^2+y_2^2 \to 0$, but

$$x_1^2 + y_1^2 = \frac{k^2}{k^2-1} \to 1.$$

**Remark.** The last equation in (52) can be written as

$$x_2^2 + y_2^2 = \frac{-1+\sqrt{1-\tan\lambda}}{2}$$

1) For the sake of convenience, we call it circle, too. Actually, only its projection on $(x_2, y_2)$ plane is a circle.

and from the first two we get

$$x_1^2+y_1^2=\frac{k^2}{k^2-1}=\frac{1+\sqrt{1-\tan\lambda}}{2}.$$

Then by means of (41), we see that the two circles in (52) satisfy the equations

$$x^2+y^2=x_1^2+y_1^2-x_2^2-y_2^2+2i(x_1x_2+y_1y_2)=1\pm i\sqrt{-\tan\lambda}=1\pm\sqrt{\tan\lambda} \qquad (55)$$

respectively, and this is just the same as (37), although $\lambda$ is negative now. Notice that the plus and minus signs in the right hand side of (55) only explain that these two circles are situated in different planes of (46), but not signify that they have different radii.

The reason we can find the disappeared limit cycles of the real plane in any other place in the above three examples is that, in every example the equation of an integral surface $f(x, y)=0$ is known. And if this integral surface intersects the real plane, the intersection curve is always a limit cycle, i. e., $f(x_1, y_1)=0$ is the equation of this cycle. On the other hand, if only the existence but not the equation of the limit cycle is known, it will still be a difficult problem to find out the disappeared limit cycle of the real plane as the parameter varies. Thus, for example the real quadratic differential system of type (I) (see [6])

$$\frac{dx}{dt}=-y+\delta x+lx^2+xy, \qquad \frac{dy}{dt}=x$$

when $\delta l>0$, and

$$\frac{dx}{dt}=-y+\delta x+xy+ny^2, \qquad \frac{dy}{dt}=x$$

when $\delta n>0$ all belong to this case. We have confidence in this problem that the same situation would appear for these systems.

## References

[1] Ye Yanqian, Some Problems in the Qualitative Theory of Ordinary Differential Equations, (to appear in Journ. Diff. Equs.).

[2] Ye Yanqian, Several Topics on Qualitative Theory of Ordinary Differential Equations, *J. Xinjiang Univ. (Nat. Sci)*, 1 (1980), 1—32.

[3] Il'iašenko, Ju. S., Global and Local Aspects of the Theory of Complex Differential Equations, *Proc. Intern. Congr. Math., Helsinki*, (1978), 821—826.

[4] Yeh, Y. C., Limit Cycles of Certain Nonlinear Differential Systems, *Sci. Rec., New Ser.*, **1**: 6 (1957), 359—361.

[5] Chin, Y. X., Integral Curves defined by Differential Equations, Sci. Press, (1959).

[6] Yeh, Y. C., Limit Cycles of Certain Nonlinear Differential Systems II, *Sci. Rec., New Ser.*, **2**: **9** (1958), 276—279.

*Chin. Ann. of Math.*
3 (4) 1982

# ON THE GROWTH AND THE DISTRIBUTION OF VALUES OF EXPONENTIAL SERIES CONVERGENT ONLY IN THE RIGHT HALF-PLANE

YU JIARONG (YU CHIA-YUNG)
(*Wuhan University*)

Dedicated to Professor Su Bu-chin on the Occasion of his 80th Birthday and his 50th Year of Educational Work

Ritt, J. F. investigated, the growth of entire functions defined by exponential series and introduced the Ritt order or order ($R$). Mandelbrojt, S. [5] and Valiron G. [10] studied the growth and the distribution of values of such functions. Blambert, M. [3], Sunyer i Balaguer, F. [7], Tanaka, C. [8], the author[13,14] and others continued to do research work in this respect,

For analytic functions defined by exponential series convergent only in the right half-plane, the author[14] introduced the order ($R$) and studied some exponential series and random exponential series. In this paper we introduce the order ($R$–$H$) of such functions. Applying an extension of an inequality and we study the growth and the distribution of values of such functions in some horizontal half-strips and obtain results similar to the case of seme entire functions defined by exponential series.[1)]

**1. The order ($R$) in the right half-plane.** Consider the exponential series

$$f(s)=\sum_{n=0}^{+\infty} a_n e^{-\lambda_n s}, \tag{1.1}$$

where $\{a_n\}$ is a sequence of complex numbers, $0=\lambda_0<\lambda_1<\cdots<\lambda_n\uparrow+\infty$, $s=\sigma+it$, $\sigma$ and $t$ being real variables. Suppose that

$$\overline{\lim_{n\to+\infty}}\frac{\log n}{\lambda_n}=\overline{\lim_{n\to+\infty}}\frac{\log|a_n|}{\lambda_n}=0. \tag{1.2}$$

Then the abscissa of convergence and that of absolute convergence of (1.1) is 0 and the series (1.1) defines a function $f(s)$ analytic in the right-half plane.

Let
$$M(\sigma)=\sup_{-\infty<t<+\infty}|f(\sigma+it)|\,(\sigma>0).$$

The quantity

$$\rho=\overline{\lim_{\sigma\to+0}}\frac{\overset{+}{\log}\,\overset{+}{\log}\,M(\sigma)}{\log(1/\sigma)} \tag{1.3}$$

is called order ($R$) of $f(s)$ in $\sigma>0$. We have the following theorem:

---

Manuscipt recived Dec. 5, 1981.

1) Some results in this paper have been announced in a Note[15].

**Theorem 1. 1.**[2)] *Suppose that* (1.2) *and*

$$\varlimsup_{n\to+\infty}\frac{n}{\lambda_n}=D_1<+\infty \tag{1.4}$$

*are verified. Then* $f(s)$ *is of order* $(R)$ $\rho$ *in* $\sigma>0$

$$\Leftrightarrow \varlimsup_{n\to+\infty}\frac{\overset{+}{\log}\,\overset{+}{\log}\,|a_n|}{\log\lambda_n}=\frac{\rho}{\rho+1}, \tag{1.5}$$

*where* $\dfrac{\rho}{\rho+1}$ *must be replaced by* (1) *in the case* $\rho=+\infty$.

Following Hiong Kin-lai [4] we introduce a proximate order $(R)$ in the case $\rho=+\infty$. Let $\rho(u)$ $(u>0)$ be a strictly increasing positive funcition such that

1) $\lim\limits_{u\to+\infty}\rho(u)=+\infty$,

2) $\lim\limits_{u\to+\infty}\dfrac{\log U(u')}{\log U(u)}=1$,

where
$$u'=u+\frac{u}{\log U(u)},\ U(u)=u^{\rho(u)}.$$

If

$$\varlimsup_{\sigma\to+0}\frac{\overset{+}{\log}\,\overset{+}{\log}\,M(\sigma)}{\log U(1/\sigma)}=1, \tag{1.6}$$

we say that $f(s)$ is of order $(R\text{-}H)$ $\rho(1/\sigma)$ in $\sigma>0$. We establish. a theorem similar to Theorem 1.1.

**Theorem 1. 2.** *Suppose that* (1.2) *and* (1.4) *are verified. Then*

$$f(s) \text{ is of order } (R\text{-}H)\ \rho(1/\sigma)\Leftrightarrow \varlimsup_{u\to+\infty} t_n=1, \tag{1.7}$$

*where*
$$t_n=\begin{cases}\log\lambda_n/\log U(\lambda_n/\log|a_n|) & (|a_n|>1),\\ 0 & (|a_n|\leqslant 1).\end{cases}$$

*Proof* First we prove that if $\varlimsup\limits_{n\to+\infty} t_n=1$, then

$$\varlimsup_{\sigma\to+0}\frac{\overset{+}{\log}\,\overset{+}{\log}\,M(\sigma)}{\log U(1/\sigma)}\leqslant 1. \tag{1.8}$$

For any $\eta>0$, there exists an integer $N>0$ such that for $n>N$ and $|a_n|>1$

$$\lambda_n<\left[U\left(\frac{\lambda_n}{\log|a_n|}\right)\right]^{1+\eta}.$$

Let $v=U(u)$ and $u=\varphi(v)$ be two reciprocally inverse functions. Then for $n>N$ and $|a_n|>1$,

$$\varphi(\lambda_n^{\frac{1}{1+\eta}})<\frac{\lambda_n}{\log|a_n|}$$

and consequently

2) This theorem was stated in [14] without the condition (1.4). But in order to prove it we must have $\sum\limits_{n=0}^{\infty}e^{-\lambda_n\varepsilon\sigma}=O\left(\dfrac{1}{\varepsilon\sigma}\right)$ $(\varepsilon>0,\ \sigma\to+0)$ for which the condition (1.2) is not sufficient ([14], p. 102). the condition (1.4) must be added in other related theorems in [15].

$$|a_n|e^{-\lambda_n\sigma}<\exp\left[\frac{\lambda_n}{\varphi(\lambda_n^{\frac{1}{1+\eta}})}-\lambda_n\sigma\right]. \tag{1.9}$$

Evidently (1.9) holds when $|a_n|\leqslant 1$. Hence we have (1.9) for $n>N$.

Fix $\sigma>0$ and take $\bar{\lambda}$ such that

$$\frac{1}{\sigma}\left(1+\frac{1}{\log U(1/\sigma)}\right)=\varphi(\bar{\lambda}^{\frac{1}{1+\eta}}).$$

Then
$$U\left[\frac{1}{\sigma}\left(1+\frac{1}{\log U(1/\sigma)}\right)\right]=\bar{\lambda}^{\frac{1}{1+\eta}}$$
and consequently $\bar{\lambda}=[U(1/\sigma)]^{1+\eta+o(1)}$ $(\sigma\to+0)$.

By (1.9), when $\lambda_n>\bar{\lambda}$, and $n>N$

$$|a_n|e^{-\lambda_n\sigma}<\exp\left[\lambda_n\left(\frac{1}{\varphi(\bar{\lambda}^{\frac{1}{1+\eta}})}-\sigma\right)\right]=\exp\left[\frac{-\lambda_n\sigma}{1+\log U(1/\sigma)}\right]<1.$$

Let
$$m(\sigma)=\max_n|a_n|e^{-\lambda_n\sigma}$$
and
$$n(\sigma)=\max_k\{k\,|\,|a_k|e^{-\lambda_k\sigma}=\max_n|a_n|e^{-\lambda_n\sigma}\}.$$

For sufficiently small $\sigma$, we have $\lambda_{n(\sigma)}<\bar{\lambda}$. Since [13]

$$\log m(\sigma)=A+\int_\sigma^a\lambda_{n(x)}\,dx\quad(0<\sigma<1),$$

where a $(>\sigma)$ and $A$ are constants, we obtain

$$\log m(\sigma)<[U(1/\sigma)]^{1+2\eta+O(1)}(\sigma\to+0).$$

On the other hand, for any $\varepsilon>0$, there exists an integer $N_1>0$ such that for any $n>N_1$,

$$\lambda_n>n/(D_1+\varepsilon).$$

Suppose that $\sigma$ is fixed such that $n(\sigma)>N_1$. Then

$$\begin{aligned}|f(s)|&\leqslant\sum_{\lambda_n\leqslant\bar{\lambda}}|a_n|e^{-\lambda_n\sigma}+\sum_{\lambda_n>\bar{\lambda}}|a_n|e^{-\lambda_n\sigma}\\&\leqslant(D_1+\varepsilon)[U(1/\sigma)]^{1+\eta+o(1)}\exp\{[U(1/\sigma)]^{1+2\eta+o(1)}\}\\&\quad+\sum_{n=1}^{+\infty}\left\{\exp\left[-\frac{\sigma}{1+\log U(1/\sigma)}\right]\right\}^{\frac{n}{D_1+\varepsilon}}\quad(\sigma\to+0).\end{aligned}$$

Hence for sufficiently small $\sigma>0$,

$$\begin{aligned}M(\sigma)&<\exp\{[U(1/\sigma)]^{1+3\eta}\}+\frac{2(D_1+\varepsilon)}{\sigma}[1+\log U(1/\sigma)]\\&<\exp\{[U(1/\sigma)]^{1+4\eta}\}\end{aligned}$$

and consequently (1.8) holds.

Now we prove that if $\overline{\lim_{n\to+\infty}}\,t_n=1$, then we cannot have

$$\overline{\lim_{\sigma\to+\infty}}\frac{\log^+\log^+ M(\sigma)}{\log U(1/\sigma)}=c<1. \tag{1.10}$$

Suppose that (1.10) would hold, Choose $\varepsilon>0$ and that $c+2\varepsilon<1$. Then there would exist $\sigma_0(0<\sigma_0<1)$ such that for $0<\sigma<\sigma_0$

$$\log M(\sigma)<[U(1/\sigma)]^{c+\varepsilon}$$

and conseqnently for $n=0, 1, 2, \cdots$,

$$\log|a_n|-\lambda_n\sigma<[U(1/\sigma)]^{c+\varepsilon}. \tag{1.11}$$

On the other hand, there exist arbitrarily large integers $n$ such that

$$\lambda_n>[U(\lambda_n/\log|a_n|)]^{1-\varepsilon} \quad (|a_n|>1). \tag{1.12}$$

Take such sufficiently large $n$ and take $\sigma(0<\sigma<\sigma_0)$ such that

$$[U(1/\sigma)]^{c+\varepsilon}=\frac{\lambda_n}{(\lambda_n/\log|a_n|)\log U(\lambda_n/\log|a_n|)}. \tag{1.13}$$

Combining (1.11) and (1.13), we see that for any $\eta>0$, there are arbitrarily large $n$ and $\sigma\in(0, \sigma_0)$ such that

$$\frac{1}{\sigma}<\frac{\lambda_n}{\log|a_n|}\left[1+\frac{1}{\log U(\lambda_n/\log|a_n|)}\right]$$

and consequently

$$U(1/\sigma)<[U(\lambda_n/\log|a_n|)]^{1+\eta} \tag{1.14}$$

By (1.13) and (1.14) we have, for those sufficiently large $n$ for which (1.12) holds.

$$\lambda_n<[U(\lambda_n/\log|a_n|)]^{(1+\eta)(c+\varepsilon)}\frac{\lambda_n}{\log|a_n|}\log U(\lambda_n/\log|a_n|)$$

and, for the other sufficiently large $n$, $t_n\leqslant 1-\varepsilon$, whence $\overline{\lim\limits_{n\to+\infty}}\, t_n<1$, contrary to the hypothesis. The sufficiency of the condition (1.7) is proved.

We can prove easily the necessity of this condition.

The proof is completed.

**2. The order in a horizontal half-strip.** Let $t_0$ be a real number and $a$ be a positive number. Denote the horizontal half-strip $\{s|\sigma>0, |t-t_0|\leqslant a\}$ by $S=S(t_0, a)$. Let

$$M_S(\sigma)<\max_{|t-t_0|\leqslant a}|f(\sigma+it)|\,(\sigma>0),$$

where $f(s)$ is defined by (1.1), (1.2) being verified. Replacing $M(\sigma)$ in (1.3) and (1.6) by $M_S(\sigma)$ we obtain the definitions of the orders ($R$) and ($R$-$H$) of $f(s)$ in $S=S(t_0, a)$. In order to study these orders we establish a lemma.

Inspired by the idea of Anderson, J. M. and Binmore, K. G. [1], we suppose that

$$\inf_{q>0}\overline{\lim_{x\to+\infty}}\frac{N((x+1)q)-N(xq)}{q}=D<+\infty, \tag{2.1}$$

and

$$\underline{\lim_{n\to+\infty}}\frac{\log(\lambda_{n+1}-\lambda_n)}{\log\lambda_n}>-\infty, \tag{2.2}$$

where $N(x)$ denotes the number of $\lambda_n$ less than $x(>0)$. Evidently if

$$\lim_{n\to+\infty}(\lambda_{n+1}-\lambda_n)=+\infty, \tag{2.3}$$

then (2.1) and (2.2) hold with $D=0$. We can prove $D_1\leqslant D$.

**Lemma 2. 1.** *Suppose that the series* (1.1) *satifies* (1.2), (2.1) *and* (2.2). *Then for any* $\varepsilon>0$, *for any real number* $t_0$ *and for any* $\sigma>0$, *we have*

$$|a_k| \leqslant A\lambda_k^B M_S(\sigma)e^{\lambda_k\sigma} \quad (k=0,\ 1,\ 2,\ \cdots), \tag{2.4}$$

*where* $S=S(t_0,\ \pi(D+\varepsilon))$, *and* $A$ *and* $B$ *are positive constants depending only on* $\varepsilon$ *and* $\{\lambda_n\}$.

The proof of this lemma is similar to that of Theorem 7 in [1]. In virtue of (2.1) for any $\varepsilon>0$ there exist a positive real number $q_0$ and a positve integer $M$ such that the number of $\lambda_j$ satisfying

$$nq_0 \leqslant \lambda_j < (n+1)q_0 \quad (n>M)$$

does not exceed $(D+\varepsilon)q_0$. Suppose that

$$\lambda_{p+1} = \min\{\lambda_l \mid \lambda_l \geqslant (M+1)q_0\}.$$

For the fixed $k$ construct $j(t)$, $J(x)$, $h_\xi(t)$, $H_\xi(x)$ as in [1]. Corresponding to $\lambda_1,\ \lambda_2, \cdots,\ \lambda_p$, there exist[2] functions $g_\xi(t)$ and $G_\xi(x)$ such that $g_\xi(t) \in L(-\infty,\ +\infty)$ and

1) $G_\xi(x) = \int_{-\infty}^{+\infty} e^{ixt} g_\xi(t)\,dt,$

2) $g_\xi(t) = 0 \left( |t| > \frac{\pi}{2}\frac{p}{\lambda_k} + p\xi \right),$

3) for the fixed value $k$

$$G_\xi(\lambda_k - \lambda_j) = 0 \quad (j=1,\ 2,\ \cdots,\ p),$$

4) $\lim_{\xi\to+0} G_\xi(0) = \prod_{j=0}^{p} \cos\frac{\pi}{2}\cdot\frac{\lambda_j}{\lambda_k},$

5) $\int_{-\infty}^{+\infty} |g_\xi(t)|\,dt \leqslant 1.$

We define $L(x) = G_\xi(x)H_\xi(x)J(x)$ and $l(t) = (g_\xi * h_\xi * j)(t)$ similar to $L(x)$ and $l(t)$ in [1] and define

$$f_N(s) = \sum_{j=0}^{N} a_j e^{-\lambda_j\sigma} e^{-i\lambda_j t} \quad (\sigma>0).$$

Then
$$\int_{-\infty}^{+\infty} l(t) e^{i\lambda_k t} f_N(s)\,dt = a_k e^{-\lambda_k\sigma} L(0),$$

whence
$$a_k e^{-\lambda_k\sigma} L(0) = \int_{-\eta}^{\eta} l(t) e^{i\lambda_k t} f_N(s)\,dt,$$

where $\eta = \pi\rho\delta + \tau_k + 2hp\xi + (\pi p/2\lambda_k) + p\xi$; $\rho,\ \delta,\ \tau_k,\ h$ and $p$ being the same as in [1]. Reasoning as in [1] and taking account of (2.2) to estimate

$$\prod_j \left(\cos\frac{\pi}{2}\cdot\frac{\lambda_{k-j}}{\lambda_k}\right)\left(\cos\frac{\pi}{2}\cdot\frac{\lambda_k}{\lambda_{k+j}}\right),$$

we come to the conclusion in Lemma 2.1.[3)]

If (2.3) holds, we have a more precise result.

$$|a_k| \leqslant AM_S(\sigma)e^{\lambda_k\sigma} \quad (k=0,\ 1,\ 2,\ \cdots), \tag{2.5}$$

where $S=S(t_0,\ \varepsilon)$ and $A$ is a positive constant depending only on $\varepsilon$ and $\{\lambda_n\}$.

The proof is similar to that of Lemma 2.1.

Now we apply Lemma 2.1 to prove the following theorem:

3) Below the formula (5.13) in [1], there is a misprint. We must have $\tau_k = 2^{-1}\pi\left(hp\lambda_k^{-1} + \sum_{j=k+1}^{k+hp} \lambda_j^{-1}\right)$.

**Theorem 2.1.** *Suppose that the series* (1.1) *satisfies* (1.2), (2.1) *and* (2.2). *Then the order* ($R$) *or* ($R$–$H$) *of* $f(s)$ *in* $\sigma>0$ *is the same as that in any half-strip* $S(t_0, a)$, *where* $t_0$ *is an arbitrary real number* $a$ *is an arbitrary number* $>\pi D$.

*Proof* 1) Suppose that $f(s)$ is of finite order ($R$) $\rho>0$. Then in any half-strip $S=S(t_0, a)$ $(a>\pi D)$, $f(s)$ is of order ($R$) $\rho_1\leqslant\rho$. If $f(s)$ were of order ($R$) $\rho^*<\rho$ in a half-strip $S^*=S(t_0^*, a^*)$ $(a^*>\pi D)$, for any $\varepsilon>0$, there would exist $\sigma_0>0$ such that for $\sigma\in(0, \sigma_0)$

$$\overset{+}{\log} M_{S^*}(\sigma)<(1/\sigma)^{\rho^*+\varepsilon}.$$

By the above inequality and Lemma 2.1 we would have, for $\sigma\in(0, \sigma_0)$ and $k\in\{0, 1, 2, \cdots\}$

$$\overset{+}{\log}|a_k|\leqslant\log A+B\log\lambda_k+(1/\sigma)^{\rho^*+\varepsilon}+\lambda_k\sigma. \tag{2.6}$$

Substituting $\left(\dfrac{\rho^*+\varepsilon}{\lambda_k}\right)^{1/(\rho^*+1+\varepsilon)}$ for $\sigma$ in (2.6), we would get

$$\varlimsup_{k\to+\infty}\frac{\overset{+}{\log}\,\overset{+}{\log}|a_k|}{\log\lambda_k}\leqslant\frac{\rho^*}{\rho^*+1}<\frac{\rho}{\rho+1},$$

contrary to (1.5). The case $\rho=0$ or $\rho=+\infty$ can be studied in the same way.

2) The case that $f(s)$ is of order ($R$–$H$) $\rho(\sigma)$ in $\sigma>0$ can be studied similarly. The proof of the Theorem is then completed.

We have similar results for proximate orders ($R$)[14]. In particular we can prove

**Theorem 2. 2.** *Suppose that the series* (1.1) *satisfies* (1.2), (2.1) *and* (2.2). *Then if*

$$\varlimsup_{\sigma\to+0}\sigma\overset{+}{\log} M(\sigma)=+\infty,^{4)} \tag{2.7}$$

*we have*

$$\varlimsup_{\sigma\to+0}\sigma\overset{+}{\log} M_S(\sigma)=+\infty \tag{2.8}$$

*for any half-strip* $S=S(t_0, a)$, *where* $t_0$ *is an arbitrary real number and* $a$ *is an arbitrary number* $>\pi D$.

**3. Picard points and Borel points.** Now we study the distribution of values of the function $f(s)$ in the vicinity of $\sigma=0$. If $f(s)$ takes, in every neighborhood of a certain point $s_0$, every finite complexvalue infinitely many times, with one possible *point* of $f(s)$. We have the following theorem: exception, then $s_0$ is called a *Picard*

**Theorem 3. 1.** *Suppose that the series* (1.1) *satisfies* (1.2), (2.1), (2.2) *and* (2.7). *Then in any interval of length* $2\pi D$ *on* $\sigma=0$, *there exists at least a Picard point of* $f(s)$.

*Proof* By Theorem 2.2, for an arbitrary real number $t_0$ and for an arbitrary number $a>\pi D$, (2.7) holds, where $S=S(t_0, a)$. Divide the horizontal half-strip $S$

4) This condition is equivalent to

$$\varlimsup_{n\to+\infty}\frac{\log|a_n|}{\sqrt{\lambda_n}}=+\infty.^{[14]}$$

into two horizontal half-strips $S^{(1)}$ and $S^{(2)}$ of the same breadth. Then

$$\overline{\lim_{\sigma\to+0}}\ \sigma \overset{+}{\log} M_{S^{(j)}}(\sigma) = +\infty \tag{3.1}$$

must be true for at least one of the half-strips $S^{(j)}(j=1, 2)$. Denote one half-strip for which (3.1) holds by $S_1$.

Divide $S_1$ into two horizontal half-strips $S_1^{(1)}$ and $S_1^{(2)}$ and repeat the above process indefinitely.

We obtain a sequence of harizental half-strips $\{S_l\}$ in $\sigma>0$ $(l=1, 2, \cdots)$ for which we have

$$\overline{\lim_{\sigma\to+0}}\ \sigma \overset{+}{\log} M_{S_l}(\sigma) = +\infty, \tag{3.2}$$

the breadth of $S_{l+1}$ being half of that of $S_l$. Hence on the line joining $i(t_0-a)$ and $i(t_0+a)$ there exists a point $it^*$ such that for any $\eta>0$,

$$\overline{\lim_{\sigma\to+0}}\ \sigma \overset{+}{\log} M_{S^*}(\sigma) = +\infty,$$

where $S^*=S(t^*, \eta)$.

Now we can prove $t^*$ is a Picard point of $f(s)$ as in the proof of Theorem 4.3 in [14]. Since the above reasoning is valid for any $a>\pi D$ and since a limiting point of Picard points is itself a Picard point, we can easily complete the proof of the Theorem.

We consider at present Borel $(R)$ points. Suppose that $f(s)$ is defined by (1.1) and that (1.2) holds. Let $it_1$ be a point on $\sigma=0$ and $\eta$ be a positive number. For any finite complex number $\zeta$, range the points $s$ to satisfy $f(s)=\zeta$, $|s-it_p|<\eta$, Re $s=\sigma>0$ in a sequence $\{s_n(\zeta, it_1, \eta)\}$, where Re $s_n(\zeta, it_1, \eta)=\sigma_n(\zeta, it_1, \eta)$ is non increasing. We have

**Theorem 3. 1.** *Suppose that the series* (1.1) *satisfies* (1.2), (2.1) *and* (2.2) *and that $f(s)$ is of order $(R)$ $\rho(0<\rho<+\infty)$ in $\sigma>0$. Then in any interval of length $2\pi D$ on $\sigma=0$, there exists at least a Borel $(R)$ point of order at least $\rho$ and at most $\rho+1$.*

That is to say, in any interval of length $2\pi D$ on $\sigma=0$, there exists a point $it_1$ such that for any sufficiently small positive number $\eta>0$ and for any finite complex number $\zeta$, the series $\sum_n[\sigma_n(\zeta, it_1, \eta)]^\tau$ converges if $\tau>\rho+1$ and that for any positive number $\eta$ and for any finite complex number $\zeta$, with a possible exception, the series $\sum_n[\sigma_n(\zeta, it_1, \eta)]^\tau$ diverges if $\tau<\rho$.

In order to prove Theorem 3.2 we show first, as in the proof of Theorem 3.1, that there exists a point $it_0$ in any interval of length greater than $2\pi D$ on $\sigma=0$ such that for any $\eta>0$

$$\overline{\lim_{\sigma\to+0}} \frac{\overset{+}{\log}\overset{+}{\log} M_S(\sigma)}{\log(1/\sigma)} = \rho,$$

where $S = S(t_0, \eta)$. Then we can complete the proof as in the proof of Thearem 4.4 in [14] and in that of the previous Theorem.

We study now the case of infinite order $(R)$. Suppose that the series $(1.1)$ satisfies $(1.2)$ and $(1.4)$ and that

$$\varlimsup_{\sigma \to +0} \frac{\overset{+}{\log}\overset{+}{\log}\overset{+}{\log} M(\sigma)}{\log(1/\sigma)} = \rho_1 \, (0 < \rho_1 < +\infty). \tag{3.3}$$

Construct[11,14] a differentiable function $\rho_1(r) \, (r > 0)$ such that

$$\lim_{r \to +\infty} \rho_1(r) = \rho_1, \quad \varlimsup_{r \to +\infty} \rho_1'(r) r \log r = 0 \tag{3.4}$$

and that

$$\varlimsup_{\sigma \to +0} \frac{\overset{+}{\log}\overset{+}{\log} M(\sigma)}{U_1(1/\sigma)} = 1, \tag{3.5}$$

where $U_1(r) = r^{\rho_1(r)}$ is a strictly increasing function. Let $k(r)$ be a continuons function defined for $r > 0$ such that

$$\lim_{r \to +\infty} k(r) = k \, (0 < k < +\infty). \tag{3.6}$$

We can show [11] thst

$$\lim_{r \to +\infty} \frac{U_1(k(r)r)}{U_1(r)} = k^{\rho_1}. \tag{3.7}$$

Put $U(r) = \exp[U_1(r)]$ and $\rho(r) = U_1(r)/\log r$. Then $U(r) = r^{\rho(r)}$ and $\rho(r)$ is evidently an order $(R\text{-}H)$ of $f(r)$ in $\sigma > 0$.

**Theorem 3.3.** *Suppose that the series $(1.1)$ satisfies $(1.2)$, $(2.1)$, $(2.2)$ and $(3.3)$ and that $f(s)$ is of order $(R\text{-}H)$ $\rho(1/\sigma)$ as defined above. Then in any interval of length $2\pi D$ on $\sigma = 0$, there exists at least a Borel $(R\text{-}H)$ point of order at least*

$$\rho\left(\frac{1}{6\sigma}\right)\left(1 + \frac{\log 6}{\log \sigma}\right) \text{ and at most } \rho\left(\frac{6}{\sigma}\right)\left(1 - \frac{\log 6}{\log \sigma}\right).$$

That is to say, in any interval of length $2\pi D$ on $\sigma = 0$. There exists a point $it_1$ such that for any sufficiently small number $\eta > 0$ and for any finite complex number $\zeta$, the series $\sum_n [U(6/\sigma_n(\zeta, it_1, \eta))]^{-\tau}$ converges if $\tau > 1$ and that for any positive number $\eta$ and for any finite complex number $\zeta$, with a possible exception, the series $\sum_n [U(1/6\sigma_n(\zeta, it_1, \eta)]^{-\tau}$ diverges if $\tau < 1$.

*Proof* As in the proofs of the previous theorems we show that there exists a point $it_0$ in any interval of length greater than $2\pi D$ on $\sigma = 0$ such that for any $\eta > 0$

$$\varlimsup_{\sigma \to +0} \frac{\overset{+}{\log}\overset{+}{\log} M_S(\sigma)}{\log U(1/\sigma)} = 1, \tag{3.8}$$

where $S = S(t_0, \eta)$.

Let $\varepsilon$ be such that $0 < \varepsilon < \pi$. The applications[9]

$$z = e^{-s+it_0}, \quad Z = z^{\pi/2\varepsilon} \text{and } w = \frac{Z^2 - 1 + 2Z}{Z^2 - 1 - 2Z}$$

transform the domain $\sigma > 0$, $|t - t_0| < \varepsilon$ into the domains $|z| < 1$, $|\arg z| < \varepsilon$; $|Z| < 1$,

$|\arg Z| < \frac{\pi}{2}$ and $|w|<1$; $s=i(t_0 \pm \varepsilon)$ and $+\infty$ correspond respectively to $z=e^{\pm i\varepsilon}$ and 0; $Z=\pm i$ and 0; and $w=\pm i, -1$.

Under the above applications we obtain

$$f(s)=f_1(z)=f_2(Z)=f_3(w).$$

Put $M_S(\sigma) = \sup_{|t-t_0|<\varepsilon} |f(\sigma+it)|$, $\overline{M}_S(\sigma) = \sup_{|t-t_0|<\varepsilon} |f(x+it)| \, (x \geqslant \sigma,\ \sigma>0)$,

$$\overline{M}_1(r) = \sup_{|z|<r} |f_1(z)| \, (|\arg z|<\varepsilon,\ 0<r<1),$$

$$\overline{M}_2(R) = \sup_{|Z|<R} |f_2(Z)| \left(|\arg Z|<\frac{\pi}{2},\ 0<R<1\right),$$

$$M_2(R) = \sup_{|Z|=R} |f_2(Z)| \left(|\arg Z|<\delta\left(<\frac{\pi}{2}\right),\ R>C\right),$$

$$M_3(S) = \max_{|w|\leqslant S} |f_3(w)| \, (0<S<1),$$

where $S=S(t_0, \varepsilon)$. We have

$$M_S(\sigma) \leqslant \overline{M}_S(\sigma) \leqslant M(\sigma).$$

Hence[12, 14]

$$1 = \overline{\lim_{\sigma\to+0}} \frac{\overset{+}{\log}\overset{+}{\log} \overline{M}_S(\sigma)}{\log U(1/\sigma)} = \overline{\lim_{\sigma\to+0}} \frac{\overset{+}{\log}\overset{+}{\log} \overline{M}_S(\sigma)}{\log U(1/(1-e^{-\sigma}))}$$

$$= \overline{\lim_{s\to+0}} \frac{\overset{+}{\log}\overset{+}{\log} \overline{M}_1(s)}{\log U(1/(1-s))} = \left(\frac{2\varepsilon}{\pi}\right)^{\rho_1} \overline{\lim_{R\to1-0}} \frac{\overset{+}{\log}\overset{+}{\log} \overline{M}_2(R)}{\log U(1/(1-R))}.$$

Since (3.8) holds for any $\eta>0$, we deduce as above

$$\overline{\lim_{R\to1-0}} \frac{\overset{+}{\log}\overset{+}{\log} M_2(R)}{\log U(1/(1-R))} = \left(\frac{\pi}{2\varepsilon}\right)^{\rho_1}.$$

By (3.11), when $|w|\leqslant S$, the corresponding $Z$ satisfies $|Z|<(S+1)/2$, By (3.10), when $Z$ satisfies $|Z|=3S-2$, $|\arg Z|<\delta$ and $C<3S-2<1$, the corresponding $w$ satisfies $|w|<S$. Consequently we have

$$M_2(3S-2) \leqslant M_3(S) \leqslant \overline{M}_2((S+1)/2)$$

and

$$\left(\frac{\pi}{6\varepsilon}\right)^{\rho_1} = \overline{\lim_{S\to1-0}} \frac{\overset{+}{\log}\overset{+}{\log} M_2(3S-2)}{\log U(3/[1-(3S-2)])} \leqslant \overline{\lim_{S\to1-0}} \frac{\overset{+}{\log}\overset{+}{\log} M_3(S)}{\log U(1/(1-S))}$$

$$\leqslant \overline{\lim_{S\to1-0}} \frac{\overset{+}{\log}\overset{+}{\log} \overline{M}_2((S+1)/2)}{\log U(1/2[1-(S+1)/2])} = \left(\frac{\pi}{\varepsilon}\right)^{\rho_1}.$$

Hence by (3.7)

$$\overline{\lim_{S\to1-0}} \frac{\overset{+}{\log}\overset{+}{\log} M_3(S)}{\log U(k_1/(1-S))} = 1, \tag{3.12}$$

where $k_1 = k\pi/2\varepsilon$ and $1/3 \leqslant k \leqslant 2$.

Let $T_3(S)$ be the Nevanlinna characteristic function of $f_3(w)$ $(|w|=S)$. By (3.12) we have

$$\overline{\lim_{S\to1-0}} \frac{\overset{+}{\log}\overset{+}{\log} T_3(S)}{\log U(k_1/(1-S))} = 1. \tag{3.13}$$

In fact, we have, by an inequality of Nevanlinna [6], for any $a>1$

$$\log M_3\left(\frac{1+(a-1)S}{a}\right)\geqslant T_3\left(\frac{1+(a-1)S}{a}\right)\geqslant\frac{1-S}{1+(2a-1)S}\log M_3(S),$$

whence

$$1=\varlimsup_{S\to1-0}\frac{\overset{+}{\log}\overset{+}{\log}M_3\left(\frac{1+(a-1)S}{a}\right)}{\log U(ak_1/(a-1)(1-S))}\geqslant\varlimsup_{S\to1-0}\frac{\overset{+}{\log}T_3\left(\frac{1+(a-1)S}{a}\right)}{\log U(ak_1/(a-1)(1-S))}$$

$$\geqslant\varlimsup_{S\to1-0}\frac{\overset{+}{\log}\overset{+}{\log}M_3(S)}{\log U(ak_1/(a-1)(1-S))}\geqslant\left(\frac{a-1}{a}\right)^{\rho_1}.$$

**Since the above inequalities hold for any $a>0$, we get (3.13).**

**Hence[4] $f_3(w)$ has a Borel point $w_1(|w_1|=1)$ of order $(H)$**

$$\rho(k_1/1-|w|)\left[1+\frac{\log k_1}{\log(1/1-|w|))}\right].^{[4]}$$

**By the applications mentioned above we can complete the proof of Theorem 3.3.**

## Referencs


[1] Anderson J. M. and Binmore K. G., *Proc. London Math. Soc.*, **18** (1968), 49—68.

[2] Binmore, K. G., *J. London Math. Soc.*, **41** (1966), 693—696.

[3] Blambert, M., *Ann. scient. Éc. Norm. Sup.*, *3e série*, **79** (1962), 353—375.

[4] Hiong Kin-lai, *J. de Math.*, *9e série*, **14** (1935), 233—308.

[5] Mandelbrojt, S. Séries de Dirichlet. Principes et méthodes, Paris, Gauthier-Villars, 1969.

[6] Nevanlinna, R., Le théarème de Picard-Borel et la théorie des fonctions méromorphes, Paris, Gauthier-Villars, 1929.

[7] Sunyer i Balaguer F., *Proc. Amer. Math. Soc.*, **4** (1953), 310—322.

[8] Tanaka, C., *Yokohama Math. J.*, **2** (1954), 151—164.

[9] Valiron, G., *Bull. Sc. Math.*, **56** (1932), 10—32.

[10] Valiron, G., *Proc. Nat. Acad. Sc., U. S. A.* **20** (1934), 211—215.

[11] Valiron, G. Fonctions entières d'ordre fini et fonctions méromorphes, Genève, L'Enseignement Mathématique, 1960.

[12] Yang Lo et Shiao Shiou-zhi, *Scientia Sinica*, **15** (1965), 1556—1573.

[13] Yu, Chia-yung, *Ann. scient, Éc. Narm. Sup.*, *3e série*, **68** (1951), 65—104.

[14] Yu, Chia-yung, *Acta Math Sinica*, **21** (1978), 97—118.

[15] Yu, Chia-Yung, *C. R. Acad. Sc. Paris*, **288** (1979), *Sér. A*, 891—893.

*Chin. Ann. of Math.*
3 (4) 1982

# ON A $\rho$-INCREASING FAMILY OF POINT-TO-SET MAPS

YUE MINYI
(*Institute of Applied Mathematics, Academia Sinica*)

Dedicated to Professor Su Bu-chin on the Occasion of His 80th Birthday and his 50th Year of Educational Work

## 1. Introduction.

Since the publication of Zangwill's book "Nonlinear Programming-A Unified Approch" in 1969, the theory of point-to-set maps has come into increasing use during the last ten or more years in papers on optimization, especially, on the convergence theory of the iteration algorithms (Cf. MP Study no. 10, 1979). One of the advantages of the approach introduced by Zangwill is that one can treat in a unified way the problems concerning the convergence properties of algorithms. Zangwill showed that a lot of methods can be viewed as applications of the fixed point method $x_{i+1}\in\Gamma(x_i)$, where $\Gamma$ is a point-to-set map depending upon the given particular algorithm.

Huard (1975) made a research on Zangwill's results. He designed two models of algorithms (Algorithms 4.2 and 5.2 in his paper), and proved that these algorithms, under some specified conditions, can be viewed as special cases of Zangwill's model. Therefore, Zangwill's results can be applied to these models. The advantage which Huard's models have over the Zangwill's is that it can be used more easily to decide whether a given algorithm belongs or not to these models, and, if it does, the convergence property follows as a consequence.

Denel (1979) introduced a $\rho$-decreasing family of point-to-set maps and extended Zangwill's approach from another point of view. In his paper (1979), Denel introduced many notions such as monotone decreasing family, uniform regularity, pseudo upper (lower) continuity, etc., and several algorithms. He proved that his algorithms have some convergence properties if certain conditions are satisfied. Denel claimed that his algorithm $A_1'$ leads to the same sequence as those produced by the algorithm 4.2 given by Huard (1975). However, we find that when the sequence $\{\gamma_n\}$ (Cf. Denel (1979)) is given, the sequence produced by $A_1'$ is only a part of those produced by the algorithm 4.2, because the parameter $\rho$ introduced in $A_1'$ has influence on the sequence. Therefore, Huard's algorithm 4.2 can not be said to be a special case of $A_1'$. However, as pointed

Manuscript received september 9, 1980.

out by Denel (1979) through two examples, with his algorithm $A_1'$, one can modelize a lot of well-known methods that can not be modelized with the classical approachs, so Denel's approach is, in a sense, a further development of the classical ones.

Huard (1979) made an extension of Zangwill's results, but he did not show what is the use of his result.

In this paper we introduce a $\rho$-increasing family of point-to-set maps, then define two algorithms and show that under certain simple conditions every accumulation point of the sequence generated by these algorithms belongs to a well-defined set. Then we prove that the results of Zangwill (1969) and Huard (1975) are direct consequences of our results. We prove also that Huard's algorithm (Proposition $D$ (1979)) is convergent if our conditions just mentioned are satisfied. To show that our approach is a real extension of the classical ones, we prove that the convergence of linearized method of centers with partial linearization, which, as Denel showed, could not be modelized with the classical approachs, is a direct consequence of our results.

## 2. Notations and Definitions.

The following notations are used throughout this paper:

$N=\{1, 2, 3, \cdots\}$;

$E\subset R^n$, a compact set;

$\mathscr{P}(E)$, set of the subsets of $E$;

$V(x)$, a relative neighbourhood of $x$, i. e., $V(x)=E\cap U(x)$, where $U(x)$ is a euclidean neighbourhood;

$E^0$, the interior of $E$;

$\{F_\rho(x)\mid\rho\geqslant 0\}$, an increasing family of point-to-set maps $F_\rho(x)$: $E\to\mathscr{P}(E)$, i. e., the family satisfying

$$F_{\rho_1}(x)\supset F_{\rho_2}(x), \quad \forall \rho_2\geqslant\rho_1\geqslant 0, \ \forall x\in E;$$

$\Omega\subset R^n$, a given set, $\Omega\sqsupseteq\{x\mid x\in F_0(x)\}$;

$P=\{X\in E\mid x\in\Omega \text{ or } F_0(x)=\emptyset\}$.

**Definition.** *A point-to-set map $F$:$E\to\mathscr{P}(E)$ is said to be upper continuous* (u. c.) *at $x\in E$, if for any given sequence $\{x_k\in E,\ k\in N\}\to x$ and sequence $\{y_k\in E,\ k\in N\}\to y$ satisfying $y_k\in F(x_k)$, $\forall k\in N$, we have $y\in F(x)$.*

$F$ is said to be upper continuous on $E$ if it is upper continuous at every point of $E$.

**Definition.** *A point-to-set map $F$: $E\to\mathscr{P}(E)$ is said to be lower continuous* (l. c.) *at $x\in E$, if for any sequence $\{x_k\in E,\ k\in N\}\to x$ and any $y\in F(x)$ there exists sequence $\{y_k\in E,\ k\in N\}\to y$ and positive integer $k_0$ such that $y_k\in E(x_k)$, $\forall k\geqslant k_0$.*

$F$ is said to be lower continuous on $E$ if it is lower continuous at every point of $E$.

**Definition.** *F is said to be continuous on E if it is both upper and lower continuous on E.*

## 3. Assumptions and a Fundamental Theorem.

We make the following assumptions:

H1: If $\rho_\nu \to 0$, $x_\nu \to x_0$, $y_\nu \in F_{\rho_\nu}(x_\nu)$, $y_\nu \to y_0$, then we have $y_0 \in F_0(x_0)$;

H2: There exists a function $h(\cdot)$ continuous on $E$ having the following property: If for some $\rho \geqslant 0$ we have $x' \in F_\rho(x)$ $x \overline{\in} F_\rho(x)$, where $x \overline{\in} P$, then we have $h(x') > h(x)$.

**Theorem 1.** *Under the assumptions* H1 *and* H2, *there exist, for any given* $x_0 \in E \backslash P$, $V(x_0)$ and $\rho_0 > 0$, *such that*

$$h(x'') > h(x_0), \quad \forall x' \in (E \backslash P) \cap V(x_0), \quad \forall x'' \in F_{\rho_0}(x').$$

*Proof* Assume that the conclusion is false. Then for any given monotonely decreasing sequence $\{\rho_i\}$ of nonnegative reals converging to zero, there exist sequences $\{x'_i\}$ and $\{x''_i\}$ with $x'_i \in \mathrm{E} \backslash P$, $x''_i \in F_{\rho_i}(x'_i)$, $\forall i \in N$, $\{x'_i\} \to x_0$, such that $h(x''_i) \leqslant h(x_0)$, $\forall i$. By assumptions, $E$ is compact, so that there exists $N' \subset N$ such that $\{x''_i,\ i \in N'\} \to x''_0$. By H1, we have $x''_0 \in F_0(x_0)$. Since $h$ is continuous on $E$, we have $h(x''_0) \leqslant h(x_0)$. On the other hand, by H2, we have $h(x''_0) > h(x_0)$, since $x_0 \overline{\in} P$. Thus we have a contradiction.

**Corollary 1.** *Under* H1 *and* H2, *for* $x_0 \in E \backslash P$, *there exist* $V(x_0)$ *and* $\rho_0 > 0$, *such that* $h(x'') > h(x_0)$, $\forall 0 \leqslant \rho \leqslant \rho_0$, $\forall x' \in (E \backslash P) \cap V(x_0)$, $\forall x'' \in F_\rho(x')$.

*Proof* This is a direct consequence of Theorem 1 and the monotonicity of

$$\{F_\rho \mid \rho \geqslant 0\}.$$

## 4. Algorithms.

**Algorithm A1.** Let $x_0 \in E$ be the starting point and $\{\rho_n\}$ be a given monotonely decreasing sequence of nonnegative reals converging to zero.

Step $n$:

if $x_n \in P$, then stop;

$$\left.\begin{array}{l}\text{if } F_{\rho_n}(x_n) = \emptyset \\ \text{or } x_n \in F_{\rho_n}(x_n)\end{array}\right\}, \text{ then } \begin{cases} x_{n+1} = x_n, \\ \rho_{n+1} \leftarrow \rho_n; \end{cases}$$

$$\left.\begin{array}{l}\text{if } F_{\rho_n}(x_n) \neq \emptyset \\ \text{and } x_n \overline{\in} F_{\rho_n}(x_n),\end{array}\right\} \text{then} \begin{cases} \text{choose } x_{n+1} \in F_{\rho_n}(x_n), \\ \rho_{n+1} \leftarrow \rho_n; \end{cases}$$

End of step $n$.

## 5. Convergence property.

**Theorem 2.** *Let* $\{\rho_n\}$ *be a given monotonely decreasing sequence converging to zero. Then, under the assumptions* H1 *and* H2, *algorithm* A1 *either constructs an infinite* *sequence having all its accumulation points in $P$, or terminates at an $x_k \in P$ in a finite number of steps.*

*Proof* Obviously we can suppose $x_n \overline{\in} P$, $\forall n \in N$. By the definition of $\{x_n\}$ and the assumption H2, it is evident that $\{h(x_n)\}$ is a monotonely increasing sequence, so that it has a limit. We have $F_{\rho_n}(x_n) \neq \emptyset$, $\forall n \in N$, otherwise there exists an $n$ for which $F_{\rho_n}(x_n) = \emptyset$. By the monotonicity of $F_\rho$, we have $F_0(x_n) = \emptyset$, this means $x_n \in P$ and contradicts our assumption.

Let $\{x_n\}_{N'} \to x^*$. Without loss of generality we assume $x_n \overline{\in} F_{\rho_n}(x_n)$, $\forall n \in N'$. If it is not so, then there are only two possibilities:

(i) From some $n_0$ onwards we have $x_n \in F_{\rho_n}(x_n)$, $\forall n \in N'$. In this case, by H1, we must have $x^* \in F_0(x^*)$, and the Theorem holds;

(ii) There is an infinite sequence $N'' \subset N'$, for which $x_n \overline{\in} F_{\rho_n}(x_n)$. In this case, we replace $N'$ by $N''$ without violating the limit of $\{x_n\}_{N'}$.

Now assume that $x^* \overline{\in} P$. By Corollary 1, there exist $V(x^*)$ and $\rho_0 > 0$, such that $h(x'') > h(x^*)$, $\forall 0 \leqslant \rho \leqslant \rho_0$, $\forall x' \in (E \backslash P) \cap V(x^*)$, $\forall x'' \in F_\rho(x')$. Therefore, there exists $n_0 > 0$ such that $h(x_{n+1}) > h(x^*)$, $\forall v \geqslant n_0 + 1$, $n \in N'$ and $x_{n+1} \in F_{\rho_n}(x_n)$. Since $\{h(x_n)\}$ is a monotonely increasing sequence, we have $h(x_n) \geqslant h(x_{n_0+2}) > h(x^*)$, $\forall n \geqslant n_0 + 1$.

By the continuity of $h(x)$, we have $h(x^*) > h(x^*)$, a contradiction. So we must have $x^* \in P$.

## 6. Applications of Theorem 2.

In this section we will prove that Zangwill's theorem and Huard's theorem (Prop. 4.1, 1975) are special cases of Theorem 2.

**Zangwill Theorem** *Let $E \subset R$ be compact, $P \subset E$ be a given solution set, $F$: $E \to \mathscr{P}(E)$ be a point-to-set map, and $f$: $E \to R$ be a continuous function. Suppose:*

( i ) $F(x) \neq \emptyset$, $\forall x \in E \backslash P$;

( ii ) *$F$ is upper continuous on $E \backslash P$;*

( iii ) $f(y) > f(x)$, $\forall y \in F(x)$, $\forall x \in E \backslash P$.

*Let $A$ be the algorithm defined on $E$:*

*Starting value: $x_0 \in E$;*

*Step n: if $x_n \in P$, then stop;*

*if $x_n \overline{\in} P$, take $x_{n+1} \in F(x_n)$.*

*Then $A$ either constructs an infinite sequence having all its accumulation points in $P$ or terminates at an $x_k \in P$ in a finite number of steps.*

**Corollary of Theorem 2.** *Zangwill's theorem holds.*

*Proof* Let, in Theorem 2, *$F_\rho(x)$ be independent of $\rho$*, i. e., $F(x) = F_\rho(x)$, $\forall \rho \geqslant 0$. Assumption (ii) implies that H1 holds. Assumption (iii) implies that H2 holds (in this case, we take $h(x) = f(x)$).

Huard (1975) proved the following Theorem (Prop. 4.1):

Let $T \subset R_+$ be compact;

$h$: $R^n \to R$ be continuous;

$\Delta$: $E \to \mathscr{P}(E)$ be a point-to-set map such that

( i ) $\Delta$ is continuous on $E$;

( ii ) $x \in \Delta(x)$, $\forall x \in E$;

g: $E \times E \to R$ be continuous;

$M_\Delta(x, \varepsilon) = \{y \in \Delta(x) \mid g(y, x) \geqslant \max_{z \in \Delta(x)} g(z, x) - \varepsilon\}$.

The function $g$ is supposed such that

$\forall (x, \varepsilon) \in E \times T$, $\forall y \in M_\Delta(x, \varepsilon)$, $g(x, x) < g(y, x) \Rightarrow h(x) < h(y)$.

The algorithm A is defined by:

Starting value: $x_0 \in E$, $\{\varepsilon_n\} \subset T$ be a given non-increasing sequence converging to zero.

Step $n$: if $x_n \in M_\Delta(x_n, \varepsilon_n)$, then $x_{n+1} = x_n$, $\varepsilon_{n+1} \leftarrow \varepsilon_n$,

if $x_n \overline{\in} M_\Delta(x_n, \varepsilon_n)$, then $x_{n+1} \in M_\Delta(x_n, \varepsilon_n)$, $\varepsilon_{n+1} \leftarrow \varepsilon_n$.

**Huard Theorem** (Prop. 4.1). *Under the assumptions given above, every accumulation point $x^*$ of the infinite sequence generated by the algorithm A belongs to* $M_\Delta(x^*, 0)$.

Now we are going to prove that the theorem above is a special case of Theorem 2.

**Corollary 3.** *Huard Theorem* (Prop. 4.1) *holds.*

*Proof* Taking $F_\rho(x) = M_\Delta(x, \rho)$, $P = \{x \in E \mid F_0(x) = \emptyset$ or $x \in F_0(x)\}$, we only need to prove that under all the assumptions given above, family $\{F_\rho(x) \mid \rho \geqslant 0\}$ satisfies all the requirements of Theorem 2.

It is easy to prove that $F_\rho(x) \neq \emptyset$, $\forall \rho \geqslant 0$, $\forall x \in E$. Namely, $\Delta(x)$ is non-empty and compact, $g(y, x)$ is continuous in $y$, so for any $x \in E$ and $\rho \geqslant 0$, there exists $y_0 \in \Delta(x)$ such that $g(y_0, x) \geqslant \max_{y \in \Delta(x)} g(y, x) - \rho$.

( i ) It is obvious that $\{F_\rho(x) \mid \rho \geqslant 0\}$ is a non-decreasing family of point-to-set maps.

( ii ) H1 holds. Let $\rho_\nu \to 0$, $x_\nu \to x_0$, $y_\nu \in M_\Delta(x_\nu \rho_\nu)$, $y_\nu \to y_0$: $g(y_\nu, x_\nu) \geqslant \max_{y \in \Delta(x_\nu)} g(y, x_\nu) - \rho_\nu$. Then, since $\Delta(x)$ is (upper) continuous, we have $y_0 \in \Delta(x_0)$. Now we are going to prove

$$g(y_0, x_0) = \max_{y \in \Delta(x_0)} g(y, x_0).$$

Namely, if $g(x_0, y_0) < \max_{y \in \Delta(x_0)} g(y, x_0)$, there exist $\varepsilon > 0$ and $y_1 \in \Delta(x_0)$ such that $g(y_0, x_0) + \varepsilon < g(y_1, x_0)$. Since $\Delta(x)$ is (lower) continuous, there exist $\bar{y}_\nu \to y_1$, $\bar{y}_\nu \in \Delta(x_\nu)$, $\forall \nu \in N$. From this we have

$$g(y_\nu, x_\nu) \geqslant g(\bar{y}_\nu, x_\nu) - \rho_\nu.$$

Since $g(y, x)$ is continuous, we have

$$g(y_0, x_0) \geqslant g(y_1, x_0),$$

a contradiction.

(iii) H2 holds. This is a direct consequence of the definition of $F_\rho(x)$.

## 7. Another Algorithm of Huard.

Huard (1979) proved the following

Proposition (Prop. 0). Let $E \subset R^n$ be closed, $Q \subset P \subset E$. Let $F_1$ and $F_0$: $(F\backslash Q) \to \mathscr{P}(E)$ be point-to-set maps such that $F_1 \supset F_0$. Suppose that for all $x \in E\backslash Q$ we have

($\alpha_0$) $F_0(x) \neq \emptyset$,

($\beta_0$) $x' \in (E\backslash Q) \cap F_1(x) \Rightarrow F_1(x') \subset F_1(x)$, and

($\gamma_0$) if $x \overline{\in} P$, then there exists $V(x)$ such that

$$x' \in (E\backslash Q) \cap V(x),\ x'' \in (E\backslash Q) \cap F_0(x') \Rightarrow x \overline{\in} \overline{F_1(x'')}.$$

Then every accumulation point of the sequence $\{x_k\}$ generated by the following algorithm A3 belongs to P.

**Algorithm A3.**

$$x_0 \in E;$$

$$x_{i+1} \in F_0(x_i), \text{ if } x_i \overline{\in} P;$$

$$x_{i+1} \in F_0(x_i) \cup \{x_i\}, \text{ if } x_i \in P\backslash Q;$$

$$x_{i+1} = x_i, \text{ if } x_i \in Q.$$

Now we are going to prove the following

**Theorem 3.** *Assume that $F_1$ is upper continuous on $E\backslash P$, and $P \supset \{x \in E \mid x \in F_1(x)\}$. Then the assumption $\gamma_0$ in the proposition mentioned above is a consequence of the other assumptions.*

*Proof* If the conclusion is false, then there exist $\{x'_\nu\} \to x$, $x'_\nu \in E\backslash Q$, $\forall \nu \in N$, and $x''_\nu \in (E\backslash Q) \cap F_0(x'_\nu)$ such that $x \in \overline{F_1(x''_\nu)}$, $\forall \nu \in N$. Therefore for a fixed $\nu$ there exists sequence $\{y_{\nu_j}\}$ with $\lim\limits_{j\to\infty} y_{\nu_j} = x$, and $y_{\nu_j} \in F_1(x''_\nu)$, $\forall\ j$. It is easy to see that there exists sequence $\{y_{n_{j_n}}\}$ with $\lim\limits_{n\to\infty} y_{n_{j_n}} = x$ and $y_{n_{j_n}} \in F_1(x''_n)$. Since $x''_n \in (E\backslash Q) \cap F_0(x'_n)$, we have, by ($\beta_0$), $y_{n_{j_n}} \in F_1(x'_n)$. Since $F_1(x)$ is upper continuous, we have $x \in F_1(x)$, a contradiction. So we must have $x \overline{\in} \overline{F(x)}$.

**Remark 1.** In the proposition mentioned above, it has been implicitly supposed that $P \supset \{x \in E \mid x \in F_0(x)\}$. Namely, if there exists $x_0$ such that $x_0 \in F_0(x_0)$, $x_0 \overline{\in} P$, then, taking $x = x' = x'' = x_0$, we have, by ($\gamma_0$), $x_0 \overline{\in} \overline{F_1(x_0)}$. Therefore we have $x_0 \overline{\in} F_0(x_0)$, a contradiction.

**Remark 2.** Since. in Algorithm A3, we have two sets $P$ and $Q$, Theorem 2 can not be applied to this case. Theorem 3 only shows that, under our assumption, every accumulation point of the sequence $\{x_\nu\}$ generated by A3, according to Huard's (Prop. 0), belongs to $P$.

### 8. Algorithm A2.

Now we introduce the following assumptions:

H3: $\{\Delta_\rho(x) \mid \rho \geqslant 0\}$ is a non-increasing (in $\rho$) family of point-to-set maps;

H4: if $\rho_\nu \to 0$, $x_\nu \to x_0$, $y_\nu \in \Delta_{\rho_\nu}(x_\nu)$, $y_\nu \to y_0$, then $y_0 \in \Delta_0(x_0)$;

H5: if $x_0 \in E$, $\Delta_0(x_0) \neq \emptyset$, then there exist $V(x_0)$ and $\rho_0 > 0$ such that $\Delta_\rho(x) \neq \emptyset$, $\forall 0 \leqslant \rho \leqslant \rho_0$, $\forall x \in V(x_0)$.

**Notations:**

$$M1 = \{x \in E \mid \Delta_0(x) = \emptyset\},$$
$$M2 = \{x \in E \mid \exists z \in \Delta_0(x): F_0(x, z) = \emptyset\},$$
$$M3 = \{x \in E \mid \exists z \in \Delta_0(x): x \in F_0(x, z)\},$$
$$M = M1 \cup M2 \cup M3,$$
$$P = M1 \cup M2 \cup \Omega, \ \Omega \supset M3.$$

**Algorithm A2.** Let $x_0 \in E$, Let $\{\rho_n\}$ and $\{\rho'_n\}$ be two given non-decreasing sequences of non-negative reals converging to zero.

Step $n$: if $x_n \in P$, stop;

Phase 1: if $\Delta_{\rho_n}(x_n) = \emptyset$, then $x_{n+1} = x_n$, $\rho_{n+1} \leftarrow \rho_n$, $\rho'_{n+1} \leftarrow \rho'_n$;

if $\Delta_{\rho_n}(x_n) \neq \emptyset$, then $z_n \in \Delta_{\rho_n}(x_n)$;

Phase 2: if $F_{\rho'_n}(x_n, z_n) = \emptyset$ or $x_n \in F_{\rho'_n}(x_n, z_n)$, then

$x_{n+1} = x_n$, $\rho_{n+1} \leftarrow \rho_n$, $\rho'_{n+1} \leftarrow \rho'_n$;

if $F_{\rho'_n}(x_n, z_n) \neq \emptyset$ and $x_n \overline{\in} F_{\rho'_n}(x_n, z_n)$, then

$x_{n+1} \in F_{\rho'_n}(x_n, z_n)$, $\rho_{n+1} \leftarrow \rho_n$, $\rho'_{n+1} \leftarrow \rho'_n$.

**Theorem 4.** *Let $\{F_\rho(x, z) \mid \rho \geqslant 0\}$ be a non-decreasing* (in $\rho$) *family of point-to-set maps ($F_\rho(x, z)$ is a map of $(x, z)$) and satisfy* H1 *and* H2; *let $\Delta_\rho(x)$ be a non-increasing (in $\rho$) family of point-to-set maps and satisfy* H4 *and* H5. *Algorithm A2 either stops at a step $k$ with $x_k \in M$ or generates a sequence $\{x_k\}$ with all its accumulation points in $M$.*

*Proof* It is obvious that we can suppose $x_n \overline{\in} M$, $\forall n \in N$. In this case, we have

(1) $$\Delta_0(x_n) \neq \emptyset, \ \forall n \in N,$$

(2) $$F_0(x_n, z) \neq \emptyset, \ \forall z \in \Delta_0(x_n), \ \forall n \in N.$$

From this we conclude that if $\Delta_{\rho_n}(x_n) \neq \emptyset$, then $F_0(x_n, z) \neq \emptyset$ $\forall z \in \Delta_{\rho_n}(x_n)$, and therefore, $F_{\rho'_n}(x_n, z_n) \neq \emptyset$.

Obviously, $\{h(x_n)\}$ is a monotonely increasing sequence. Let $\{x_n\}_{N'} \to x^*$, $N' \subset N$. If $x^* \overline{\in} M$, then $\Delta_0(x^*) \neq \emptyset$. By H5 and (2) there exists $n_0$ such that $\Delta_{\rho_n}(x_n) \neq \emptyset$, $z_n \in \Delta_{\rho_n}(x_n)$, $F_{\rho'_n}(x_n, z_n) \neq \emptyset$, $\forall n \geqslant n_0$, $n \in N'$. As in the proof of Theorem 2, we can assume $x_n \overline{\in} F_{\rho'_n}(x_n, z_n)$, $\forall n \in N'$. Namely, if it is not so, then there are two possibilities:

(i) From some $n$ onwards, we have $x_n \in F_{\rho'_n}(x_n, z_n)$, $\forall n \in N'$. Since $E$ is compact, by H4, there exists $N''\subset N'$ such that $\{z_n\}_{N''}\to z^*\in \Delta_0(x^*)$; then by H1, we have $x^*\in F_0(x^*, z^*)$. Thus we have $x^*\in M3\subset M$, and the theorem holds.

(ii) There is an infinite sequence $N''\subset N'$ such that $x_n\overline{\in} F_{\rho_n'}(x_n, z_n)$, $\forall n\in N''$. In this case, we replace $N'$ by $N''$ without violating the limit of $\{x_n\}_{N'}$.

Since $E$ is compact, we can assume $\{z_n\}_{N'}\to \bar{z}$. By H4, we have $\bar{z}\in \Delta_0(x^*)$. Since $x^*\overline{\in} M$, by Corollary 1, there exist $V(x^*, \bar{z})$ and $\rho_0>0$ such that $h(x'')>h(x^*)$, $\forall (x', z')\in V(x^*, \bar{z})$, $\forall x''\in F_\rho(x', z')$, $\forall 0\leqslant\rho\leqslant\rho_0$. Thus there exists $n_0$ such that $h(x_{n+1})>h(x^*)$, $\forall x_{n+1}\in F_{\rho_n'}(x_n, z_n)$, $\forall n\geqslant n_0+1$, $n\in N'$. Since $\{h(x_n)\}$ is a non-decreasing sequence, we have $h(x_n)\geqslant h(x_{n_0+2})>h(x^*)$, $\forall n>n_0+1$, $n\in N'$. Since $h(x)$ is continuous, we have $h(x^*)>h(x^*)$, a contradiction. Therefore we must have $x^*\in M$.

## 9. Applications of Theorem 4.

Huard (1975) proved the following theorem (Prop. 5.1):

**Theorem.** *Let $A\subset R^n$ be closed and $B\subset R^n$ be convex compact such that $A\cap B\neq\emptyset$. Let $T\subset R_+$ be compact, $f$: $R^n\to R$ continuous, $\Delta$: $B\to\mathscr{P}(B)$ upper continuous on $B$, $\Delta(x)\neq\emptyset$, $\forall x\in A\cap B$, $g$: $B\times B\to R$ continuous*

$$M(x, z, \varepsilon)=\{y\in[x, z]\mid g(y, x)\geqslant \max_{w\in[x,z]} g(w, x)-\varepsilon\}.$$

*Suppose that*

$$(*)\qquad M(x, z, \varepsilon)\cap A\cap B\neq\emptyset,\ \forall x\in A\cap B,\ \forall z\in\Delta(x),\ \forall \varepsilon\geqslant 0, \text{ and that}$$

$$\forall x\in A\cap B,\ \forall z\in B,\ \forall \varepsilon\in T,\ \forall y\in M(x, z, \varepsilon),\ g(x, x)<g(y, x)\Rightarrow f(x)<f(y).$$

*Then the following algorithm either terminates at a finite step $k$ with $x_k\in M$ or generates an infinite sequence having all its accumulation points in $M$, where*

$$M=A\cap B\cap M(x, z, 0).$$

**Algorithm A4.** $x_0\in A\cap B$ is arbitrarily given.

Step k: We have $x_k\in A\cap B$, and choose: $z_k\in\Delta(x_k)$, $\varepsilon_k\in T$

$$x_{k+1}\in A\cap B\cap M(x_k, z_k, \varepsilon_k), \text{ if } x_k\overline{\in} M(x_k, z_k, \varepsilon_k).$$

$$x_{k+1}=x_k, \text{ if } x_k\in M(x_k, z_k, \varepsilon_k).$$

Now we are going to prove the following

**Theorem 5.** *Huard's theorem above* (Prop. 5.1) *is a special case of Theorem 4.*

*Proof* Set

$$E=A\cap B,\ \Delta_\rho(x)=\Delta(x),\ F_\rho(x, z)=A\cap B\cap M(x, z, \rho).$$

Let $\{\rho_\nu\}$ be a non-increasing sequence of non-negative reals converging to zero. Then we want to prove that, under the assumptions given in the above Theorem, all the conditions of Theorem 4 are satisfied, and that Algorithm A4 is a special case of Algorithm A2.

( i ) It is obvious that $F_\rho$ $(x, z)$ is a non-decreasing family (in $\rho$).

( ii ) From $(*)$ we have $F_\rho(x, z)\neq\emptyset$, $\forall\rho\geqslant 0$. Thus we have $M1=M2=\emptyset$.

(iii) Obviously $[x, z]$ is a continuous map of $(x, z)$, so the part (ii) in the proof of Corollary 3 is also effective in this case. Therefore H1 holds.

(iv) H2 follows immediately from the definition of $M(x, z, \varepsilon)$.

From the fact that $F_\rho(x, z) \neq \emptyset$ and $\Delta(x) \neq \emptyset$, $\forall x \in A \cap B$, we see immediately that Algorithm A4 is a special case of Algorithm A2.

Denel (1979) has cited two examples to show that his algorithms have more applications than those given by Huard (1975). Now we show that Theorem 4 is also applicable to these examples. As the arguments are quite similar, we only discuss the first one in detail.

**Example** Linearized method of centers with partial linearization (Huard(1978)).

The problem to be solved is

$$\begin{aligned} &\max f(x) \\ &\text{s. t. } g_i(x) \geqslant 0, \ i=1, \cdots, m, \\ &x \in B, \end{aligned}$$

where the functions are coucave, continuously differentiable and $B$ is a compact polyhedron.

Denote

$$A = \{x \mid g_i(x) \geqslant 0, \ i=1, \cdots, m\}.$$

We suppose that $A^0 \cap B \neq \emptyset$.

For a given $\varepsilon > 0$, denote

$$d'_\varepsilon(z, x) = \min \{f'(z, x) - f(x), \ g'_i(z, x), \ i \in I_\varepsilon(x)\}, \text{ where}$$

$$f'(z, x) = f(x) + \nabla f(x) \cdot (z - x),$$

$$g'_i(z, x) = g_i(x) + \nabla g_i(x) \cdot (z - x), \ i=1, \cdots, m,$$

$$I_\varepsilon(x) = \{i \in \{1, \cdots, m\} \mid g_i(x) < \varepsilon\},$$

$$d(t, f(x)) = \min \{f(t) - f(x), \ g_i(t), \ i=1, \cdots, m\}.$$

By "linearized method of centres with partial linearization" we mean the following two-phase algorithm:

(1) For a given $x_\nu$, find $z_\nu$ such that

$$d'_\varepsilon(z_\nu, x_\nu) = \max_{z \in B} d'_\varepsilon(z, x_\nu);$$

(2) For given $x_\nu$ and $z_\nu$, find $x_{\nu+1}$ such that

$$d(x_{\nu+1}, f(x_\nu)) = \max_{t \in [x_\nu, z_\nu]} d(t, f(x_\nu)).$$

For fixed $\rho > 0$ we define

$$\Delta_\rho(x) = \{z \in B \mid d'_\varepsilon(z; x) \geqslant d'_\varepsilon(t; x), \ \forall t \in B\},$$

$$\Delta_0(x) = \{z \in B \mid d'_\varepsilon(z; x) \geqslant d(t, f(x)), \ \forall t \in B\},$$

$$F_\rho(x, z) = \{y \in [x, z] \mid d(y, f(x)) \geqslant \max_{t \in [x, z]} d(t, f(x)) - \rho\},$$

$$F_0(x, z) = \{y \in [x, z] \mid d(y, f(x)) \geqslant \max_{t \in [x, z]} d(t, f(x))\}.$$

Now we are going to prove that families $\{\Delta_\rho(x) \mid \rho \geqslant 0\}$ and $\{F_\rho(x, z) \mid \rho \geqslant 0\}$ satisfy all the conditions in Theorem 4:

(i) $\{\Delta_\rho(x)\,|\,\rho\geqslant 0\}$ is non-increasing in $\rho$. In fact, for any concave function $\varphi(x)$ we have

$$\varphi(t)\leqslant\varphi(x)+\nabla\varphi(x)\cdot(t-x),$$

and therefore

$$\begin{aligned}d(t, f(x))&=\min\{f(t)-f(x),\ g_i(t),\ i=1,\cdots,m\}\\&\leqslant\min\{f'(t, x)-f(x),\ g_i'(t, x),\ i=1,\cdots,m\}\\&\leqslant\min\{f'(t, x)-f(x),\ g_i'(t, x),\ i\in I_s(x)\}\\&=d_s'(t;\ x).\end{aligned}$$

From this we have $\Delta_\rho(x)\subset\Delta_0(x)$.

(ii) $\{\Delta_\rho(x)\,|\,\rho\geqslant 0\}$ satisfies H4.

*Proof* Let $x_\nu\to\bar{x}$, $z_\nu\to\bar{z}$ and $d_s'(z_\nu, x_\nu)=\max\limits_{t\in B} d_s'(t, x_\nu)$. Now we are going to prove that $\bar{z}=\Delta_0(\bar{x})$, or

$$d_s'(\bar{z}, \bar{x})\geqslant\max_{t\in B} d(t, f(\bar{x})).$$

It is known that $d_s'(z, x)$ is an upper semicontinuous function of $(z, x)$ and that max $\{d(t, f(x))\,|\,t\in B\}$ is a continuous function of $x$ (Cf. Denel (1979), p. 64). From this we have

$$\overline{\lim_{\nu\to\infty}}\, d_s'(z_\nu, x_\nu)\leqslant d_s'(\bar{z}, \bar{x}).$$

So we have only to show

$$\overline{\lim_{\nu\to\infty}}\, d_s'(z_\nu, x_\nu)\geqslant d(t, f(\bar{x})),\ \forall t\in B.$$

By definition of $z_\nu$, for all $t\in B$ we have

$$\begin{aligned}d_s'(z_\nu, x_\nu)&\geqslant d_s'(t, x_\nu)\\&\geqslant\min\{\nabla f(x_\nu)\cdot(t-x_\nu),\ g_i(x_\nu)+\nabla g_i(x_\nu)\cdot(t-x_\nu), i=1,\cdots,m\}.\end{aligned}$$

Since $\{f(x_\nu)\}$ is a non-decreasing family, we have

$f(x_\nu)\leqslant f(\bar{x})$, therefore the above expression

$$\geqslant\min\{f(t)-f(\bar{x}),\ g_i(t),\ i=1,\cdots,m\}=d(t;\ f(\bar{x})).$$

This is what we want to prove.

(iii) $F_\rho(x, z)$ is obviously non-decreasing in $\rho$.

(iv) H1 is a direct consequence of Theorem A15 in Huard (1975).

(v) H2 follows directly from the definition of $F_\rho(x, z)$.

(vi) Evidently $\Delta_0(x)\neq\emptyset$, $\Delta_\rho(x)\neq\emptyset$, $\forall x\in A\cap B$, so we have H5.

(i)—(vi) indicates that conditions H1—H5 in Theorem 4 all hold. Therefore all the accumulation points of the sequence $\{x_\nu\}$ generated by the algorithm belong to

$$\Gamma=\{x\in A\cap B\,|\,x\in F_0(x, z),\ z\in\Delta_0(x)\}.$$

Now we are going to prove that, if $x\in\Gamma$, then $x$ is an optimal solution of the given problem (following the way of Huard (1975), p. 325).

Since $x\in F_0(x;\ z)$, we have

$$d(x, f(x))\geqslant d(t, f(x)),\ \forall t\in[x, z],\ \text{or}$$

$$\min\{0,\ g_i(x),\ i=1, \cdots, m\} \geqslant \min\{f(t)-f(x),\ g_i(t),\ i=1, \cdots, m\},$$
$$\forall t \in [x, z].$$

From this we have, $\forall t \in [x, z]$

$$\min\{f(t)-f(x),\ g_i(t),\ i=1, \cdots, m\} \leqslant 0.$$

Therefore, we have

$$\min\{f(t)-f(x),\ g_i(t),\ i=I_s(x)\} = \min\{f(t)-f(x),\ g_i(t),\ i=1, \cdots, m\}$$
$$\leqslant 0, \quad \text{if } t \text{ is sufficiently near to } x. \tag{1}$$

Now we are going to prove that

$$\min\{\nabla f(x)\cdot(z-x),\ \nabla g_i(x)\cdot(z-x)+g_i(x),\ i \in I_s(x)\} \leqslant 0. \tag{2}$$

If, on the contrary, (2) is false, then there exists $b=b(x, z)>0$ such that

$$\nabla f(x)\cdot(z-x) \geqslant b,\ g_i(x)+\nabla g_i(x)\cdot(z-x) \geqslant b,\ \forall i \in I_s(x).$$

Let $t=x+\lambda(z-x)$, $0 \leqslant \lambda \leqslant 1$. Since $f(t)$, $g_i(t)$, $i=1, \cdots, m$ are all continuously differentiable, then, by mean-value theorem, there exist $0<\theta$, $\theta_i<1$, such that

$$f(t)-f(x) = \lambda \nabla f(x+\theta\lambda(z-x))\cdot(z-x),$$
$$g_i(t) = g_i(x)+\lambda \nabla g_i(x+\theta_i\lambda(z-x))\cdot(z-x),\ \forall i \in I_s(x).$$

Thus, if $\lambda$ is small enough, we have

$f(t)-f(x)>0$, $g_i(t)>0$, $\forall i \in I_s(x)$, which contradicts (1).

Since $z \in \Delta_0(x)$, by (2), we have

$$\min\{f(t)-f(x),\ g_i(t),\ i=1, \cdots, m\}$$
$$\leqslant \min\{f'(z, x)-f(x),\ g_i'(z, x),\ i \in I_s(x)\} \leqslant 0,\ \forall t \in B.$$

From the inequality, it follows immediately that

$$f(t) \leqslant f(x),\ \forall t \in A^0 \cap B.$$

Since $A$ is convex, $f(x)$ is continuous, we have

$$f(t) \leqslant f(x),\ \forall t \in A \cap B.$$

This is what we want to prove.

## References

[1] Denel, J., Adaptation and Performance of the Linearized Method of Centers, *Cahiers du CERO (Bruxelles)* **16**, (1974), 447—458.

[2] Denel, J., Extentions of the Continuity of Point-to-set Maps: Applications to Fixed Point Algorithms, *Mathematical Programming Study*, **10** (1979), 48—68.

[3] Huard, R., Optimization Algorithms and Point-to-set Maps, *Mathematical Programmieg*, **8** (1975), 308—331.

[4] Huard, R., Implementation of Gradient Methods by Tangential Discretization, *Journal of Optimization, Theory and Applications*, **28** (1978).

[5] Huard, R., Extentions of Zangwill Theorem, *Mathematical Programming Study*, **10** (1979), 98—103.

[6] Zangwill, W. I., Nonlinear Programming, A Unified Approach (Prentice Hall, Englewood Cliff, RI), (1969).

*Chin. Ann. of Math.*
3 (4) 1982

# ON $A_2^2$-POLYHEDRA

ZHANG SUCHENG (CHANG S. C.)
*(Institute of Mathematics, Academia Sinica)*

Dedicated to Professor Su Bu-chin on the Occasion of his 80th Birthday and his 50th Year of Educational Work

A finite simply connected 4-dimensional polyhedron is denoted for brevity by $A_2^2$-polyhedron. To such a polyhedron there is an associated cohomology ring consisting of cohomology groups and Pontrjagin square[1]. A one-one correspondence between the homotopy types of this sort of polyhedra and the properly isomorphic classes of cohomology rings has been well established by Whitehead, J. H. C.[1]. However, the proof seems to be very difficult. The purpose of this note is to introduce $A_2^2$-homology co-ring and give simple proofs of the following two Theorems:

**Theorem 1.** *If an $A_2^2$-homology co-ring $H$ is given, there exists an $A_2^2$-polyhedron whose homology co-ring is properly isomorphic to $H$.*

The polyhedron in Theorem 1 is said to realize the given $A_2^2$-homology co-ring.

Let $H$ and $H'$ be two $A_2^2$-homology co-rings realized by $K$ and $K'$ respectively. We have

**Theorem 2.** *If $h$: $H \to H'$ is a proper homomorphism, there is a continuous map $\phi$: $K \to K'$ such that $\phi$ induces $h$.*

These two Theorems imply a one-one correspondence between the homotopy types of $A_2^2$-polyhedra and the proper isomorphism classes of $A_2^2$-homology co-rings.

## §1. Algebraic Preliminaries.

Let $H_i (i=2, 3, 4)$ be abelian groups of finite generators, where $H_4$ is free. To each integer $m \geqslant 2$, there is a natural projection $\mu_{m,0}$: $H_i \to (H_i)_m = H_i/mH_i (i=2, 3, 4)$. As usual, ${}_mH_i$ denotes a subgroup of $H_i$ such that $a \in {}_mH_i$ if and only if $ma=0$. Construct a group $H_i(m)$ as a direct sum

$$H_i(m) = \mu_{m,0} H_i + \Delta_m^* H_{i-1}, \tag{1}$$

$\Delta_m^*$ being an isomorphism of ${}_mH_{i-1}$ into $H_i(m)$. For brevity, $\Delta_m^*$ is written as $\Delta^*$. Let

$$\Delta: H_i(m) \to H_{i-1}$$

be a homomorphism such that

Manuscript received December 18, 1980

$$\Delta^{-1}(0)=\mu_{m,0}H_i, \tag{2}$$
$$\Delta\Delta^*|_m H_{i-1}=1.$$

Let a homomorphism

$$\mu_{l,m}: H_i(m)\to H_i(l)$$

be defined by

$$\mu_{l,m}(\mu_{m,0}x+\Delta_m^* y)=\mu_{l,0}\Big(\frac{l}{(l,\ m)}x\Big)+\Delta_l^*\Big(\frac{m}{(l,\ m)}y\Big),\ x\in H_i,\ y\in {}_mH_{i-1}, \tag{3}$$

where $(l, m)$ denotes the greatest common divisor of $l$ and $m$.

Suppose $H_2=\sum\limits_{\alpha\in A} Z_\alpha(\sigma_\alpha)$, where $A$ is a finite ordered set and $Z_\alpha(\sigma_\alpha)$ is a cyclic group of order $\sigma_\alpha$, which may be zero. An abelian group (see Whitehead, J. H. C.[3]), $\Gamma(H_2)$ is so constructed that it consists of the elements of the form

$$\sum_{\alpha<\beta,\alpha,\beta\subset A} a_{\alpha,\beta}e_{\alpha,\beta},$$

where $e_{\alpha,\beta}$ is a generator of order $(\sigma_\alpha, \sigma_\beta)$ and $a_{\alpha,\beta}\in Z_\alpha((\sigma_\alpha, \sigma_\beta))$, a cyclic group of order $(\sigma_\alpha, \sigma_\beta)$ if $\alpha\neq\beta$, while $e_{\alpha,\alpha}$ a generator of order $(\sigma_\alpha^2, 2\sigma_\alpha)$ and $a_{\alpha,\alpha}\in Z_\alpha((\sigma_\alpha^2, 2\sigma_\alpha))$, a cyclic group of order $(\sigma_\alpha^2, 2\sigma_\alpha)$ if $\alpha\in A$. In case $\alpha\neq\beta$, we assume $e_{\alpha,\beta}=e_{\beta,\alpha}$. The group $\Gamma(H_2)$ is called the homology module of $H_2$.

Let $f$: $H_2\to H_2'$ be a homomorphism, where $H_2'=\sum\limits_{\beta\in A'} Z_\beta(\sigma_\beta')$, $A'$ being also a finite ordered set. Then

$$f(g_\alpha)=\sum_{\beta\in A'} b_{\alpha,\beta}g_\beta',$$

in which $g_\alpha$ and $g_\beta'$ denote generators of $H_2$ and $H_2'$ respectively, and $b_{\alpha,\beta}$ are integers satisfying

$$\sigma_\alpha b_{\alpha,\beta}\equiv 0,\ \mathrm{mod}\ \sigma_\beta'.$$

Let $\Gamma(H_2')$ be the homology module of $H_2'$. There is a homomorphism

$$\Gamma(f): \Gamma(H_2)\to\Gamma(H_2')$$

defined, in correspondence with $f$: $H_2\to H_2'$, by

$$\Gamma(f)\Big(\sum_{\alpha,\beta\subset A} a_{\alpha,\beta}e_{\alpha,\beta}\Big)=\sum_{\alpha,\beta\subset A} a_{\alpha,\beta}\Gamma(f)e_{\alpha,\beta}, \tag{4}$$

and

$$\Gamma(f)e_{\alpha,\beta}=\sum_{\mu,\nu\subset A'} b_{\alpha,\mu}b_{\beta,\nu}e_{\mu,\nu}'+\sum_{\mu\in A'} b_{\alpha,\mu}b_{\beta,\mu}e_{\mu,\mu}', \tag{5}$$
$$\Gamma(f)e_{\alpha,\alpha}=\sum_{\beta\in A'} b_{\alpha,\beta}^2 e_{\beta,\beta}'+\sum_{\gamma<\delta,\gamma,\delta\subset A'} b_{\alpha,\gamma}b_{\alpha,\delta}e_{\gamma,\delta}',$$

where the occurring coefficients are well defined according to the following Lemma or its similar arguments

**Lemma 1.** *If $a_{\alpha,\beta}$ is a reduced* mod $(\sigma_\alpha, \sigma_\beta)$ *integer, then $a_{\alpha,\beta}b_{\alpha,\mu}b_{\beta,\nu}$ is a well defined* mod $(\sigma_\mu', \sigma_\nu')$ *integer.*

*Proof* The integer $a_{\alpha,\beta}$ may vary by a multiple of $(\sigma_\alpha, \sigma_\beta)$. The integers $b_{\alpha,\mu}$ and $b_{\beta,\nu}$ may be written as $\Big(\frac{r\sigma_\mu'}{(\sigma_\alpha,\ \sigma_\mu')}+l\sigma_\mu'\Big)$ and $\Big(\frac{s\sigma_\nu'}{(\sigma_\beta,\ \sigma_\nu')}+k\sigma_\nu'\Big)$ respectively, where $r$, $s$, $l$, $k$ are integers. Then

$$(a_{\alpha,\beta}+t(\sigma_\alpha,\sigma_\beta))\Big(\frac{r\sigma'_\mu}{(\sigma_\alpha,\sigma'_\mu)}+l\sigma'_\mu\Big)\Big(\frac{s\sigma'_\nu}{(\sigma_\beta,\sigma'_\nu)}+k\sigma'_\nu\Big)$$
$$\equiv(a_{\alpha,\beta}+t(\sigma_\alpha,\sigma_\beta))\Big(\frac{r\sigma'_\mu}{(\sigma_\alpha,\sigma'_\mu)}\Big)\Big(\frac{s\sigma'_\nu}{(\sigma_\beta,\sigma'_\nu)}\Big),\ \text{mod}(\sigma'_\mu,\sigma'_\nu).$$

Since $\sigma_\alpha/(\sigma_\alpha,\sigma_\beta)$ and $\sigma_\beta/(\sigma_\alpha,\sigma_\beta)$ are relatively prime to each other, there are integers $p$ and $q$ such that

$$p\frac{\sigma_\alpha}{(\sigma_\alpha,\sigma_\beta)}+q\frac{\sigma_\beta}{(\sigma_\alpha,\sigma_\beta)}=1.$$

Therefore

$$t(\sigma_\alpha,\sigma_\beta)\Big(\frac{r\sigma'_\mu}{(\sigma_\alpha,\sigma'_\mu)}\Big)\Big(\frac{s\sigma'_\nu}{(\sigma_\beta,\sigma'_\nu)}\Big)$$
$$=t(\sigma_\alpha,\sigma_\beta)\Big(p\frac{\sigma_\alpha}{(\sigma_\alpha,\sigma_\beta)}+q\frac{\sigma_\beta}{(\sigma_\alpha,\sigma_\beta)}\Big)\Big(\frac{r\sigma'_\mu}{(\sigma_\alpha,\sigma'_\mu)}\Big)\Big(\frac{s\sigma'_\nu}{(\sigma_\beta,\sigma'_\mu)}\Big)\equiv 0,\ \text{mod}(\sigma'_\mu,\sigma'_\nu),$$

bacause $\sigma_\alpha/(\sigma_\alpha,\sigma'_\mu)$ and $\sigma_\beta/(\sigma_\beta,\sigma'_\mu)$ are integers. Hence $a_{\alpha,\beta}b_{\alpha,\mu}b_{\beta,\nu}$ are determined independently of the different choices of the representatives of $a_{\alpha,\beta}$, $b_{\alpha,\mu}$ and $b_{\beta,\nu}$.

Similar arguments for other cases are omitted.

Though $\Gamma(f)$ is a homomorphism, yet $\Gamma(f+g)\neq\Gamma(f)+\Gamma(g)$. we have

**Lemma 2.** *$\Gamma(f)$ is a covariant functor.*

This is evident after computation or geometric reasoning by keeping in mind that $e_{\alpha,\beta}(\alpha\neq\beta)$ actually means Whitehead product and $e_{\alpha,\alpha}$ a map of Hopf invariant 1 (See § 2).

Give a set of homomorphisms

$$\gamma_m\colon H_4(m)\to(\Gamma(H_2))_m,\ m=0,2,3,\cdots,$$

so that

$$\gamma_p\mu_{p,q}|H_4(q)=\bar{\mu}_{p,q}\gamma_q, \tag{6}$$

in which

$$\bar{\mu}_{p,q}\colon(\Gamma(H_2))_q\to(\Gamma(H_2))_p$$

means the projection of $\frac{p}{(p,q)}x'$ into $(\Gamma(H_2))_p$, if $x'$ represents an element of $(\Gamma(H_2))_q$. Evidently it is independent of the choice of $x'$ in its class. A homology co-ring[1)] contains groups, which are $H_i(i=2,3,4)$, $H_i(m)$ $(m=2,3,\cdots)$ and homology module $(\Gamma(H_2))_m$, and homomorphisms among these groups, which are $\mu_{p,q}$: $H_i(q)\to H_i(p)$ $(p=2,3,4,\cdots;q=0,2,3,\cdots)$, $\bar{\mu}_{p,q}$: $(\Gamma(H_2))_q\to(\Gamma(H_2))_p(q=0,2,3,\cdots,p=2,3,\cdots)$, $\Delta_m$: $H_i(m)\to H_{i-1}(m=2,3,\cdots)$, and $\gamma_m$: $H_4(m)\to(\Gamma(H_2))_m$ $(m=0,2,3,\cdots)$, the latter satisfying (6). Hereafter we use $\mu_{p,q}$ to denote both $\mu_{p,q}$ and $\bar{\mu}_{p,q}$.

Let homomorphisms $f_i$: $H_i\to H'_i$ $(i=2,3,4)$ be given. They induce homomorphisms

$$f_i(m)\colon H_i(m)\to H'_i(m)$$

by

$$f_i(m)\,|\,\mu_{m,0}H_i=\mu_{m,0}f_iH_i, \tag{7}$$
$$f_i(m)\,|\,\Delta_m^*({}_mH_{i-1})=\Delta_m^*f_{i-1}\,{}_mH_{i-1} \tag{8}$$

1) Homology co-ring is dual to cohomology ring in [1].

**Lemma 3.**

$$f_r(p)\mu_{pq}=\mu_{p,q}f_r(q), \tag{9}$$
$$f_{r-1}\Delta=\Delta f_r(p) \tag{10}$$
$$(r=2,\ 3,\ \cdots,\ p=2,\ 3,\ \cdots;\ q=0,\ 2,\ 3,\ \cdots).$$

This Lemma follows directly from (2), (3), (7), and (8).

Let $H$ and $H'$ be two homology co-rings. A set of homomorphisms $\{f_r(m)\}$ ($m=0, 2, 3, \cdots$) between the corresponding homology groups of $H$ and $H'$ with integer coefficients or with integer coefficients reduced mod $m$ ($m=0, 2, 3, \cdots$) is called a $(\mu, \Delta)$–homomorphism of $H$ into $H'$ if (9) and (10) are all satisfied. The homomorphism $\Gamma(f_2)$ of $\Gamma(H_2)$ into $\Gamma(H'_2)$ is induced by $f_2$: $H_2\to H'_2$ according to (4) and (5), which carries ${}_m\Gamma(H_2)$ into ${}_m\Gamma(H'_2)$ and produces a homomorphism

$$\Gamma(f_2)_m:\ (\Gamma(H_2))_m\to(\Gamma(H_2))_m.$$

If $f_2$ is an isomorphism onto, $f_2^{-1}$ exists, and by Lemma 2

$$\Gamma(f_2)\Gamma(f_2^{-1})=1_{\Gamma(H'_2)},\quad \Gamma(f_2^{-1})\Gamma(f_2)=1_{\Gamma(H_2)}.$$

Both $\Gamma(f_2)$ and $\Gamma(f_2^{-1})$ are isomorphisms onto. Hence $(\Gamma(f_2))_m$ and $(\Gamma(f_2^{-1}))_m$ are also isomorphisms onto.

If, besides (6), the following diagrams

$$\begin{array}{ccc} H_4(m) & \xrightarrow{\gamma_m} & (\Gamma(H_2))_m \\ \downarrow f_4(m) & & \downarrow \Gamma(f_2)_m \\ H'_4(m) & \xrightarrow{\gamma'_m} & (\Gamma(H'_2))_m, \end{array} \tag{11}$$
$$m=0,\ 2,\ 3,\ 4,\ \cdots,$$

commute, then the $(\mu, \Delta)$ homomorphism $\{f_i(m)\}$ ($m=0, 2, 3, \cdots, i=2, 3, 4$) and $\Gamma(f_2)$ are said to constitute a proper homomorphism of the homology co-ring $H$ into another $H'$. No doubt, proper isomorphism is an equivalence relation.

## § 2. Geometric Interpretation of homology Co-ring.

§ 2.1. The three dimensional skeleton of an $A_2^2$–polyhedron is able to be reduced to

$$K^3=(S_1^2\cup e_1^3(\sigma_1))\vee\cdots\vee(S_{t_2}^2\cup e^3(\sigma_{t_2}))\vee S_{t_2+1}^2\vee S_{t_2+2}^2\vee\cdots\vee S_{t_2+p_2}^2$$
$$\vee S_1^3\vee\cdots\vee S_{p_3+t_3}^3. \tag{12}$$

An $A_2^2$–polyhedron may be obtained within its homotopy type by attaching four dimensional cells to $K^3$. Since $\Pi_3(S_1^2\cup e_1^3(\sigma_1))$ is the image of

$$i:\ \Pi_3(S_1^2)\to\Pi_3(S_1^2\cup e^3(\sigma_1)),$$

it is necessary to compute the kernel of the homomorphism $i$ which is, however, known (See Chang, S. C.[4]). to be generated by Whitehead product, $[S_1^2, \sigma_1 S_1^2]$ and the composition element $\sigma_1\cdot h$, where $h$ denotes the Hopf map, $S^3\to S^2$, and $\sigma_1$ is a map of $S^2$ to $S_1^2$ of degree $\sigma_1$. Hence $\Pi_3$ $(S_1^2\cup e_1^3(\sigma_1))$ is a cyclic group of order

$(\sigma_l^2, 2\sigma_1)$. Evidently

$$\Pi_3(K^3)=\sum_{l=1}^{t_2+p_2} Z_l((\sigma_l^2, 2\sigma_l))+\sum_{\alpha<\beta,1}^{t_2+p_2} Z_{\alpha,\beta}((\sigma_\alpha, \sigma_\beta))+\{b_1, \cdots, b_{p_3+t_3}\},$$

where $\sigma_l=0$, if $l=t_2+1, \cdots, t_2+p_2$, and $\{b_1, \cdots, b_{p_3+t_3}\}$ is a free module of $p_3+t_3$ generators. Now

$$\Pi_3(K^3)=\Gamma(H_2(K))+\{b_1\cdots, b_{p_3+t_3}\},$$

"+" being the direct sum. Let

$$\boldsymbol{p}: \Pi_3(K^3)\to\Gamma(H_2(K))$$

be a projection. Let $f:A_2^2\to A_2'^2$ be a cellular map, then $f$ induces the homomorphisms

$$f_2: H_2(A_2^2)\to H_2(A_2'^2),$$
$$f_*: \Pi_3(K^3)\to\Pi_3(K'^3),$$

where $K^3$ and $K'^3$ are the three skeletons of $A_2^2$ and $A_2'^2$ respectively, so that

$$f_*|\Gamma(H_2(A_2^2)): \Gamma(H_2(A_2^2))\to\Gamma(H_2(A_2^2))$$

coincides with $\Gamma(f_2)$.

The group $i\Pi_3(K^2)\,(\subset\Pi_3(K^3))$ is determined by $t_2, \sigma_1, \cdots, \sigma_{t_2}$ and $p_2$, which are known if we know $H_2(A_2^2)$. Hence $i\Pi_3(K^2)$ is preferably written as $\Gamma(H_2(A_2^2))$.

§ 2.2. An $A_2^2$-polyhedron $K$ may be constructed within its homotopy type by attaching 4-dimensional cells to $K^3$ given by

$$\begin{aligned}\beta e_l^4&=\bar\sigma_l S_l^3+\sum_{\alpha<\gamma} a_{l,\alpha,\gamma}e_{\alpha,\gamma},\quad l=1, \cdots, t_3,\\ \beta e_{t_3+r}^4&=\sum_{\alpha<\gamma} a_{t_3+r,\alpha,\gamma}e_{\alpha,\gamma},\quad r=1, \cdots, p_4,\end{aligned}\tag{13}$$

$\beta e_*^4$ being the homotopy boundary of $e_*^4$, which implies

$$\partial e_l^4=\bar\sigma_l S_l^3,\quad l=1, \cdots, t_3,$$
$$\partial e_{t_3+r}^4=0,\qquad r=1, \cdots, p_4.$$

For convenience, $S_l^3$ and $e_l^4$ are also used to denote their homology classes with integers as coefficients or with integers reduced mod $\bar\sigma_l$ as coefficients[2]. If $a=\sum_{i=1}^{p_3+t_3}\alpha_i S_i^3\in{}_mH_3$, then $m\alpha_i\equiv0$, mod $\bar\sigma_i$, $i=1, \cdots, p_3+t_3$, whence

$$\alpha_i=\frac{r_i\bar\sigma_i}{(m, \bar\sigma_i)},\quad i=1, \cdots, p_3+t_3,\tag{14}$$

where $r_i=1, \cdots, (m, \bar\sigma_i)$. The group ${}_mH_3$ is generated by $\bar\sigma_l S_l^3/(m, \bar\sigma_l)$ $(l=1, \cdots, t_3)$ of order $(m, \bar\sigma_l)$. Though $\Delta^*$ is not unique, yet a special one may be defined by[3]

$$\Delta_m^*\left(\frac{\bar\sigma_l S_l^3}{(m, \bar\sigma_l)}\right)=\frac{m}{(m, \bar\sigma_l)}e_l^4,\tag{15}$$

because

2) Hereafter $\frac{m}{(m, \bar\sigma_l)}e_l^4$ also denotes a homology class mod $m$ and $\frac{\bar\sigma_l S_l^3}{(m, \bar\sigma_l)}$ a homology class mod $(m, \bar\sigma_l)$.

3) It is evident that $e_l^4$ may vary by a linear expression $\sum_{r=1}^{p_4} u_{l,r}e_{t_3+r}^4$, where $u_{l,r}$ are integers if (16) is required to be valid. It tells that $\Delta^*$ is not unique. See § 2 of Whitehead, J. H. C. [1] for its cohomology version.

$$\Delta_m\Delta_m^*\Big(\frac{\bar{\sigma}_l S_l^3}{(m,\ \bar{\sigma}_l)}\Big)=\Delta_m\Big(\frac{me_l^4}{(m,\ \bar{\sigma}_l)}\Big)=\frac{\bar{\sigma}_l}{(m,\ \bar{\sigma}_l)}S_l^3,\quad l=1,\ \cdots,\ t_3. \tag{16}$$

From (15) it also leads to

$$\Big(\mu_{p,m}\Delta_m^*\Big(\frac{\bar{\sigma}_l S_l^3}{(m,\ \bar{\sigma}_l)}\Big)=\mu_{p,m}\frac{m}{(m,\ \bar{\sigma}_l)}e_l^4=\frac{p}{(p,\ m)}\cdot\frac{m}{(m,\ \bar{\sigma}_l)}e_l^4$$

$$=\Delta_p^*\Big(\frac{m}{(p,\ m)}\cdot\frac{\bar{\sigma}_l}{(m,\ \bar{\sigma}_l)}S_l^3\Big), \tag{17}$$

in which I remark that $p\Big(\frac{m}{(p,\ m)}\Big)\Big(\frac{\bar{\sigma}_l}{(m,\ \bar{\sigma}_l)}\Big)\equiv 0$, mod $\bar{\sigma}_l$. From (17) and

$$\mu_{p,m}\mu_{m,0}H_i=\mu_{p,0}\Big(\frac{p}{(p,\ m)}H_i\Big), \tag{18}$$

it is shown that (3) is valid if $i=4$. To complete the verification of (3) it is still necessary to consider the cases that $i=3$ and $i=2$.

A formula similar to (15) is

$$\Delta_m^*\Big(\frac{\sigma_l}{(m,\ \sigma_l)}S_l^2\Big)=\frac{m}{(m,\ \bar{\sigma}_l)}e^3(\sigma_l), \tag{19}$$

because it is easy to verify $\Delta_m\Delta_m^*=1$ and

$$\mu_{p,m}\Delta_m^*\Big(\frac{\sigma_l}{(m,\ \sigma_l)}S_l^2\Big)=\Delta_p^*\Big(\Big(\frac{m\sigma_l}{(p,\ m)(m,\ \sigma_l)}\Big)S_l^2\Big).$$

If $i=3$. formula (3) is able to be verified in the same way as the case when $i=4$ by use of (19) in place of (15). The case $i=2$ is simple, because

$$H_2(A_2^2,\ Z_p)=\mu_{p,0}H_2(A_2^2,\ Z).$$

This finishes the verification[4] of (3).

**Lemma 3.** *Let a four-dimensional simply connected CW-complex $K$ be given by* (12) *and*

$$\beta e_\lambda^4=\bar{\sigma}_\lambda S_\lambda^3+\sum_{\alpha<\gamma}a_{\lambda,\alpha,\gamma}e_{\alpha,\gamma},\quad \lambda=1,\ \cdots,\ t_3, \tag{13}$$

*then $a_{\lambda,\alpha,\gamma}$'s are determined* mod $\bar{\sigma}_\lambda$ *within the homotopy type of $K$.*

*proof* Let $g$: $S_\lambda^3\to S_\lambda^3\vee S^3$ be a comultiplication. Map $S^3$ into $K$ by $\sum c_{\alpha,\gamma}e_{\alpha,\gamma}$, $c_{\alpha,\gamma}$ being arbitrary integers. After these two processes $\beta e_\lambda^4=\bar{\sigma}_\lambda S_\lambda^3+\sum_{\alpha<\gamma}a_{\lambda,\alpha,\gamma}e_{\alpha,\gamma}$ becomes $\bar{\sigma}_\lambda S_\lambda^3+\sum_{\alpha<\gamma}a_{\lambda,\alpha,\gamma}e_{\alpha,\gamma}+\bar{\sigma}_\lambda\sum_{\alpha<\gamma}c_{\alpha,\gamma}e_{\alpha,\gamma}$. Construct a new CW-complex $L$, which contains, besides all the other cells of $K^4$ except $e_\lambda^4$, a new cell $e_\lambda'^4$, that is attached to $K^3$ by $\bar{\sigma}_\lambda S_\lambda^3+\sum_{\alpha<\gamma}(a_{\lambda,\alpha,\gamma}e_{\alpha,\gamma}+\bar{\sigma}_\lambda c_{\alpha,\gamma}e_{\alpha,\gamma})$. Let $E^4$ be a euclidean four dimensional unit cell containing the points $\{(x_1,\ x_2,\ x_3,\ x_4)\}$, $0\leqslant x_i\leqslant 1$ $(i=1,\ 2,\ 3,\ 4)$. Let $\phi$: $E^4\to\bar{e}_\lambda^4$ and $\psi$: $E^4\to\bar{e}_\lambda'^4$ be the characteristic maps of $\bar{e}_\lambda^4$ and $\bar{e}_\lambda'^4$ respectively, where identification topology is used. Then $\psi\phi^{-1}$ is a homomorphism in the interior of $\bar{e}_\lambda^4$. If $\beta e_\lambda^4$ is

4) The verification has been carried out for reduced complexes. But it is true for general $A_2^2$-polyhedron, because the latter is of the same homotopy type as a reduced. Or we may choose suitable basis for chain groups of an $A_2^2$-polyhedron, so that (12) and (13) are valid.

considered as a point set, then $\phi^{-1}(\beta\bar{e}_\lambda^4 \cap S_\lambda^3) = \phi^{-1}S_\lambda^3$ is able to be considered as a three dimensional unit cell $E^3$, containing the points $\{(x_1, x_2, x_3, 0)\}$, $0 \leqslant x_i \leqslant 1$, $(i=1, 2, 3)$. Divide $E^3$ into $E_l^3 (l=1, \cdots, \bar{\sigma}_\lambda)$ such that $(x_1, x_2, x_3, 0) \in E_l^3$ if $(l-1) \cdot \frac{1}{\bar{\sigma}_\lambda} \leqslant x_3 \leqslant l \cdot \frac{1}{\bar{\sigma}_\lambda}$, $1 \leqslant l \leqslant \bar{\sigma}_\lambda$. Then $\phi | E_l^3 : (E_l^3, \dot{E}_l^3) \to (S^3, 0)$ may be supposed to represent $1 \in \Pi_3(S_l^3)$. Divide $E_l^3$ again into $E_{l,1}^3$ and $E_{l,2}^3$ such that a point $\mathscr{P}(x_1, x_2, x_3, 0)$ of $E_l^3$ will belong to $E_{l,1}^3$ or $E_{l,2}^3$, if $0 \leqslant x_1 \leqslant \frac{1}{2}$ or $\frac{1}{2} \leqslant x_1 \leqslant 1$. Assume that $\psi | E_{l,1}^3 : (E_{l,1}^3, \dot{E}_{l,1}^3) \to (S_\lambda^3, 0)$ represents $1 \in \Pi_3(S_\lambda^3)$ and $\psi | E_{l,2}^3 : (E_{l,2}^3, \dot{E}_{l,2}^3) \to (K^3, 0)$ represents $\sum (c_{\alpha,\gamma} e_{\alpha,\gamma})$ in $\Pi_3(K^3, 0)$. Then

$$\begin{aligned} \psi\phi^{-1} | \beta e_\lambda^4 &= \psi\phi^{-1}(\beta e_\lambda^4 \cap (K^3 - S_\lambda^3) + \beta e_\lambda^4 \cap S_\lambda^3) \\ &= 1 | \beta e_\lambda^4 \cap (K^3 - S_\lambda^3) + (1 \vee \sum c_{\alpha,\gamma} e_{\alpha,\gamma}) \circ g | \bar{\sigma}_\lambda S_\lambda^3, \end{aligned}$$

which is single valued. From Lemma 3 in [2], $\psi\phi^{-1}$ is continuous. Define a continuous map

$$\xi : K \to L$$

so that $\xi | K - e_\lambda^4 = 1$, $\xi | \bar{e}_\lambda^4 = \psi\phi^{-1}$. Evidently $\xi$ induces isomorphisms of the homology groups of $K$ with integer coefficients onto the corresponding groups of $L$. Hence $\xi$ is a homotopy equivalence. However, the homotopy boundaries of the 4-dimensional cells of $K$ and $L$ are equal except $\beta e_\lambda^4$ in $K$ and $\beta e_\lambda'^4$ in $L$. They differ by $\bar{\sigma}_\lambda \sum_{\alpha<\gamma} c_{\alpha,\gamma} e_{\alpha,\gamma}$. Repeating this procedure at most $t_3$ times to alternate the homotopy boundaries of $e_\lambda^4 (\lambda = 1, \cdots, t_3)$ succesively finishes the proof of this Lemma.

The elements of $H_4(m)$ are $\mu_{m,0} x + \Delta_m^* \sum_l \frac{r_l \bar{\sigma}_l}{(m, \bar{\sigma}_l)} S_l^3 (r_l = 1, \cdots, (m, \bar{\sigma}_l))$, where $x \in H_4$. From (15) they are actually represented by

$$\mu_{m,0}\left(x + \sum_l \frac{r_l \cdot m}{(m, \bar{\sigma}_l)} e_l^4\right).$$

Here $x (\in H_4)$ has unique representative, which is written as $x$ for brevity. Now

$$\mu_{m,0}\beta\left(x + \sum_l \frac{r_l \cdot m}{(m, \bar{\sigma}_l)} e_l^4\right)$$

is uniquely determined for each element $\mu_{m,0}\left(x + \sum_l \frac{r_l \cdot m}{(m, \bar{\sigma}_l)} e_l^4\right)$ of $H_4(m)$. Define

$$\gamma_m : H_4(m) \to (\Gamma(H_2))_m, \quad m = 0, 2, 3, \cdots,$$

according to Whitehead, J. H. C.[3] that

$$\begin{aligned} \gamma_m \mu_{m,0}\left(x + \sum_l \frac{r_l \cdot m}{(m, \bar{\sigma}_l)} e_l^4\right) &= \mu_{m,0}\beta\left(x + \sum_l \frac{r_l \cdot m}{(m, \bar{\sigma}_l)} e_l^4\right) \\ &= \mu_{m,0} p\beta\left(x + \sum_l \frac{r_l \cdot m}{(m, \bar{\sigma}_l)} e_l^4\right), \end{aligned} \tag{20}$$

because of (13) and $\frac{m}{(m, \bar{\sigma}_l)} \cdot \bar{\sigma}_l \equiv 0$, mod $m$. Hence

$$\gamma_p\mu_{p,m}\mu_{m,0}\Big(x+\sum_l \frac{r_l\cdot m}{(m,\bar\sigma_l)}e_l^4\Big)=\gamma_p\Big\{\mu_{p,0}\Big(\frac{p}{(p,m)}x+\sum_l \frac{p\cdot r_l\cdot m}{(p,m)\cdot(m,\bar\sigma_l)}e_l^4\Big)$$
$$=\mu_{p,0}\mathbf{p}\beta\Big(\frac{p}{(p,m)}x+\sum_l \frac{p\cdot r_l\cdot m}{(p,m)\cdot(m,\bar\sigma_l)}e_l^4\Big),$$

while

$$\mu_{p,m}\gamma_m\mu_{m,0}\Big(x+\sum_l \frac{r_l\cdot m}{(m,\bar\sigma_l)}e_l^4\Big)=\mu_{p,0}\mathbf{p}\beta\Big(\frac{p}{(p,m)}x+\sum_l \frac{p\cdot r_l\cdot m}{(p,m)\cdot(m,\bar\sigma_l)}e_l^4\Big).$$

Consequently the following diagram

$$\begin{array}{ccc} H_4(m) & \xrightarrow{\mu_{p,m}} & H_4(p) \\ \gamma_m\downarrow & & \downarrow\gamma_p \\ (\Gamma(H_2))_m & \xrightarrow{\mu_{p,m}} & (\Gamma(H_2))_p \end{array} \tag{6}$$

commutes. This verifies (6) in § 1, if $\gamma_m$ is definod by (20).

Let

$$\Delta_p\colon H_i(A_2^2, Z_p)\to H_{i-1}(A_2^2)$$

be defined by $\frac{1}{p}\partial$, then $\Delta\Delta^*=1$ follows from (15) and (16). We have the groups $H_i(A_2^2)$ $(i=2, 3, 4)$, $H_i(A_2^2, Z_m)$ $(m=2, 3, \cdots,)$ and $\Gamma(H_2(A_2^2))$ along with homomorphisms $\mu_{p,q}$, $\Delta$ and $\gamma_m$. They together constitute a homology co-ring of an $A_2^2$–polyhedron. This explains the geometrio meaning of an abstract homology co-ring associated to a given $A_2^2$–polyhedron.

## § 3. Realizability.

§ 3.1. If $f\colon A_2^2\to A_2'^2$ is a cellular map, it induces homomorphisms of the homology groups $H_i(A_2^2, Z_m)$ into the corresponding groups $H_i(A_2^2, m)$, $i=2, 3, 4$, $m=0, 2, 3, \cdots$. These homomorphisms, denoted by $f_*$, commute with $\mu_{p,q}$ and $\Delta$, i. e., (9) and (10) are satisfied. If $x\in H_4(A_2^2, m)$ and if $x'$ is a mod$m$ cycle belonging to $x$, then

$$\begin{array}{ccc} x' & \xrightarrow{\gamma_m} & \Gamma(H_2(A_2^2))_m \\ \Big\downarrow f_4(m) & & \Big\downarrow \Gamma(f_2)_m \\ f_4(m)x & \xrightarrow{\gamma_m} & \Gamma(H_2(A_2'^2)) \end{array} \tag{21}$$

commute if $m=0,2,3,\cdots$. Therefore the homomorphism $\gamma_m\colon H_4(A_2^2, m)\to\Gamma(H_2(A_2^2))_m$ satisfies

$$(\Gamma(f_2))_m\cdot\gamma_m=\gamma_m\cdot f_4(m), \tag{22}$$

where $(\Gamma(f_2))_m$ is determined by $f_2\colon H_2(A_2^2)\to H_2(A_2'^2)$. It is worth while to remark that $\gamma_m$ is able to be considered as a homomorphism

$$\gamma_m\colon H_4(A_2^2, m)\to(i\Pi_3(K^2))_m,$$

then $\gamma_m$ satisfies

$$f_* \cdot \gamma_m = \gamma_m \cdot f_4(m), \tag{23}$$

where $f_*$: $(i\Pi_3(K^2))_m \to (i\Pi_3(K^2))_m$ is induced by $f|K^2$. Of course (22) and (23) are equivalent. All these homomorphisms induced by a cellular map $f$: $A_2^3 \to A_2'^3$ constitute a proper homomorphism of the homology co-ring $H(A_2^3)$ into the homology co-ring $H(A_2'^3)$.

If $H_i (i=2, 3, 4)$ are given abelian groups, where $H_4$ is free, then $H_i(m)$, $(m=0, 2, 3, \cdots)$, $\Delta$, $\mu_{p,q}$ can be defined by (1), (2), (3), and $\Gamma(H_2)$ can, moreover, be constructed by its definition in § 1. Furthermore, $\gamma_m (m=0, 2, 3, \cdots)$ may be arbitrarily given homomorphisms if (6) is satisfied. Then an abstract homology co-ring is completely determined with a lot of arbitrary factors. Is it able to be realized by an $A_2^3$-polyhedron? If so, is there a CW-map to realize a proper homomorphism of a homology co-ring into another? The answers are contained in Theorem 1 and Theorem 2.

Before going to prove Theorem 1 and Theorem 2, we need to prove

**Lemma 4.** *The homomorphisms*

$$\gamma_m: H_4(m) \to (\Gamma(H_2))_m, \quad m=0, 2, 3, \cdots$$

*are determined if*[5)]

$$\gamma_{\bar\sigma_l} | \Delta_{\bar\sigma_l}^* S_l^3: \Delta_{\bar\sigma_l}^* S_l^3 \to (\Gamma(H_2))_{\bar\sigma_l}, \quad l=1, \cdots, t_2 \tag{24}$$

*and*

$$\gamma | H_4: H_4 \to \Gamma(H_2) \tag{25}$$

*are known.*

*Proof* Each element of $H_4(m)$ is able to be written as $\mu_{m,0}x + \Delta_m^* \sum k_l S_l^3$, where $k_l = \dfrac{r_l \cdot \bar\sigma_l}{(m, \bar\sigma_l)}$, $r_l = 1, \cdots, (m, \bar\sigma_l)$. Hence $\gamma_m(H_4(m))$ is determined if $\gamma_m | \mu_{m,0} H_4$ and $\gamma_m \Delta_m^* k_l S_l^3$ are known. By means of (15) we find

$$\begin{aligned}\gamma_m \Delta_m^* k_l S_l^3 &= \gamma_m\left(\frac{k_l \cdot m}{\bar\sigma_l} e_l^4\right) = \gamma_m \cdot \mu_{m,\bar\sigma_l}\left(\frac{k_l \cdot (m, \bar\sigma_l)}{\bar\sigma_l}\right) e_l^4 \\ &= \gamma_m \mu_{m,\bar\sigma_l} \Delta_{\bar\sigma_l}^*\left(\frac{k_l \cdot (m, \bar\sigma_l)}{\bar\sigma_l} S_l^3\right) \\ &= \mu_{m,\bar\sigma_l} \gamma_{\bar\sigma_l} \Delta_{\bar\sigma_l}^* (r_l S_l^3).\end{aligned} \tag{26}$$

Besides this we have from (6) that

$$\gamma_m \mu_{m,0} | H_4 = \mu_{m,0}(\gamma | H_4).$$

Hence (24) and (25) provide the lemma.

*Proof of Theorem 1.* Let $K$ be the CW-complex given by (12) and (13). Then the other boundary relations are

$$\begin{aligned}\partial e_i^3(\sigma_i) &= \sigma_i S_i^2 \quad (i=1, \cdots, t_2), \\ \partial S_j^3 &= 0 \qquad (j=1, \cdots, p_3+t_3), \\ \partial S_h^2 &= 0 \qquad (h=1, \cdots, p_2+t_2).\end{aligned}$$

5) By $S_l^3$ we mean a generator of $H_3$ of order $\bar\sigma_l$.

Therefore $H_i(K)\approx H_i$, $i=2, 3, 4$. Let these three isomorphisms be denoted by $h$. Because

$$H_i(K,\ m)=\mu_{m,0}H_i(K)+\Delta_m^*({}_mH_{i-1}(K))$$

and

$$H_i(m)=(H_i)_m+\Delta_m^*({}_mH_{i-1}),$$

we may define

$$h\colon H_i(K,\ m)\to H_i(m)$$

by (7) and (8), i. e.,

$$h(\mu_{m,0}x+\Delta_m^*y)=\mu_{m,0}(hx)+\Delta_m^*(hy), \tag{27}$$

where $x\in H_i(K)$, $y\in {}_mH_{i-1}(K)$.

Now $H_n(K,\ Z_m)$ and $H_n(K,\ Z_p)$ may be computed from the boundary relations in $K$. From (27) it is evident that

$$\begin{array}{ccc} H_n(K,\ Z_m) & \xrightarrow{\mu_{p,m}} & H_n(K,\ Z_p) \\ \downarrow h & & \downarrow h \\ H_n(m) & \xrightarrow{\mu_{p,m}} & H_n(p) \end{array}$$

is commutative, where $\mu_{p,m}\colon H_n(K,\ Z_m)\to H_n(K,\ Z_p)$ is defined by

$$\mu_{p,m}[c]=\left[\frac{p}{(p,\ m)}c\right],$$

$c$ being a cycle mod$m$ belonging to a homology class mod $m$ of $H_n(K,\ Z_m)$. Moreover, $h\Delta=\Delta h$, where $\Delta$ in the right hand side is that in the abstract homology co-ring $H$, while in the left hand side is that in $H(K)$. From (27) we know that $h\colon H_n(K,\ Z_p)\to H_n(p)$ is isomorphism for all $n$ and $p$. Lemma 2 assures $[\Gamma(h)]_m$ is an isomorphism onto. By the definition of $\mu_{m,p}$ we see that

$$[\Gamma(h)]_m\mu_{m,p}=\mu_{m,p}[\Gamma(h)]_p. \tag{28}$$

Let $\mu_{m,0}x+\Delta_m^*k_lS_l^3$ denote an element of $H_4(A_2^2)$, where $x\in H_4(A_2^2)$, $k_l=\dfrac{r_l\bar\sigma_l}{(m,\ \bar\sigma_l)}$, $r_l=1, \cdots, (m,\ \bar\sigma_l)$. Then

$$\begin{aligned}[\Gamma(h)]_m\gamma_m(\mu_{m,0}x+\Delta_m^*k_lS_l^3)&=[\Gamma(h)]_m(\mu_{m,0}\gamma x)+[\Gamma(h)]_m\mu_{m,\bar\sigma_l}\gamma_{\bar\sigma_l}\Delta_{\bar\sigma_l}^*r_lS_l^3\\&=\mu_{m,0}[\Gamma(h)]\gamma x+\mu_{m,\bar\sigma_l}[\Gamma(h)]_{\bar\sigma_l}\gamma_{\bar\sigma_l}\Delta_{\bar\sigma_l}^*r_lS_l^3,\end{aligned} \tag{29}$$

from (6), (26) and (28). In case

$$\begin{gathered}\Gamma(h)\cdot\gamma=\gamma\cdot h\,|\,H_4,\\ [\Gamma(h)]_{\bar\sigma_l}\cdot\gamma_{\bar\sigma_l}=\gamma_{\bar\sigma_l}\cdot h\,|\,\Delta_{\bar\sigma_l}^*S_l^3\quad(l=1,\ \cdots,\ t_3)\end{gathered} \tag{P}$$

are satisfied, we find from (29) that

$$\begin{aligned}[\Gamma(h)]_m\gamma_m(\mu_{m,0}x+\Delta_m^*k_lS_l^3)&=\mu_{m,0}\cdot\gamma\cdot hx+\mu_{m,\bar\sigma_l}\cdot\gamma_{\bar\sigma_l}\cdot h\Delta_{\bar\sigma_l}^*\cdot r_lS_l^3\\&=\gamma_m(\mu_{m,0}hx)+\gamma_m\cdot\mu_{m,\bar\sigma_l}\Delta_{\bar\sigma_l}^*r_lhS_l^3\\&=\gamma_m(\mu_{m,0}hx)+\gamma_m\Delta_m^*\left(\frac{r_l\cdot\bar\sigma_l}{(m,\ \bar\sigma_l)}\right)hS_l^3\\&=\gamma_m\cdot h(\mu_{m,0}x+\Delta_m^*k_lS_l^3)\quad(l=1,\ \cdots,\ t_3).\end{aligned}$$

Because any element of $H_4(A_2^2, Z_m)$ is a linear expression of $(\mu_{m,0}x+\Delta_m^* k_l S_l^3)$, it is concluded that $(\Gamma(h))_m\cdot\gamma_m=\gamma_m\cdot h$ for all $m$. Now $\Gamma(h)$ and $(\Gamma(h))_{\bar\sigma_l}$ $(l=1, \cdots, t_3)$ are isomorphisms onto. The condition (P) becomes

$$\begin{aligned}\gamma&=[\Gamma(h)]^{-1}\cdot\gamma(H)\cdot h|H_4(K),\\ \gamma_{\bar\sigma_l}&=[\Gamma(h)]_{\bar\sigma_l}^{-1}\cdot\gamma_{\bar\sigma_l}(H)\cdot h|\Delta_{\bar\sigma_l}^* S_l^3\quad(l=1, \cdots, t_3),\end{aligned}\tag{P'}$$

which means $h$: $H(K)\to H$ is a proper isomorphism if the attaching mappings of the cells $e_l^4$, $(l=1, \cdots, t_3+p_4)$ are

$$\begin{aligned}\beta e_l^4&=\bar\sigma_l S_l^3+[\Gamma(h)]_{\bar\sigma_l}^{-1}\cdot\gamma(H)\cdot h(e_l^4(\bar\sigma_l)),\quad l=1, \cdots, t_3,\\ \beta e_{t_3+u}^4&=[\Gamma(h)]^{-1}\gamma(H)\cdot h(e_{t_3+u}^4),\quad u=1, \cdots, p_4,\end{aligned}\tag{P''}$$

where $e_l^4(\bar\sigma_l)$ denotes a mod $\bar\sigma_l$ homology class represented by $e_l^4(\bar\sigma_l)$, and $e_{t_3+u}^4$ an integral homology class represented by $e_{t_3+u}^4$. The conditions (P'') can be assigned to $K$. This completes the proof of Theorem 1.

§ 3.2. Let $K$ and $K'$ be two simply connected CW-complexes and let $K^r$ (or $K'^r$) be the $r$-dimensional skeleton of $K$ (or $K'$). By a chain group $C_n(K)$ we mean $\Pi_n(K^n, K^{n-1}$, $n=2, 3, \cdots$. Let $\beta$: $\Pi_n(K^n, K^{n-1})\to\Pi_{n-1}(K^{n-1})$ denote the homotopy boundary operator and $j$: $\Pi_{n-1}(K^{n-1})\to\Pi_{n-1}(K^{n-1}, K^{n-2})$ the injection. Then $j\beta=\partial$ is a homology boundary operator. A chain map $g_n$: $C_n(K)\to C_n(K')$, $n=2, 3, \cdots$ induces a $(\mu, \Delta)$ homomorphism of the homology co-ring $H(K)$ into the homology co-ring $H(K')$. On the other hand, if a $(\mu, \Delta)$-homomorphism $h$: $H(K)\to H(K')$ is given, we have

**Lemma 5.** *There is a chain map*

$$g_n: C_n(K)\to C_n(K'),\quad n=0, 1, 2, \cdots$$

*realizing* $h$.

Let $a\in H_n(K)$ or $H_n(K, Z_m)$. If $a'\in a$ and $j_m g a'=h j_m a'$ for all $a$, where $j_m$ denotes the injection of $a'$ or $ga'$ into its homology class, then $h$ is said to be realized by $g$. This Lemma is the dual of Lemma 4 in [1]. We sketch the proof of Lemma 5 as follows: The chain map $g$: $C_n(K)\to C_n(K')$ realizes $h$ if and only if $hj_{\sigma_l^n}a_l^n=j_{\sigma_l^n}ga_l^n$ $(l=1, \cdots, q_n)$ for all $n$, where a basis $a_1^n, \cdots, a_{q_n}^n$ for each group $C_n(K)$ is so chosen that $\partial a_i^n=\sigma_i^n a_i^{n-1}$ $(i=1, \cdots, q_n)$, where $\sigma_i^n$ may be zero, and $a_i^{n-1}$ are basis elements of $C_{n-1}(K)$. Then the required chain map $g$ is able to be constructed by induction.

*Proof of Theorem 2.* Through the procedure of the proof of Theorem 1, it is shown that a reduced $A_2^2$-polyhedron exists so that its homology co-ring is properly isomorphic to a given homology co-ring $H$. Let $K$ and $K'$ be two reduced $A_2^2$-polyhedra which realize two homology co-rings $H$ and $H'$ respectively. Let $g$: $C_n(K)\to C_n(K')$ for all $n$ be a chain map realizing a given proper homomorphism $h$: $H(K)\to H(K')$. To prove Theorem 2 it becomes necessary and sufficient to show the existence of a CW-map $\phi$: $K\to K'$ realizing the chain map $g$: $C_n(K)\to C_n(K')$. Because $K^2$ and $K'^2$ are

6) $C_0(K)=Z$, $C_1(K)=\phi$.

bouquets of 2–dimensional spheres, $g: C_2(K)\to C_2(K')$ is easily realized by a CW-map $\phi: K^2\to K'^2$. Now $K^3$ takes its fashion as (12). Analogously $K'^3$ is written as

$$K'^3=\bigvee_{n=1}^{p_2'+t_2'}(S_n'^2\cup e_u'^3(\bar{\sigma}_u'))\vee\bigvee_{v=1}^{p_3'+t_3'}S_v'^3, \qquad (12')$$

where $\bar{\sigma}'_u$ is zero if $u>t'_2$. Then

$$gS_l^3=\sum_{u=1}^{t_2'}a_{l,u}e_t'^3(\bar{\sigma}_u')+\sum_{v=1}^{p_3'+t_3'}b_{l,v}S_v'^3 \quad (l=1, \cdots, p_3+t_3),$$

$$ge_h^3(\sigma_h)=\sum_{u=1}^{t_2'}a_{hu}e_u'^3(\bar{\sigma}_u')+\sum_{v=1}^{p_3'+t_3'}b_{l,v}S_v'^3 \quad (h=1, \cdots, t_2).$$

Because $\partial g=g\partial$, it follows that

$$a_{lu}=0 \quad (l=1, \cdots, p_3+t_3, u=1, \cdots, t_2'),$$

$$a_{hu}\bar{\sigma}_u'=\sigma_h r_{h,u} \quad (h=1, \cdots, t_2, u=1, \cdots, t_2'),$$

in ich we have assumed $gS_h^2=\sum r_{hu}S_u^2$. Since

$$\beta g S_l^3=\phi=\phi\beta S_l^3,$$

$$\beta ge_h^3(\sigma_h)=\sum_{u=1}^{t_2'}a_{hu}\bar{\sigma}_u' S_u'^2=\sum_{u=1}^{t_2'}\sigma_h r_{hu}S_u'^2=\sigma_h\phi S_h^2=\phi\beta e_h^3(\sigma_h),$$

the map $\phi$ is sble to be extended to a map of $K^3$ into $K'^3$ so that $\phi$ induces a chain map $\phi_*: \Pi_3(K^3, K^2)\to\Pi_3(K'^3, K'^2)$ satisfying

$$\phi_* e_h^3(\sigma_h)=ge_h^3(\sigma_h),$$

$$\phi_* S_l^3=gS_l^3,$$

if $e_h^3(\sigma)$ and $S_l^3$ are considered as elemento of $\Pi_3(K^3, K^2)$.

Let $e_h^4(\bar{\sigma}_h)$ be a cell of $K^4$. It is evident that $j_{\bar{\sigma}_h}e_h^4(\bar{\sigma}_h)$ is an element of $H_4(K, \bar{\sigma}_h)$ with unique representativo, which is $e_h^4(\bar{\sigma}_h)$ itself. Let $\hat{j}$ denote an isomorphism of the group of spherical homology classes of $K^3$ generated by $S_1^3, \cdots, S_{p_3+t_3}^3$ in (12) onto the direct summand $\{b_1, \cdots, b_{p_3+t_3}\}$ in

$$\Pi_3(K^3)=i\Pi_3(K^2)+\{b_1, \cdots, b_{p_3+t_3}\}.$$

In other words

$$\Pi_3(K^3)=\Gamma(H_2(K^3)+\hat{j}\{S_1^3, \cdots, S_{p_3+t_3}^3\},$$

where $\{S_1^3, \cdots, S_{p_3+t_3}^3\}$ is a free module generated by the 3–spheres $S_i^3 (i=1, \cdots, p_3+t_1)$ attached at a point. Let $i: \Gamma(H_2(K))\to\Pi_3(K^3)$ be an injection such that $\boldsymbol{pi}=1_{\Gamma(H_2(K))}$. Then $\Pi_3(K^3)$ becomes

$$\Pi_3(K^3)=\boldsymbol{ip}\Pi_3(K^3)+(1-\boldsymbol{ip})\Pi_3(K^3). \qquad (30)$$

Hence

$$\beta ge_h^4(\sigma_h')=\boldsymbol{ip}\beta ge_h^4(\sigma_h')+(1-\boldsymbol{ip})\beta ge_h^4(\sigma_h').$$

Because

$$\cdots\xrightarrow{\beta}\Pi_3(K^2)\xrightarrow{i}\Pi_3(K^3)\xrightarrow{j}\Pi_3(K^3, K^2)\xrightarrow{\beta}\cdots$$

is exact, the homomorphism

$$j(1-\boldsymbol{ip})\beta g: C_4(K, Z_{\bar{\sigma}_h})\to\{S_1^3, \cdots, S_{p_3+t_3}^3\}$$

becomes $\partial g=g\partial: C_4(K, Z_{\bar{\sigma}_h})\to\{S_1^3, \cdots, S_{p_3+t_3}^3\}$ since $\boldsymbol{ip}\Pi_3(K^3)=i\Pi_3(K^3)$ and $ji=0$.

Furthermore, $j|(1-\boldsymbol{ip})\beta gC_4(K, Z_{\bar{\sigma}_h})$ is an isomorphism because the kernel of $j$ is $\boldsymbol{ip}\Pi_3(K^3)$ which is a direct summand in (30). On the other hand

$$\phi\beta e_h^4(\bar{\sigma}_h)=\phi\boldsymbol{ip}\beta e_h^4(\bar{\sigma}_h)+\phi(1-\boldsymbol{ip})\beta e_h^4(\bar{\sigma}_h).$$

Now

$$j\phi(1-\boldsymbol{ip})\beta e_h^4(\bar{\sigma}_h)=\phi j(1-\boldsymbol{ip})\beta e_h^4(\bar{\sigma}_h)=\phi\partial e_h^4(\bar{\sigma}_h)=g\partial e_h^4(\bar{\sigma}_h),$$

because $\phi$ is cellular and realizes $g$ in $K^3$. Hence

$$j\phi(1-\boldsymbol{ip})\beta e_h^4(\sigma)=j(1-\boldsymbol{ip})\beta ge_h^4(\bar{\sigma}_h).$$

Because $j|(1-\boldsymbol{ip})\Pi_3(K^3)$ is isomorphism into, we have

$$\phi(1-\boldsymbol{ip})\beta e_h^4(\bar{\sigma}_h)=(1-\boldsymbol{ip})\beta ge_h^4(\bar{\sigma}_h).$$

Consequently

$$\beta ge_h^4(\bar{\sigma}_h)-\phi\beta e_h^4(\bar{\sigma}_h)=\boldsymbol{ip}\beta ge_h^4(\bar{\sigma}_h)-\phi\boldsymbol{ip}\beta e_h^4(\bar{\sigma}_h)=\boldsymbol{i}(\boldsymbol{p}\beta g-\phi\boldsymbol{p}\beta)e_h^4(\bar{\sigma}_h). \tag{31}$$

By definition

$$\mu_{\bar{\sigma}_h,0}(\boldsymbol{p}\beta ge_h^4(\bar{\sigma}_h)-\phi\boldsymbol{p}\beta e_h^4(\bar{\sigma}_l))=\gamma_{\bar{\sigma}_h}[ge_h^4(\bar{\sigma}_h)]-\Gamma(\phi)\gamma_{\bar{\sigma}_h}[e_h^4(\bar{\sigma}_h)],$$

in which $[ge_h^4(\bar{\sigma}_h)]$ is the same homology class as $h[e_h^4(\bar{\sigma}_h)]$ owing to the realization of $h$ by $g$, while $\Gamma(\phi)=\Gamma(h)$ owing to the realization of $h|H_2(K)$ by $\phi$. It follows from (22) that

$$\mu_{\bar{\sigma}_h,0}(\boldsymbol{p}\beta ge_h^4(\bar{\sigma}_h)-\phi\boldsymbol{p}\beta e_h^4(\bar{\sigma}_h))=0$$

in the homology module $(\Gamma(H_2))_{\bar{\sigma}_h}$, which means that

$$\boldsymbol{p}\beta(ge_h^4(\bar{\sigma}_h))-\phi\boldsymbol{p}\beta e_h^4(\bar{\sigma}_h)=\bar{\sigma}_h(\sum a'_{\mu,\nu}e'_{\mu,\nu}).$$

From (31) we have

$$\beta ge_h^4(\bar{\sigma}_h)-\phi\beta e_h^4(\bar{\sigma}_h)=\boldsymbol{i}\bar{\sigma}_h(\sum a'_{\mu,\nu}e'_{\mu,\nu}), \tag{32}$$

where $\sum a'_{\mu,\nu}e'_{\mu,\nu}\in i\Pi_3(K'^2)$. We remark

$$\beta e_h^4(\bar{\sigma}_h)=\bar{\sigma}_hS_h^3+\sum a_{\alpha,\gamma}e_{\alpha,\gamma}.$$

Define a CW-map $\phi'$: $K^3\to K'^3$ such that

$$\phi'|K^3-S_h^3=\phi|K^3-S_h^3,$$

but $\phi'|S_h^3$ is the composition of the following maps:

$$S_h^3\xrightarrow{\omega}S_1^3\vee S_2^3\xrightarrow{\mu\vee\nu}S_h^3\vee K'^3\xrightarrow{\phi\vee 1}K'^3,$$

where $\omega$ clutches $S_h^3$ by its equator to obtain $S_1^3\vee S_2^3$, $\mu$ maps $S_1^3$ onto $S_h^3$ of degree 1 and $\nu$ maps $S_2^3$ into $K'^3$ so that $\nu$ represents ${}_+\sum a'_{\mu,\nu}e'_{\mu,\nu}$. The map $\phi'$ still realizes the chain map $g$: $K^3\to K'^3$. But (32) becomes

$$\beta ge_h^4(\bar{\sigma}_h)-\phi'\beta e_h^4(\bar{\sigma}_h)=0.$$

It means that $\phi'$ can be extended over to realize $ge_h^4(\bar{\sigma}_h)$. If $\bar{\sigma}_h=0$, we have

$$\beta ge_{i_s+u}^4=\boldsymbol{p}\beta ge_{i_s+u}^4=\gamma h[e_{i_s+u}^4],$$

$$\phi\beta e_{i_s+u}^4=\Gamma(h)\cdot\gamma\cdot[e_{i_s+u}^4].$$

From (11) with $m=0$ we have $\gamma\cdot h=\Gamma(h)\cdot\gamma$, which means $\beta ge_{i_s+u}^4=\phi\beta e_{i_s+u}^4$. This guarantees the extendability of $\phi$ to realize $ge_{i_s+u}^4$, $u=1, \cdots, p_4$. In short, the chain map $g$ is able to be realized by a cellular map. Q. E. D.

## References

[1] Witehead, J. H. C., On simply connected 4-dimensional polyhedra, *Comm. Math. Helvetici*, **22** (1949) 48—92.

[2] Witehead, J. H. C., On a Theorem due to Borsuk, *Bull. Amer. Math. Soc.*, **54** (1948), 1125—1132.

[3] Witehead, J. H. C., A certain exact sequence, *Annals of Math.*, **52** (1950), 51—110.

[4] Chang, S. C., Some Suspension Theorems, *Quarterly, J. Math.* (*Oxford*), **2**: **1**. (1950), 310—316.